CW01281745

THE BEES OF SUSSEX

James Power

with mapping by Bob Foreman
and the Sussex Biodiversity Record Centre

with support from

west sussex county council East Sussex County Council SOUTH DOWNS NATIONAL PARK TRUST South Downs National Park Authority Sussex Wildlife Trust

Proceeds from the sale of this book will be donated to the
Bees, Wasps and Ants Recording Society and to the Sussex Biodiversity Record Centre

BWARS — Bees, Wasps & Ants Recording Society Sussex Biodiversity Record Centre

pisces publications

For
Juliette and Wilf
Richard and Eileen

After William Kirby:

To whom can I inscribe this little work, such as it is, with more propriety than to those whose partiality first urged me to undertake it; and whose kind assistance and liberal communications have contributed so largely to bring it to a conclusion.

Accept it, therefore, as a small token of esteem of many virtues, my dear family, and of gratitude for many favours conferred upon.

Your obliged and affectionate friend
The Author

(Monographia Apum Angliae, 1802)

Published 2024 by Pisces Publications

Copyright © James Power (2024)
Copyright © of the photographs remains with the photographers
Maps © Sussex Biodiversity Record Centre and include data supplied by the Bees, Wasps and Ants Recording Society
National Character Areas data reproduced with the permission of Natural England. © Natural England 2024. Contains Ordnance Survey data © Crown copyright and database right 2024.

All rights reserved. No part of this publication may be reproduced, stored in a retrieval system or transmitted, in any form or by any means electronic, mechanical, photocopying, recording or otherwise, without the prior permission of the publishers.

First published 2024

British-Library-in-Publication Data
A catalogue record for this book is available from the British Library

ISBN 978-1-913994-11-2

Designed and published by Pisces Publications

Pisces Publications is the imprint of NatureBureau,
2C The Votec Centre, Hambridge Lane, Newbury, Berkshire RG14 5TN
www.naturebureau.co.uk – Visit our bookshop **www.naturebureau.co.uk/bookshop/**

Printed and bound in the UK by Gomer Press Ltd

FSC MIX Paper from responsible sources **FSC® C114687** www.fsc.org

Front cover: *Halictus eurygnathus* [Paul Brock]
Back cover: Ground-ivy at Seaford Head with the Seven Sisters in the background [Steven Falk]

Contents

- iv   **Acknowledgements**
- vi   **Foreword by Mike Edwards**
- 1   **Introduction**
- 2   **The state of Sussex's bees**
- 5   **An overview of Sussex**
- 10   **Assembling the data**
- 13   **The recording community**
- 17   **Explanation of the species accounts**
- 18   Glossary

Species accounts
- 20   *Dasypoda* – Pantaloon Bees
- 22   *Macropis* – Yellow Loosestrife Bees
- 24   *Melitta* – Blunthorn Bees
- 30   *Anthophora* – Flower Bees
- 40   *Apis* – Honey Bees
- 42   *Bombus* – Bumblebees
- 80   *Ceratina* – Carpenter Bees
- 82   *Epeolus* – Variegated Cuckoo Bees
- 86   *Eucera* – Long-horned Bees
- 88   *Melecta* – Mourning Bees
- 90   *Nomada* – Nomad Bees
- 144   *Xylocopa* – Large Carpenter Bees
- 146   *Anthidium* – Wool Carder Bees
- 148   *Chelostoma* – Scissor Bees
- 152   *Coelioxys* – Sharp-tail Bees
- 164   *Heriades* – Resin Bees
- 166   *Hoplitis* – Lesser Mason Bees
- 168   *Megachile* – Leafcutter Bees
- 182   *Osmia* – Mason Bees
- 196   *Stelis* – Dark Bees
- 206   *Andrena* – Mining Bees
- 320   *Panurgus* – Shaggy Bees
- 324   *Halictus* – End-banded Furrow Bees
- 332   *Lasioglossum* – Base-banded Furrow Bees
- 386   *Sphecodes* – Blood Bees
- 418   *Colletes* – Plasterer Bees
- 434   *Hylaeus* – Yellow-face Bees
- 456   Extinct species

- 462   **Making bees welcome—a personal journey**
- 464   **Gazetteer**
- 466   **How to make a record**
- 467   **References**
- 471   **Index**
- 474   **About the author**

Acknowledgements

Among the joys of exploring nature are not just the things you see but the people you meet. Studying natural history is most definitely a journey best shared with others. It is also true that this book could not have been written without the efforts of the hundreds of people who have made records of bees over the last 179 years. These range from W. Walcott who made the very first county records (of *Bombus cullumanus* and *B. subterraneus*) in 1844, to S.C. Morgan who sent in a record (of *Xylocopa violacea*) just before the cut-off date in spring 2023. Thank you.

Underpinning *The Bees of Sussex* is a dataset of 59,153 records, each of which has come from an area of land that is managed by someone—perhaps a landowner, a farmer, a gardener, a conservation ranger or a volunteer. So, if you make space for nature, thank you.

The records themselves have been assembled and curated by the Bees, Wasps and Ants Recording Society (BWARS) and the Sussex Biodiversity Record Centre (SxBRC), and I am equally grateful to each of them. SxBRC, led by the unflappable Clare Blencowe, is a model of good data management and always supportive of the recording community.

Bob Foreman is a member of the team at SxBRC and a well-known figure in the Sussex biological recording community. Not only is he the creator of the maps used throughout this book, but he is also the technical genius behind the mapping tool that made the interrogation of the original dataset, containing 96,200 unchecked records, achievable. Throughout the period of the production of *The Bees of Sussex*, he has been endlessly patient and cheerful, no matter how many times I have asked for his help.

There are many people who may have had more emails, phone calls and visits from me than they would have wished for, but who, nonetheless, have been so generous with their time and with their expertise. Foremost among these is Mike Edwards, who couldn't be a better mentor and inspiration—to me and so many others. Thank you also to his wife Sue and son Rowan who I have enjoyed getting to know over my many visits. With his pivotal role in so much that relates to bees, wasps and ants in Britain, Mike is the ideal person to provide the foreword for this book, while Rowan has devised and developed many of the systems that underpin the BWARS database.

Ian Beavis offered to undertake the enormous task of fact-checking the book, and I can't thank him enough for the care and thoroughness with which he undertook this task. Any errors that remain are, of course, mine. Ian also supplied a number of notable recent records, and he is the author of two reports on the bees and wasps on the Kent/Sussex border. Both of these contain a wealth of information on behaviour that has not otherwise been reported.

Clear and crisp writing is an attribute that I have always strived for, but never fully achieved under my own steam. Happily for me, Juliette Mitchell has read every single word and probed every sentence structure, pushing me to achieve a standard that my English teachers would have thought impossible.

Under-resourced and under-funded, museum curators and their volunteers are the unsung heroes of the nature conservation movement. So much new information has come to light thanks to the collections they maintain, and *The Bees of Sussex* is a much more complete endeavour as a result of their efforts. I have spent time in—or in some cases corresponded with—Bexhill Museum, Booth Museum of Natural History, Chilcomb House Museum, Cliffe Castle Museum, Hastings Museum, Herbert Art Gallery and Museum, Museum Wales, Natural History Museum, Oxford University Museum of Natural History and Portsmouth Museum. So, Richard Dickson, Phil Fossil, Philip Hadland, Joanne Hatton, James Hogan, Lee Ismail, Rebecca Lean, Gearóid Mac a' Ghobhainn, Stephen Miles, Joseph Monks, Bill Montgomery, Mark Pavett, Julian Porter, Wolfgang Ritter, Christine Taylor, Ross Turle and Ali Wells—thank you.

George Else is co-author with Mike Edwards of the *Handbook of the Bees of the British Isles* which has remained open on my desk for much of the last three years. It is scrupulously referenced and offers up a seemingly endless source of avenues to explore. Close reading of the handbook resulted in many questions being sent George's way, each of which was answered promptly and with even more observations. He has also supplied many new and important records.

A large volume of new data was supplied by Steven Falk, including many records from under-surveyed locations such as the Pevensey Levels. He has also very helpfully shared his extensive knowledge, particularly of *Nomada glabella* and *N. panzeri*, and has given permission for many of his wonderful photographs to be used to illustrate this book. His *Field Guide to the Bees of Great Britain and Ireland* and his Flickr pages are excellent identification resources and have been a constant source of information.

As well as supplying valuable records, Thomas Wood has enabled me to confidently include up-to-date understandings of *Andrena afzeliella* and *A. ovatula* as well as *A. scotica* and *A. trimmerana*. He also kindly shared his insights into the subtleties of identification.

Many knowledgeable friends and colleagues have helpfully clarified records, filled in gaps, supplied additional data or even double-checked specimens. Thank you, Geoff Allen, Kieran Anderson, John Badmin, Josh Baum, Chris Bentley, Rosie Bleet, Clare Boyes, Pete Brash, Jonty Denton, Scotty Dodd, Jacob Everitt, Andy Foster, Nikki Gammans, Chris Glanfield, Andrew Grace, Peter Greenhalf, Janine Griffiths-Lee, Mark Gurney, Matt Hamer, Flo Harman, Grant Hazlehurst, Peter Hodge, Martin Jenner, Richard Jones, Andy Jukes, Graeme Lyons, Amanda Millar, S.C. Morgan, Roger Morris, Glenn Norris, Nick Ostler, Nick Owens, Alice Parfitt, Mark Parsons, Andy Phillips, Colin Plant, Matt Smith, Marc Taylor, Andrew Warrington, Paul Westrich, Keith Wheeler, Adam Wright and Barry Yates. Thank you also to Martin Buckland, a Wiltshire botanist, for carefully checking the use of plant names—this was no small undertaking.

From the very beginning of this project, I resolved to enhance the text by sourcing the very best photography I could find. Thank you, then, to the photographers whose work you will find in these pages, listed here with the initials used to identify them within the text: Paul Adams [PA], Ian Beavis [IB], Owen Beckett [OB], Chris Bentley [CB], Ted Benton [TB], Matthew Berriman [MB], Rachel Bicker [RB], Rosie Bleet [RBl], Paul Brock [PB], Robin Crane [RC], Peter Creed [PC], Miles Davies [MD], Jeremy Early [JE], Mike Edwards [ME], Tim Faasen [TF], Steven Falk [SF], Nick Feledziak [NF], Michael Fogden [MF], Chris Glanfield [CG], David Gould [DG], Peter Greenhalf [PG], Peter Harvey [PH], Louise Hislop [LH], Finn Hopson [FH], Michael Howard [MH], Evan Jones [EJ], Nigel Jones [NJ], Martin Kalaher [MK], Chris Kirby-Lambert [CKL], Lukas Large [LL], Graeme Lyons [GL], Mel Mason [MM], Kevin McGee [KMcG], Mark Monk-Terry [MMT], Glenn Norris [GN], Liam Olds [LO], Nick Owens [NO], Andy Phillips [AP], Ed Phillips [EP], Nigel Phillips [NP], Andrew Philpott [APh], Wilf Power [WP], Colin Pratt [C.R. Pratt collection], Anne Katrin Purkiss [AKP], Sam Roberts [SR], Paddy Saunders [PS], Nigel Symington [NS], Ian Tew [IT], Henk Wallays [HW], Paul Westrich [PW], Robin Williams [RW], Thomas Wood [TW] and Barry Yates [BY].

Thank you also to Derek Binns, Millie Brand, John Dominick, Krisz Fekete, Daniel Greenwood, Nicki Kent, Sam Moore, Richard Moyse, Jonathan Mycock and Nigel Powell, all of whom generously offered images which, if space had allowed, I would have loved to include.

Andy Daw, Pete Hughes, Jan Knowlson and Lois Mayhew each pointed me in the right direction—and to the right photographer—when needed. Emma Chaplin generously gave me access to Sussex Wildlife Trust's image library, and any photographs from their library are identified as [SWT]. The glossary of technical terms has been enhanced immeasurably by artwork produced by Geoff Allen and Nick Owens.

Thank you to seven organisations—BWARS, East Sussex County Council, South Downs National Park Authority, South Downs National Park Trust, SxBRC, Sussex Wildlife Trust, and West Sussex County Council—for recognising the importance of bees and getting fully behind *The Bees of Sussex*. It is all the better for their backing, their generosity, and their seal of approval.

Peter Creed, most importantly, is the clever and creative mastermind behind the editing, design and layout of the book. Thank you, Peter.

And last, but in this case definitely not least, thank you, Juliette and Wilf, for giving me the time and space to disappear into the world of bees. I'll be coming back to you soon, I promise.

James Power, October 2023

Foreword

In 1978, a small group of keen entomologists interested in ants, bees and wasps met up in South Devon for the first ever Bees, Wasps and Ants Recording Society field meeting. Some of those present—the older generation or 'greybeards'—are now long gone. The younger ones—Geoff Allen, George Else, Jeremy Field and me—are now the greybeards (or greying beards).

This group could not believe their ears when Malcolm Spooner, the supreme greybeard where county recording was concerned, announced to us at the end of the week that we had added more information about the ants, bees and wasps of Devon than he had ever managed in all the years he had lived in the county. There were just more of us, and we swapped our 'finds' each time we met up during the week.

The dedication of James, admittedly aided by improvements in communications, has done the same for my knowledge of the bees of Sussex. The power of the final work is worth more than the individual parts. However, it needs the conductor to make it play and James has been an excellent conductor for this project. Read the book, think about it, and enjoy the learning; but above all, go out and add to the sum of our knowledge.

Mike Edwards, October 2023

Introduction

I'm a lucky author. There are more species of bee recorded in Sussex than in any other county in Britain apart from Kent. It's 229 currently, but—because we don't know which of the three species of 'White-tailed Bumblebee' might be here—it could be as high as 231.

With a total of 274 species for Britain as a whole (not counting the extra 11 from the Channel Islands), that's a very respectable tally.

The high number for Sussex reflects three local characteristics: the richness and diversity of the habitats here, Sussex's status as one of the warmest and driest counties in the country, and the proximity of the continent. Sussex is also fortunate to have a strong tradition of biological recording which stretches back many years.

Two personal themes run through *The Bees of Sussex*: what information would I have wanted when I first set out to record bees in Sussex (because isn't this the information new recorders will be glad to have too?), and what additional information can I source for my own interest?

I've been challenged in my research by the many significant gaps in our collective knowledge. But observations made by amateur naturalists are busy plugging these, complementing and building on the research of more established academics. It's been hard to avoid technical language completely given the nature of the subject, and the emphasis is on scientific names as these are universally accepted, whereas common names are not (in fact, most common names for solitary bees were only invented within the last decade). I trust this is no barrier to your enjoyment, however, and my hope is that this book will encourage even more people to pick up the challenge of recording bees here—and, crucially, to share their observations on behaviour.

The Bees of Sussex is not intended as an identification guide, but rather as an introduction to what species are here, how they behave and how to find them. While this book does include some pointers on how to identify the different species, either in the field or through the microscope, two excellent publications—the *Field Guide to the Bees of Great Britain and Ireland* by Steven Falk and Richard Lewington and the *Handbook of the Bees of the British Isles* by George Else and Mike Edwards—cover this topic in great depth. For anyone wanting to dig down even deeper into behaviours, *Solitary Bees* by Ted Benton and Nick Owens is an invaluable resource.

The study of bees is an evolving science, and I've dug into a wide range of sources, from nineteenth-century journals to the very latest observations by recorders in the field. I've also had to range beyond our county borders, and even across the English Channel, to fill in local understanding and give insight into behaviour. But although this gives us a fuller picture for now, it's no substitute for local, Sussex-based observations and study, because behaviour can—and does—change from place to place. With climate change driving the movement of new species from the continent, and established species also changing their range, there is ample scope for new discoveries.

I've expanded my own knowledge thanks to my work on *The Bees of Sussex*, and it has been a steep but rewarding learning curve. So, if you learn as much from using it as I have in writing it, this book will have done a fine job. And, whatever your expertise, I very much hope it meets your expectations, personal enthusiasms, and level of interest.

The state of Sussex's bees

The distribution, density and abundance of different bee species is constantly in flux, their populations responding to a multitude of factors. Among the most significant are changes to the extent and quality of habitat, evolving land-management practices, increasing urbanisation and shifting weather patterns. As these have altered over time, so have the bee populations found in Sussex.

For much of the past 100 years or so, the broad pattern for many species has been one of decline, but that does not give the full picture, which is actually more complex. Some species have maintained broadly stable populations and others have expanded their range, in many cases establishing themselves in Sussex for the first time—with a proportion of these new species being recent arrivals in Britain.

These different responses to the changing environment reflect differences in behaviour or ecology of the individual species, with those that have more generalist behaviour—for example, foraging from a wide suite of plant species—better able to adapt and many of those that have more specialist behaviours—such as collecting pollen from a narrow suite of plant species—often proving to be more vulnerable.

In the case of those species that have declined, the major shifts in range and abundance have coincided with changes in the way much of our countryside is farmed, starting with the importation of guano as fertiliser in the late nineteenth century.

The most significant adverse impacts on bee populations occurred in the period between the end of the First World War and the late 1950s. Concerns about food security underpinned an effort to intensify land-management practices and, during the early years of World War Two alone, 3,200 hectares of turf-covered downland in east Sussex were converted to cereal production.

The process of intensification, including the widespread adoption of pesticides, has continued in subsequent decades, in tandem with the continuing abandonment of large areas of marginal land (as on the steep scarp slopes of the South Downs) or their conversion to other uses. It is estimated that there were just over 7,500 hectares of heathland in Sussex in 1813, which had reduced by over 90 per cent to just 644 hectares by 1991. Much of this has been left to naturally regenerate as woodland, or has been converted to conifer plantation or permanent pasture.

As a result, there have been dramatic improvements in food production at the expense of floristic and structural diversity across much of the countryside. This has had a significant effect on many bee populations and, while a precise cause and effect cannot always be ascribed to each extinction, the intensification of land management practices has coincided with the loss of the following ten species from Sussex:

- *Osmia xanthomelana* – last record prior to 1876
- *Melecta luctuosa* – last record prior to 1887
- *Nomada sexfasciata* – last record 1920
- *Eucera nigrescens* – last record 1941
- *Bombus subterraneus* – last record 1986
- *Halictus maculatus* – last record 1879
- *Andrena rosae* – last record 1919
- *Bombus cullumanus* – last record 1923
- *Bombus distinguendus* – last record 1971
- *Bombus sylvarum* – last record 1986

Others that have seen significant population declines, particularly in the latter half of the twentieth century, include the bumblebees *Bombus barbutellus*, *B. bohemicus*, *B. campestris*, *B. humilis*, *B. muscorum*, *B. ruderarius*, *B. ruderatus* and *B. soroeensis*, as well as species such as *Anthophora retusa*, *Nomada obtusifrons*, *Osmia pilicornis*, *Andrena labiata* and *A. tarsata*. One of these, *B. bohemicus*, may now be lost from the county altogether, with records showing that its population has retreated northwards.

Climate change in Sussex is already bringing more extreme heat, longer periods of drought, and more intense flooding—and all of these are set to occur more frequently. The effects of these different events will vary from species to species, but they are likely to have played a role in the decline of many of them, particularly the cool-adapted bumblebees. In the future, populations of all species will be significantly affected.

Cereal farming [PA]

Conifer plantation on former heathland at Graffham Common [MD/SWT]

The effects of climate change are complex and include shifts in the geographic range of different species, in some cases reducing or fragmenting populations, as well as impacts on the quantity, quality and timing of pollen and nectar production by flowering plants. The changing climate also affects the capability of bees to fly, forage and overwinter.

Increasing urbanisation is significant too, particularly through the direct loss of countryside to house construction and road building. Currently about 1.6 million people live in Sussex, more than double the number living here in the years before the Second World War. Much of this population growth has taken place along the coast as well as in and around Crawley and Horsham. The impact along the coast has been especially striking, with much of the c.40 miles of coastline between Pagham in the west and Cuckmere Haven in the east now developed.

The expansion of Sussex's human population, and the associated urbanisation of large areas of the county, has had a significant impact on the county's bee populations, although the effects are not always clearcut. While many populations have been lost completely, a number of species have adapted to these new urban environments and others have been able to take advantage of pockets of suitable habitat that have survived within urban areas (Lewes Cemetery being a good example).

Coastal development at Peacehaven in 1937 [C.R. Pratt collection]

The Bees of Sussex

Within this changing environment, nine new species have been recorded in Sussex since 2000—at a rate of about one every two years—many of which are likely to have shifted their range in response to climate change or because of the opportunities created by new urban areas. These are:

- *Bombus hypnorum* – first record 2002
- *Nomada lathburiana* – first record 2011
- *Nomada ferruginata* – first record 2016
- *Nomada alboguttata* – first record 2017
- *Stelis odontopyga* – first record 2020
- *Colletes hederae* – first record 2004
- *Colletes cunicularius* – first record 2015
- *Lasioglossum sexstrigatum* – first record 2016
- *Nomada zonata* – first record 2018

Overall, the rate of colonisation means that Sussex may soon support more species than it did 100 years ago.

Two other species, *A. vaga* and *Halictus eurygnathus*, have also been recorded in Sussex since 2000, after having been last seen here in the 1940s and classified as extinct in Britain. While the other species lost from the county are unlikely to return, there are a number that have experienced periods of decline but which more recently have recovered some of their former range. Among these is *Eucera longicornis*, a species once again encountered—with increasing frequency—on legume-rich grasslands across the High Weald.

The fluctuating status of Sussex's different bee species shows that the picture is more complex than is often portrayed. But what is clear is the significance of evidence-based efforts to secure better outcomes for nature. We need data—good data—to drive the recovery of nature, and the fact that so much is known about Sussex's bees is a tribute to the recording efforts of the many naturalists who have patiently compiled biological records over the years. As the populations of Sussex's bees continue to ebb and flow in the years ahead, the role of the biological recording community will remain every bit as important.

Four species that might be coming soon

While it is difficult to anticipate which species will arrive in Sussex next, these four are possible candidates:

- *Nomada bifasciata* takes over *Andrena gravida* nests and was first recorded in Britain in 2014. It has been recorded close to the Sussex border in Kent.
- *Osmia cornuta* has spread widely across southern England since it was first recorded in 2012.
- *Andrena ventralis* was recorded for the first time in Britain in Hampshire in 2023.
- *Sphecodes albilabris* is a cuckoo of a recent arrival in Sussex, *Colletes cunicularius*, which was first recorded here in 2015 and is now widely distributed in suitable habitat. In Britain, *S. albilabris* was first recorded in Suffolk in 2020 and has subsequently also been found in Oxfordshire.

Nomada bifasciata ♀ [CG]

Osmia cornuta ♀ [PW]

Andrena ventralis ♀ [TF]

Sphecodes albilabris ♂ [CKL]

An overview of Sussex

The seven landscapes of Sussex

Sussex is a diverse county encompassing distinct and contrasting landscapes—the low-lying Coastal Plain, the rolling hills of the South Downs, the densely wooded Low and High Wealds, the flat expanses of the Pevensey Levels and Romney Marsh, and the mixed geology of the Wealden Greensand. These seven different landscapes support rich habitats, which in turn support an important diversity of bee species.

As our subject is natural history, I am not concerned here with political boundaries, apart from the outer limits of West and East Sussex County Councils which define my area of interest. Nor have I concerned myself with the Watsonian division of the county into two Vice-Counties. If I talk about west and east Sussex, then, it is with a small 'w' and 'e', and my references are purely geographical.

To dive a little deeper into the characteristics of each of our seven landscapes:

- **The Coastal Plain** sits between the dip slope of the South Downs and the coast, narrowing as it runs from Chichester Harbour in the west to Brighton in the east. It is an area dominated by towns and cities—Chichester, Bognor Regis, Littlehampton, Worthing, Shoreham-by-Sea, Hove and Brighton. It is also an area of rich, fertile farmland.

    Within this landscape are a number of important sites for bees, including East Head and Thorney Island (both within Chichester Harbour) as well as Pagham Harbour, Medmerry and the sand dunes near Littlehampton. Notable species found within this landscape include *Colletes cunicularius*, *C. marginatus*, *Megachile leachella* and *Hylaeus pectoralis*.

The Coastal Plain: Chichester Harbour [PA]

The Bees of Sussex

- **The South Downs** forms a series of gently undulating high hills and dry valleys that run for some 60 miles from West Marden and Harting in the west as far as the sea cliffs at Beachy Head in the east. They are thinly settled, with Arundel, Steyning and Lewes the main settlements. The underlying geology is chalk, although deposits of sand occur in places along the boundary with the Coastal Plain, marking the location of an ancient coastline, and on the tops of the downs.

  The steep scarp slope is generally north-facing and is where most of the surviving areas of flower-rich chalk grassland are found—just under 5,200 hectares survive, occupying less than 5 per cent of this landscape. Much of the dip slope, which rises steadily from the coast, is cultivated for crops.

  The South Downs is an important landscape for bees, with the numerous areas of flower-rich chalk grassland, chalk heath, soft-rock cliffs, vegetated shingle, saltmarsh, woodland and cultivated field margins supporting many notable species, including *Anthophora retusa*, *Bombus humilis*, *B. muscorum*, *Nomada argentata*, *Osmia pilicornis*, *Andrena niveata*, *A. marginata*, *Halictus eurygnathus*, *Colletes cunicularius* and *C. halophilus*. Formerly, this landscape also supported such species as *B. cullumanus*, *B. distinguendus*, *B. subterraneus* and *B. soroeensis*.

  Places such as Heyshott Down and Escarpment, Kingley Vale, Rewell Wood, Castle Hill, Malling Down, Lullington Heath, Seaford Head and Beachy Head are all exceptional and feature regularly in the pages that follow. A number of these sites support species associated with sandy soils, as well as those typically found on chalk grassland.

The South Downs: Firle [FH]

The South Downs: distant view of Seaford Head from cliffs above West Quay, Newhaven [WP]

- **The Wealden Greensand** is the broad swathe of Upper and Lower Greensand and Gault Clay that runs from Hampshire, past Midhurst, Petworth and Pulborough, to just past Storrington in the east. This landscape is bisected by the River Rother.

The Bees of Sussex

Of greatest significance for the county's bees are the heathlands and acid grasslands of the Lower Greensand, a narrow ribbon of sandy soils running through the middle of this landscape. Among the important locations here are Iping and Stedham Commons, Graffham Common, Lord's Piece, Ambersham Common and Wiggonholt Common.

Many of the heaths on the Lower Greensand are hot and dry and support a large number of noteworthy species, including the Heather specialists *Andrena argentata*, *A. fuscipes* and *Colletes succinctus*, as well as their cuckoos *Nomada baccata*, *N. rufipes* and *Epeolus cruciger* respectively. Other species adapted to this environment are *Dasypoda hirtipes*, *Bombus jonellus*, *Coelioxys conoidea*, *Megachile maritima*, *Halictus confusus* and *Lasioglossum prasinum*.

Wetlands include Amberley Wildbrooks and Burton Mill Pond, locations that support populations of the wetland specialists *Macropis europaea* and *Hylaeus pectoralis*.

Wealden Greensand: Iping Common [SWT]

- **The Low Weald** cuts diagonally across the centre of Sussex, from its north-west boundary with Surrey down to the Pevensey Levels. The main settlements are Crawley, Horsham, East Grinstead, Billingshurst, Haywards Heath, Burgess Hill, Henfield, Uckfield and Hailsham. This area is largely underlain by stiff, sticky clay soils. The soil is hard to cultivate and is a major reason why so much woodland has survived here. Historically, much of the woodland was coppiced or grazed as wood-pasture commons, but woodland management has now largely ceased. Between the woods is a patchwork landscape of small fields—once species-rich meadows and pasture—and hedgerows.

The Low Weald: Furnace Meadow, Ebernoe Common [MMT/SWT]

The Bees of Sussex

Dotted through the landscape are important locations for bees, including isolated sandpits (in locations where the sandy belt of the Lower Greensand reappears at the surface) and areas of woodland, acid grassland and heathland. Among the significant species found are *Bombus ruderatus*, *Eucera longicornis*, *Nomada alboguttata*, *Andrena synadelpha*, *A. helvola*, *A. falsifica*, *Sphecodes scabricollis* and *Colletes cunicularius*.

Key sites include Ebernoe Common, Chailey Common, Ditchling Common, Abbot's Wood and Rowland Wood.

- **The Pevensey Levels** is a large expanse of low-lying land sitting between Eastbourne, Hailsham and Bexhill, and was originally a shallow tidal bay. A flat landscape criss-crossed by ditches and drains, much of it is now coastal grazing marsh protected from the sea by a high bank of vegetated shingle. Behind the shingle there are small pockets of saltmarsh. While this landscape is well surveyed for other groups, it has been only patchily surveyed for bees.

  Areas of species-rich grassland, vegetated shingle and saltmarsh support species such as *Dasypoda hirtipes*, *Macropis europaea*, *Bombus muscorum*, *Eucera longicornis*, *Andrena proxima*, *Colletes halophilus* and *Hylaeus pectoralis*.

The Pevensey Levels [EJ]

- **The High Weald** is an undulating landscape of woodland, small fields, hedgerows and scattered farms, and holds strong reminders of the medieval history of this part of Sussex. It has a very diverse geology composed of alternating bands of sand and clay, with deeply incised stream valleys known as gills. It rises from the coast all the way to the heathland and woodland of Ashdown Forest, and even a little beyond. Facing the sea are extensive areas of vegetated shingle and soft-rock cliffs sitting either side of the towns of Bexhill and Hastings. Inland, the principal settlements are Battle, Hawkhurst, Wadhurst, Crowborough and Heathfield.

The High Weald: Old Lodge, Ashdown Forest [NS]

The landscape supports an important bee fauna, including populations of *Andrena thoracica*, *A. vaga* and *Colletes cunicularius* along the coast, and *Nomada guttulata*, *N. integra*, *Eucera longicornis*, *Osmia pilicornis*, *A. labiata* and *A. fulvago* inland. Thin bands of base-rich soils running through the landscape also support small outlying populations of *Osmia aurulenta* and *O. bicolor*—this is at some distance from the South Downs where they are more commonly found.

Key locations for bees include Fore Wood, Hargate Forest, Broadwater Warren, Old Lodge and Isle of Thorns (both in Ashdown Forest), the low coastal cliffs between Bulverhythe and Bexhill, plus the soft-rock cliffs, coastal grasslands and woodland at Hastings Country Park.

- **Romney Marsh**, in the south-east corner of the county, is a flat low-lying area that was formerly a shallow bay. Over time, this filled with shingle and silt, and it was subsequently drained to allow cultivation and grazing. The area between Pett and Rye Harbour is dominated by the most extensive area of vegetated shingle in the county, most of which is managed as Rye Harbour Nature Reserve. The town of Rye, just north of the nature reserve, is the area's main settlement. To the east is Camber Sands, an extensive mobile dune system. The tidal margins of the rivers support areas of saltmarsh, as does the nature reserve.

This is an important landscape for bees. At Rye Harbour Nature Reserve there are populations of the scarce bumblebees *Bombus humilis*, *B. muscorum*, *B. ruderarius* and *B. ruderatus*, as well as other bees including *Dasypoda hirtipes*, *Nomada alboguttata*, *N. ferruginata*, *Megachile leachella*, *M. maritima*, *Stelis phaeoptera*, *Andrena gravida*, *A. vaga*, *Colletes cunicularius*, *C. halophilus*, *C. marginatus* and *Hylaeus annularis*. The extinct bumblebees *B. subterraneus* and *B. sylvarum* were both last recorded in the county here, while *B. soroeensis* is still known from Lydd Ranges, just over the county boundary in Kent.

Romney Marsh: Camber Sands [BY]

These seven landscapes, each with their own geology, topography and patterns of settlement, underpin and reflect the diversity of bees across the county. But it is also worth taking a moment to focus on one of the unifying influences that stretches across them all.

This is of course the climate, which across Sussex is dominated by the weather systems that blow up the English Channel and by the proximity of the continent. The result is one of the warmest areas of the country—as well as one of the sunniest, especially along the coast—and one of the most fruitful places in all of Britain to study bees.

The Bees of Sussex

Assembling the data

The Bees of Sussex is underpinned by the combined efforts of all those who over the years have collected specimens, written up their observations in journals, or submitted biological records to one or other of the recording schemes. As a result, there are 59,153 records for the county, and these form the basis of the species accounts.

Arriving at this point has not been without its challenges. At various points during the project, the total number of possible records ranged from a peak of 96,200 to a low of 41,620. This is because existing datasets contained a number of duplications and errors, while there were also many records that had never been shared with any of the official recording schemes.

There have therefore been two main phases to the process:

1. Cleaning up the data

The core data for *The Bees of Sussex* was supplied by the Bees, Wasps and Ants Recording Society (BWARS) and the Sussex Biodiversity Record Centre (SxBRC). Merging these created a dataset of 96,200 records, which was reduced to 46,500 by removing matches so close that they were clearly duplicated records.

Thereafter, it was a question of reviewing the data species-by-species, line-by-line, to identify records that were clearly also duplicates but where the information underpinning the records was more ambiguous. This was not a straightforward process but was made possible with the aid of a mapping tool developed by SxBRC. This enabled all records for a given grid reference to be interrogated simultaneously. In all cases, the record with the most precise information (i.e. with the best grid reference/date/place name/recorder name) was retained and the others set aside. This process resulted in 4,880 records being set aside, with 41,620 records accepted as 'good data'.

In several cases, there were also question marks over the identification of a species and, while I was often able to contact the recorder and ask them to check a specimen, this, for obvious reasons, was not always possible. As a result, a number of records were set aside because they could not be trusted.

2. Seeking out new data

Large numbers of potential records from before the advent of digital databases have never been captured in an accessible form. These might exist as a data label on a pinned museum specimen or as a note or observation in a journal. There is also often, even in our digital age, a time-lag between the recorder making a record and sharing it with one or other of the recording schemes.

Capturing these records has been a time-consuming process that has involved visits to museums, careful reading of numerous journals and books, and contact with many of the people actively recording in the county. To address these in turn:

- **Museum collections**. Among the museum collections reviewed, the most notable are those held by the Booth Museum of Natural History in Brighton and by Portsmouth Museum. The Booth Museum holds a very extensive collection of Sussex specimens, with material collected by entomologists such as William Unwin, Alfred Brazenor, Alfred Jones and John Felton. The 3,106 records from this museum, few of which had been catalogued, cover the period running from the mid nineteenth century up to the late twentieth century. Portsmouth Museum yielded 1,426 records, almost all the work of Henry Guermonprez. At both museums, the identification of the specimens was painstakingly checked and the information on the data labels entered into the growing database.

  Other museums that have generated records are Cliffe Castle Museum (Keighley), Chilcomb House Museum (Winchester) and Hastings Museum. Equally helpful were Bexhill Museum, Museum Wales, Oxford University Museum of Natural History and Herbert Art Gallery and Museum (Coventry), all of whom helped to resolve conundrums, even if they did not yield new records.

  Beyond these smaller collections, there has been a targeted approach. This has involved seeking out examples of species that could have been found in the county but for which there was no evidence. This resulted in the discovery of specimens of *Stelis phaeoptera* and *Andrena rosae*, which had not otherwise been recorded from Sussex, as well as additional examples of *Nomada sexfasciata*, all in the Natural History Museum.

In total, museum collections yielded 5,000 or so new records.

- **Literature review**. Reviewing the literature to locate records has entailed delving back into some of the earliest natural history publications, as well as those that have been published as recently as 2023. The earliest publications date from the 1840s, and many of these are available to view at no cost on the Biodiversity Heritage Library website.

  Among the key sources scoured for records were *The Entomologist's Monthly Magazine*, *The Hastings and East Sussex Naturalist*, *The British Journal of Entomology and Natural History*, *The Victoria History of the County of Sussex*, *The Handbook of the Bees of the British Isles* and *The Hymenoptera Aculeata of the British Isles*.

  Reviewing the literature has resulted in 570 additional records, including noteworthy references to *Bombus cullumanus*, *Melecta luctuosa*, *Nomada glabella*, *N. sexfasciata* and *Andrena simillima*.

  There are particular challenges associated with the interpretation of records gleaned from the literature, and early records in particular have had to be treated with great caution. This is because scientific names have changed significantly over the years, while many species have since been split into two or more species, and it was not always possible to state which species the recorder was referring to. There is also the question of the reliability of the recorder.

  For these reasons, a large number of historic records have been set aside and excluded, and only those from readily identifiable species, reported from a suitable location at the right time of year, and/or from a reputable recorder, have been accepted.

  While this approach will have excluded many of what could in time prove to be 'good' records, the intention has been to achieve a high degree of confidence in the records that have been included.

- **Modern records**. At the time of writing, a number of modern records had not yet been passed on to a recording scheme, and, through direct contact with recorders, an additional 11,900 records were located. These include records for *Nomada argentata*, *Megachile circumcincta*, *Osmia pilicornis*, *Stelis breviuscula*, *Halictus eurygnathus*, *Lasioglossum semilucens*, *Colletes cunicularius* and *Hylaeus pectoralis*.

In large part, *The Bees of Sussex* has been inspired by other local atlases such as *The Flora of Sussex* and *The Butterflies of Sussex*, but more especially by publications focused on the bees, wasps and ants of other counties and regions. My hope is that, in turn, *The Bees of Sussex* will reach beyond the county's borders and inspire others to record bees and, perhaps, to produce an atlas for their own county—with the process set down in detail above serving as a useful guide.

The total number of Sussex bee records—59,153—compares very favourably with other county aculeate atlases. *The Bees of Norfolk*, for example, is based on *c.*18,000 records; *A provisional atlas of the bees, wasps and ants of Shropshire* is based on *c.*14,000 records; the *Bees of Surrey* is based on *c.*45,000 records; and *Bees, Wasps and Ants of Kent* is based on *c.*55,000 records.

A challenge in researching and writing *The Bees of Sussex* has been ongoing work examining the DNA of certain species. This included research revealing that two new species—*Nomada glabella* and *Andrena afzeliella*—were present in Sussex and that two others—*A. scotica* and *A. trimmerana*—were much more challenging to distinguish than had been believed to be the case.

Work to investigate the DNA of certain species continues, and, given the differences in behaviour that have been observed in species such as *Epeolus cruciger* and *E. variegatus*, it is thought highly likely that additional, hidden species are present in the county. These two species alone may prove to be four distinct species.

Although there is a high level of interest in bees, most species are challenging to identify, and this has limited the number of people who have moved beyond a small number of species. All told, just 18 people have ever submitted 400 or more records and strayed beyond the readily identifiable.

Inevitably, this has resulted in gaps in coverage, particularly in locations such as the Low Weald, the Pevensey Levels, and the coast running west from Brighton. There are also differences in coverage within relatively well-recorded landscapes such as the South Downs, with much of the effort focused on the best habitats and with other areas such as arable field margins frequently neglected. As naturalists, we are all drawn to our favourite places.

Although recording effort has improved significantly since his day, this is the same issue that was flagged by Edward Saunders, writing in 1905. He stated that "the list of the Aculeata of this county hardly compares well with that of either Surrey or Kent" and went on to say that although "the coast has been visited and worked by numerous entomologists, the centre and north, judging from the paucity of records, have been much neglected". He was confident that Sussex would ultimately compare well with these other counties, once better coverage had been achieved.

While the level of coverage across Sussex has indeed increased in the intervening years, significant gaps remain, and it is hoped that *The Bees of Sussex* will be a spur to greater recording effort—both in Sussex and beyond.

The recording community

Recording bees close to Gatwick Airport (l–r Josh Baum, Amanda Millar, James Power, Alice Parfitt and Krisz Fekete) [RB]

The recording community in Sussex is active, welcoming and generous. You may notice, however, a lack of diversity in the roll call of notable recorders below. Those who have generated the most records —or been pioneers in the field—have been, until now, all men. But there are many women, and people of all backgrounds, making great strides in the field, with Rosie Bleet, Krisz Fekete, Nikki Gammans and Alice Parfitt immediately springing to mind. The aspiration is that by the time the next iteration of *The Bees of Sussex* is published this will be a very different list.

William Unwin (1811–1887, 79 records) is one of the earliest known hymenopterists who worked in Sussex. He lived in Lewes and died there in straightened circumstances. He was a well-known local naturalist who collected a wide range of specimens, including birds ("both shooting and stuffing them", as was the way at this time), flowering plants, flies, beetles, wasps and bees.

In 1858, he contributed a paper entitled "A list of the insects observed in the southern parts of the county of Sussex" to *The Naturalist* magazine. In this, he produced part one of the first list of bees for the county. Unfortunately, this article concludes with the words "To be continued" but wasn't—the magazine folded before the next edition and the concluding section of the list was never produced.

Equally frustratingly, much of his collection, which is in Brighton's Booth Museum, does not have date or location labels attached. Despite this, he has contributed a number of significant records.

Henry Guermonprez (1858–1924, 1,126 records) was a significant Sussex naturalist at the end of the nineteenth century and beginning of the twentieth century. He collected widely, and his collection includes birds, fossils, plants, molluscs, marine invertebrates and bees. He was, apparently, a fine taxidermist.

He collected mostly from near his home in Bognor Regis and from the west Sussex countryside, and a large proportion of his aculeate specimens were collected in his garden. This supported a remarkable diversity of bumblebees, including *Bombus muscorum*, *B. subterraneus* and *B. sylvarum* which are now either extinct or very scarce. His collection is in Portsmouth Museum, and, happily, nearly all have a location and date label attached.

Edward Saunders (1848–1910, 137 records) was a renowned hymenopterist mostly based in Surrey. He is the author of several important publications, including the very influential *The Hymenoptera Aculeata of the British Islands*, published in 1896. He also wrote the chapter on bees, wasps and ants in *The Victoria History of the County of Sussex*, published in 1905. This includes an early list of the species recorded in the county.

He was a frequent visitor to Sussex, especially to Hastings, but also to Bognor Regis, Littlehampton and Worthing. These visits generated important records, including one of the few British records for *Halictus maculatus*. His collection is in the Natural History Museum.

Brazenor Bros (888 records) was established as a family business in Brighton by Robert Brazenor in the late nineteenth century, and their shop sold preserved and mounted wildlife specimens, many from the local area. Brazenor's style of taxidermy was described as "more enthusiastic than strictly accurate" and, on one occasion, extended to a 20 metre-long Rorqual, a species of whale. This was allowed to decay on Race Hill, Brighton, before the skeleton was displayed on Boscombe Pier, Bournemouth.

When the business closed in 1967, the business's large collection of bees and wasps was acquired by the Booth Museum. Many of these specimens had been collected by his son Alfred, and these include species that are now extinct or very scarce. Helpfully, as well as the location and the date, the labels often give observations on behaviour.

Gerald Dicker (1913–1997, 2,369 records) was an old-school gentleman naturalist who was known to protect his insect collections from the depredations of the museum beetle with DDT. He was also happy to blow smoke from his cigarette through a biro onto tree bark to see what specimens would emerge, especially on rainy days when other opportunities to collect specimens were limited.

For much of his life, he worked at the East Malling Research Station, where he became Head of Zoology. Weather permitting, he would collect specimens most days and, although based in Kent, he made many important Sussex records. His collection is in the World Museum in Liverpool.

Alfred Jones (1929–2014, 1,230 records), known as 'Wilberforce' to many, moved to Newhaven in 1965. First and foremost an expert botanist, he always carried an insect net and found the time to collect a large number of insect specimens, including bees, supporting each record with a detailed note in one of his notebooks. Fortunately, these notebooks are with his well-labelled insect collection in the Booth Museum. One specimen (labelled "*A. clarkella*, number 3879") is even accompanied by the note "PS, caused a broken arm".

John Felton (1930–1994, 415 records) was known mainly for recording aculeates in Kent and on the continent. He worked as an entomologist at Shell Research, but in his free time devoted many hours to recording and studying bees and wasps. He became County Aculeate Recorder for the Kent Field Club until his work took him to The Hague. He authored many papers, including an account of the European species of wasp in the genus *Nitela* published after his death.

He moved to Brighton upon retirement and, although he only lived in Sussex briefly, he still had time to record a good number of aculeates here, especially near his home in Patcham. His collection is in the Booth Museum.

George Else (1,895 records) has been an entomologist since his days at primary school in Portchester, Hampshire. It was not until he started work at the Natural History Museum in 1969, however, that his interests moved from butterflies and the larger moths to bees. Although it was intended that he would join the museum's Lepidoptera team, the then head of Hymenoptera was desperate for someone to assist him on aculeates and, as soon as he arrived, he was drafted to the Hymenoptera Section to take up these duties, particularly working on bees. He was told on day one that as the bee collection was so vast further staff would be recruited. Nearly 40 years later he was still the sole bee curator.

Alongside his work for the museum, he started work on a book on Britain's bees. The *Handbook of the Bees of the British Isles*, co-authored with Mike Edwards, was published in 2018. This two-volume book is the definitive account of the bees found in the British Isles.

Mike Edwards (13,581 records) is responsible for more than a fifth of all of the county's bee records. He has been at the forefront of the study of bees, wasps and ants since the 1970s, working closely with government, nature conservation charities, landowners and universities, both in Britain and abroad. He is a founder member of BWARS, has written numerous articles and publications, and is co-author with George Else of the *Handbook of the Bees of the British Isles*. His first collection of aculeate specimens is in Liverpool's World Museum.

Peter Hodge (773 records) started his working life as an engineer for British Telecom before switching to become an ecological consultant. This was after a childhood interest in entomology had been rekindled by a cousin who had asked him where to find butterflies. He is particularly known for his work on beetles and has accumulated a collection of some 18,000 British specimens, many of which are new to Britain, plus some 11,000 from countries across Europe. In between times, he has found time to work on many other insect groups, including bees.

Andrew Grace (1,199 records) is based in Bexhill. He developed his passion for aculeates by attending courses that were organised by BWARS, and with support from George Else, Mike Edwards, Chris O'Toole and others. He has spent time in Greece, working on the ALARM Pollinator Decline Programme, and as a volunteer in Asia and the south of France. In 2010 he authored the book *Introductory Biogeography to Bees of the Eastern Mediterranean and Near East*. Much of his recording is now centred on east Sussex and Kent. His collection is in Bexhill Museum.

Ian Beavis (2,132 records) has had an interest in the insects of the Tunbridge Wells area since his primary-school days. Like many, he started by studying butterflies and moths before settling on aculeates in the late 1980s. He studied Classics at Exeter University and, in 1988, published his doctoral thesis *Insects and other Invertebrates in Classical Antiquity*. In 1985 he joined Tunbridge Wells Museum and Art Gallery, where he is now Research Curator.

Most of his recording is focused on sites in the High Weald in Kent and east Sussex, with the results of his surveys summarised in two important reports published in 2002 and 2007. These include numerous first-hand observations on the behaviour of the different species recorded, including many behaviours that have not otherwise been published.

Steven Falk (3,824 records) is responsible for a remarkably high proportion of the county's bee records, including many that are notable, even though he has only ever been a visitor to the county. He first came to prominence because of the quality of his artwork, and at 15 he had already started work on the colour plates for the ground-breaking *British Hoverflies*. His passion for nature led to a career with the Nature Conservancy Council, and then Coventry and Warwickshire Museums, Buglife and, latterly, as a freelance ecologist.

He has published a number of books and journal articles, including *A Review of the Scarce and Threatened Bees, Wasps and Ants of Great Britain* in 1991 and the hugely influential *Field Guide to the Bees of Great Britain and Ireland* in 2015. Many of his superb photographs of Britain's bees can be viewed on his Flickr account, supported by information on key aspects of behaviour and key identification features.

Peter Greenhalf (459 records) has been interested in nature all his life. His particular interest in bees developed in 2010 when he attended a talk by Nikki Gammans, and he then began volunteering with the Short-haired Bumblebee Reintroduction Project, undertaking survey work across Romney Marsh. As his interest has expanded to include solitary bees and wasps, he has also become enthusiastic about recording at Rye Harbour Nature Reserve and nearby. He and his wife, S. C. Morgan, currently give talks and teach bumblebee identification courses.

James Power (3,691 records), the author of this book, first became entranced by bees and wasps when a Hornet landed at his feet while drinking a cup of tea and eating a biscuit at Sussex Wildlife Trust's offices at Woods Mill in 2010. His role as Director of Land Management was limiting the time that he had to enjoy nature, so the arrival of this Hornet was the perfect trigger for a new obsession.

His career as a nature conservationist has lasted almost 40 years and has included spells with two different Wildlife Trusts, Defra, the Severn Gorge Countryside Trust, the National Trust, and even the National Parks department in Malawi.

Chris Bentley (1,073 records) grew up near Middlesborough and became hooked on invertebrates as a small child when his aunt showed him her school's butterfly collection. He went on to study zoology and invertebrate ecology at university. He has subsequently worked in a variety of nature conservation roles across Britain, including a spell on the Skerries and, for many years, leading the biological monitoring and recording programmes at Rye Harbour Nature Reserve. His interest in invertebrates encompasses beetles, moths, spiders and, especially, flies. He is currently working as a freelance entomologist and is the county Diptera recorder.

Graeme Lyons (3,208 records) started keeping a record of the wildlife he saw around him as an eleven-year-old growing up in Staffordshire. Although he studied astrophysics at university, he has pursued his passion for nature as a career ever since. He has worked as an ecologist for both the RSPB and the Sussex Wildlife Trust, and is now an ecological consultant based in Brighton. He is a recording phenomenon, recording some 8,345 species across all groups and generating a total of 183,000 biological records. He is the county recorder for Heteroptera and spiders, and has also found time to record 163 bee species in Sussex, contributing the fourth most records for this group.

Andy Phillips (919 records) is an ecologist and naturalist based in Hastings. He currently works as an ecological consultant, and he previously worked for Hastings Borough Council as the Local Nature Reserve Officer. As well as recording bees and wasps, he was the County Spider Recorder and is now Regional Co-ordinator for the British Arachnological Society.

Thomas Wood (1,015 records) became interested in aculeates in 2012, learning the skills of solitary-bee identification from David Baldock and Mike Edwards. He then put his knowledge to good use with a PhD at the University of Sussex on farmland bees. This entailed extensive fieldwork across Sussex and Hampshire and meant that he was able to collect material from sites that are largely inaccessible to the public. He is now based at the University of Mons in Belgium, and his particular area of interest is bee taxonomy across the West Palaearctic, with a strong focus on the hyper-speciose genus *Andrena*.

Explanation of the species accounts

The following accounts cover all the species that are known to have been present in Sussex. These are laid out alphabetically within their genera, which are set down according to the most recent thinking on ancestral relationships between them, mirroring the approach taken in the *Handbook of the Bees of the British Isles*.

An introduction to each genus summarises common aspects of behaviour, and the species accounts then follow, with each one split into five sections:

- **Geography and history** sets out the distribution of each species in Sussex and, for noteworthy species, their national distribution (usually in the context of Great Britain—or 'Britain'—rather than the whole of the British Isles). This section also suggests likely locations to search and highlights significant records as well as wider aspects of their geography and history.
- **Out in the field and under the microscope** summarises a few of the fieldcraft techniques which can be used to locate a species on site, as well as pointers on how to identify each species (though this book does not aspire to be a comprehensive identification guide).

  It should be said that, unlike for birds, butterflies and plants, identifying most species of bee without the aid of a microscope is challenging. Key features to look for are explained—whether in the field for the more recognisable species, or through the microscope for the majority.

  The use of some technical terms has been unavoidable, and these are explained in the glossary below. For ease of use, the meanings of 'thorax' and 'abdomen' differ from what is strictly correct: one segment, the propodeum, is excluded from the abdomen (where it really belongs) and added to the thorax (where it does not).
- **Behaviour and interactions** describes key aspects of behaviour. This is an evolving science, and there is ample scope for new discoveries in this area.
- **On the wing** explains the flight period for each species, where possible based on the earliest and latest dates in the Sussex data since the year 2000. Where there are too few records, the dates are taken from the *Handbook of the Bees of the British Isles*, which is also the source for the end date of the first brood and the start date of the second brood in double-brooded species.
- **Feeding and foraging** is a summary of known pollen and nectar preferences. English names are used for both plant families and species. Family names are capitalised, as in 'Daisy family', and where the specific species is known, names are also capitalised, as in 'Common Ragwort'. Where the species is not known, the names are given in lower case, as in 'ragwort', and could encompass a number of possible species.

Most species accounts include two maps showing distribution—one for the period up to and including 1999, and a second covering the period from 2000 up to and including April 2023. These are based on data gathered by BWARS and SxBRC and shared in line with their data management policies. In all cases, distribution is plotted against the seven broad landscape areas found in Sussex, with records displayed as monads (one-kilometre squares).

In a number of cases, an additional layer of information is provided to highlight a key relationship underpinning the distribution of a species. This could be a close relationship between a species of bee and a particular species of plant, or a specific relationship between a parasite and its host. Examples are *Andrena hattorfiana* which forages for pollen largely from Field Scabious (where Field Scabious is also mapped), and *Coelioxys conoidea* which parasitises *Megachile maritima* (where *M. maritima* is also mapped).

The species accounts close with eleven species no longer found in Sussex, one of which was probably never here at all. These shorter accounts focus on just 'Geography and history', and include a single distribution map covering the period up to the end of 1999.

Glossary

Abdomen (figure 1) Third major body segment often with hair bands running across the body.

Antenna (figure 3) One of a pair of segmented appendages attached to the front of the head. Females have 12 antennal segments and males 13.

Basitarsus (figure 5) First of five segments that make up the tarsus on each leg. Attached to the tibia.

Basitibial plate (figure 5) Small plate at the top of the hind tibia in both sexes (but not in all species).

Caste Two forms of a female of the same species, an egg-laying queen and a worker.

Clypeus (figure 3) A plate that is on the front of the head below the eyes and above the labrum.

Daughter-queen Female bumblebee offspring that overwinters after mating and emerges in the spring to establish a new colony.

Eusocial A species that has a founding female and worker(s).

Femur (figure 5) Third segment of each leg.

Galea (figure 3) Blade-like structure that flanks the mouthparts.

Gena (figure 2) The sides of the head behind the eyes.

Genital capsule (figure 4) Male genitalia (the subtleties of which are beyond the scope of this book).

Labrum (figure 3) Hinged plate below the clypeus.

Mandible (figure 3) Jaw below the clypeus.

Marginal area (figure 1) Rear surface of each tergite.

Ocelli (figure 3) One of three simple eyes near the top of the head, arranged in a triangular pattern.

Pronotum (figure 2) First segment of the thorax, attached to the front of the scutum.

Propodeum (figure 1) Rear section of the thorax.

Propodeal triangle (figure 2) Central area at the top of the propodeum, often triangular or semi-circular in shape.

Pygidium (figure 2) A narrow plate at the tip of the abdomen, most often in females but also in some males.

Scape (figure 3) First segment of an antenna, attached directly to the head.

Scutellum (figure 2) Third plate on top of the thorax, behind the scutum.

Scutum (figure 2) Second plate on top of the thorax, behind the pronotum and before the scutellum.

Sternite (figure 4) One of a number of plates making up the underside of the abdomen—six in females (numbered 1–6), seven in males (figure 4, numbered 1–7).

Stigma (figure 6) Darkened, thickened area on the front edge of the forewing.

Sub-marginal cell (figure 6) One of two or three cells towards the front edge of the forewing.

Tarsus (figure 5) Fifth segment of each leg, itself made up of five segments, the first of which is the basitarsus.

Tergite (figures 1 & 4) One of a number of plates making up the upper surface of the abdomen—six plates in females (figure 1, numbered 1–6), seven in males (figure 4, numbered 1–7).

Thorax (figure 1) Second major body segment, between the head and the abdomen.

Tibia (figure 5) Fourth segment of each leg, attached to the end of the femur.

Tibial spur (figure 5) Spine attached to a tibia, often in pairs.

Trochanter (figure 5) Second segment of each leg, attached to the femur.

Figure 1: Female

- Head
- Thorax
- Propodeum
- Abdomen
- Marginal area
- Tergite
- 1, 2, 3, 4, 5, 6

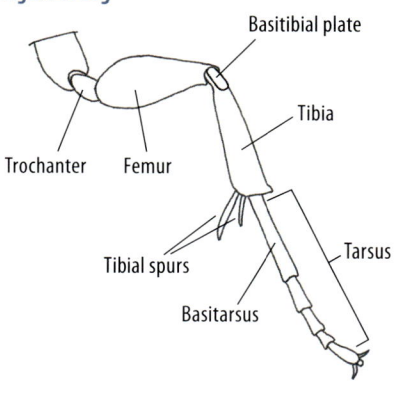

Figure 2: Female

- Pronotum
- Gena
- Scutum
- Scutellum
- Propodeal triangle
- Pygidium

Figure 3: Female head

- Ocelli
- Eye
- Scape
- Clypeus
- Labrum
- Mandible
- Galea
- Antenna
- Antennal segments 1, 2, 3, 4, 5, 6, 7, 8, 9, 10, 11, 12

Figure 4: Male abdomen

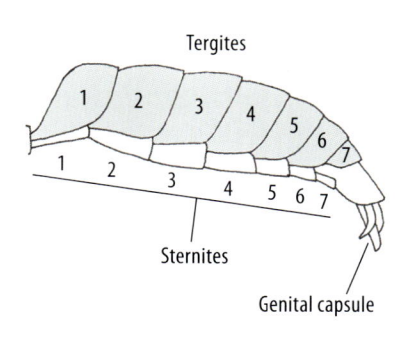

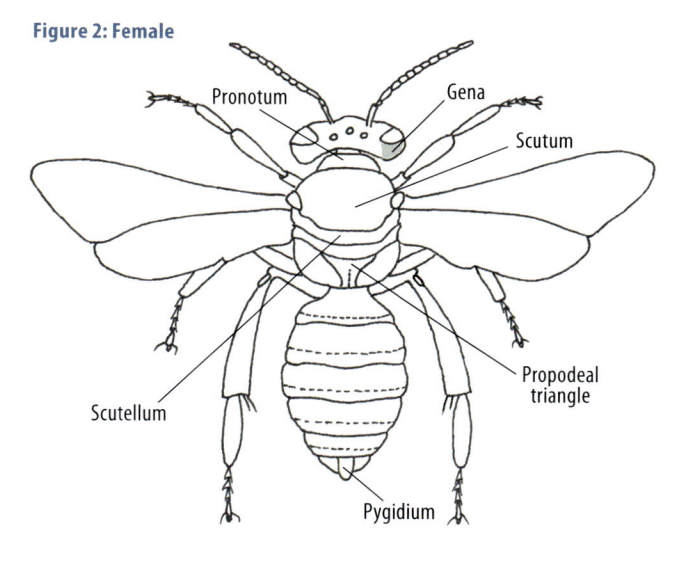

- Tergites 1, 2, 3, 4, 5, 6, 7
- Sternites 1, 2, 3, 4, 5, 6, 7
- Genital capsule

Figure 5: Leg

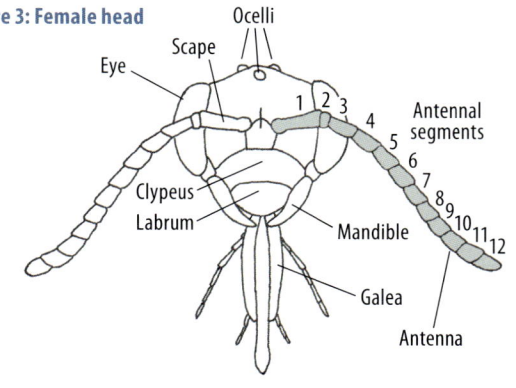

- Basitibial plate
- Tibia
- Trochanter
- Femur
- Tibial spurs
- Basitarsus
- Tarsus

Figure 6: Wings

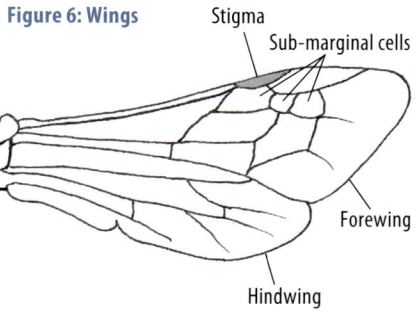

- Stigma
- Sub-marginal cells
- Forewing
- Hindwing

The Bees of Sussex

Dasypoda – Pantaloon Bees

Dasypoda are ground-nesting solitary bees found across Europe, Asia, North Africa and the Middle East. Fifteen species are known from Europe, just one of which occurs in Britain and is found in Sussex, *Dasypoda hirtipes*. This genus, unlike many others, does not appear to be attacked by cuckoo bees.

Dasypoda hirtipes Pantaloon Bee
FIRST RECORD 1876, Littlehampton (Frederick Smith) TOTAL RECORDS 135

Geography and history
Since it only thrives on light, sandy soils, *Dasypoda hirtipes* is most frequently recorded on heathlands and sandpits on the Lower Greensand, as well as areas along the coast. It has also been found inland in other areas with sandy soils, such as Ashburnham Place where it was found in 2022.

Iping and Stedham Commons and Rye Harbour Nature Reserve are reliable places to search for this bee, although any locations with sandy soils are also worth a look. It has, for example, been recorded in a sandpit not far from Lewes.

Iping Common supports a strong population of *Dasypoda hirtipes*. [NS]

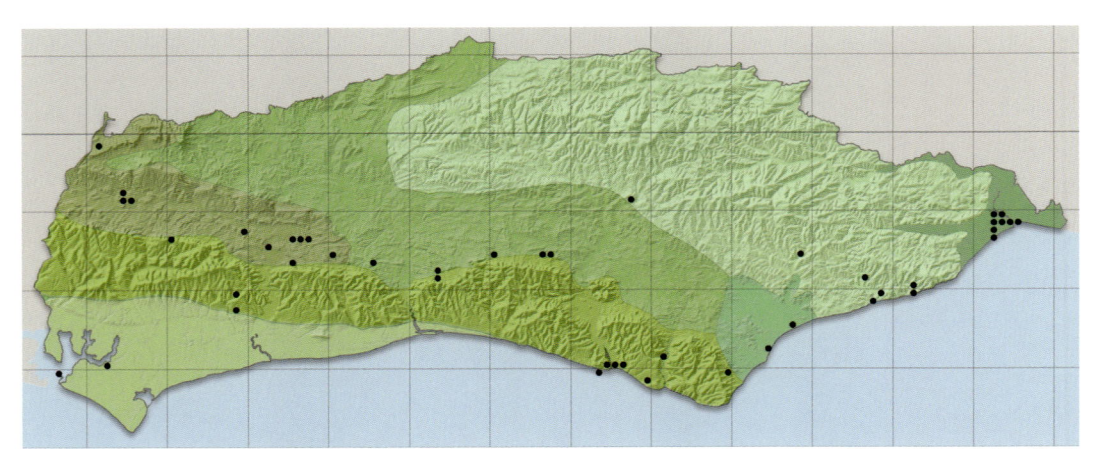

Large map: distribution 2000–2023
Small map: distribution 1844–1999

20 *The Bees of Sussex*

Dasypoda hirtipes ♀ [SF]

Dasypoda hirtipes ♂ [SF]

Out in the field and under the microscope
D. hirtipes is best searched for by checking yellow-flowered plants in the Daisy family, although it will also use other plants in this family. This is quite a large bee, and the combination of its size and the long, orange, plumose hairs on the hind legs means that the females are very distinctive. The males also have long hairs on the hind legs, but their colouring is duller overall and soon fades to grey.

While *D. hirtipes* is frequently found by checking foodplants for foraging bees, it can also be seen close to its nest sites in areas of sandy and thinly vegetated level ground. In 2011, a female was swept on a closely mown road verge near Iping Common in an area of sandy soil with 12 fresh burrows nearby.

The females frequently establish large, obvious aggregations. In using the long pollen-collecting hairs on the hind legs (as well as the pygidium) to clear the sand from the nest, they create nest entrances that are often excavated at an angle to the surface. This can result in distinctive, fan-shaped deposits of sand around the nest entrances.

Behaviour and interactions
Males fly rapidly just above the surface of a nesting aggregation while the females excavate and provision their brood cells. Females returning with a full pollen load are very quick to head down their tunnel entrance. Several females may share the same nest entrance and seal the entrance overnight or when the weather is poor.

The long hairs on the hind legs of the females enable them to transport large quantities of pollen. When they deposit the pollen loads in their brood cell, the females shape these into a 'loaf' with three conical projections. This is to reduce contact with the sides of the unlined cell, presumably to minimise the possibility of fungal infection or dampness affecting the provisions. Completed nests may have between seven and eight brood cells.

On the wing
Sussex records run from mid June until the beginning of September.

Feeding and foraging
Females collect pollen from a narrow range of plants in the Daisy family, and are reputed to principally forage from those with yellow flowers such as Cat's-ear, Common Ragwort, Common Fleabane and Smooth Hawk's-beard. However, they have also been observed foraging from purple-flowered plants in this family such as Black Knapweed and Spear Thistle, as at Ashburnham Place in 2022.

Dasypoda hirtipes ♀ on Field Scabious [MK]

Macropis – Yellow Loosestrife Bees

These ground-nesting solitary bees are found across Europe, Asia and North America. The larvae need floral oils plus pollen collected from Yellow Loosestrife for their development. Females nest either singly or in small aggregations and line their brood cells with a waterproofing substance derived from loosestrife oils—this protects the cell contents from dampness in the soil and enables them to withstand some flooding. The oil is mixed with pollen to form a 'loaf' for the larvae to feed on. Just one species occurs in Britain and is found in Sussex, *Macropis europaea*.

Macropis europaea Yellow Loosestrife Bee
FIRST RECORD 1972, Sidlesham (George Else) TOTAL RECORDS 85

Geography and history
This is one of three species of bee that are closely linked to wetlands (the other two are *Colletes halophilus* and *Hylaeus pectoralis*). It is thinly distributed across the county with small clusters of records around wetland locations such as Arundel, Burton Mill Pond and Amberley Wildbrooks. The small number of records from the Pevensey Levels is almost certainly because of under-recording. It often occurs around garden ponds with Yellow Loosestrife.

Burton Mill Pond [SR/SWT]
BELOW **Yellow Loosestrife** [PC]

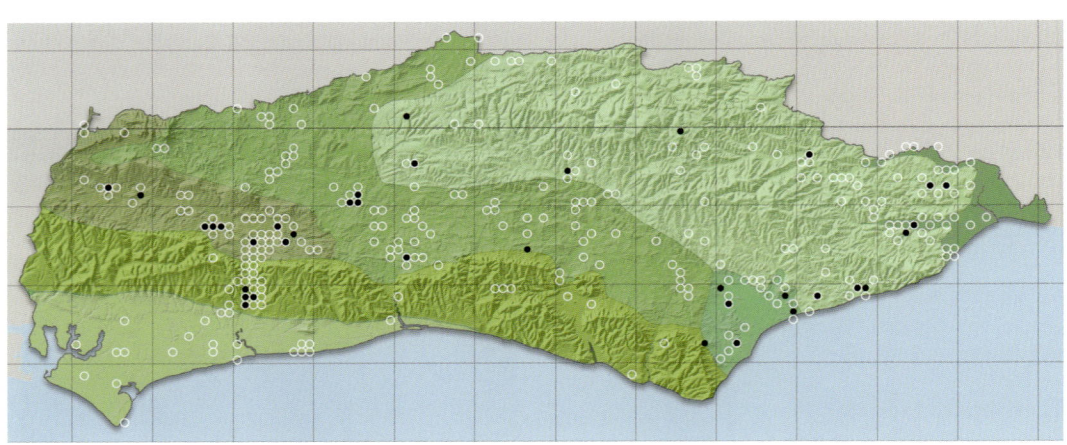

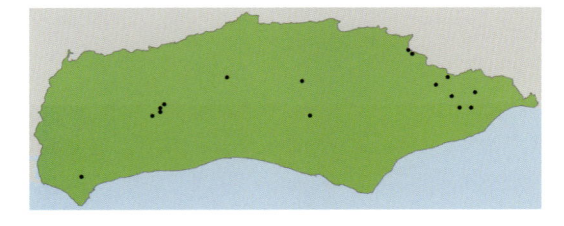

Large map: distribution 2000–2023, with white circles showing distribution of Yellow Loosestrife
Small map: distribution 1844–1999

The Bees of Sussex

Macropis europaea ♀ [BY]

Macropis europaea ♂ [PB]

Out in the field and under the microscope
As *Macropis europaea* is restricted to locations where its larval foodplant Yellow Loosestrife is found (places with shallow water and wet mud), it is worth checking any pond, fen, bog or river for this plant. When these are in flower, males and females might be seen flying rapidly around the flower heads or sheltering in and among the individual flowers.

These predominantly black bees have two sub-marginal cells on the forewings. Both sexes have very shiny abdomens and bands of white hairs on the hind margins of tergites three and four. Females have white pollen-collecting hairs on the hind tibiae, which contrast markedly with the black hairs on the very broad basitarsus. Males have extensive yellow markings on the face, yellow spots at the base of each mandible and very broad, bulging hind tibiae.

Behaviour and interactions
On emergence, males seek out patches of Yellow Loosestrife and start to patrol around the flower heads searching for females. Several males may establish themselves in the same place, competing for the chance to mate as females arrive to forage. The females have special adaptations to transport floral oils, including short, dense, velvety hairs on the front legs. These take up the protein-rich oils through capillary action. *M. europaea* can seek out Yellow Loosestrife over considerable distances.

Its close relative *M. fulvipes* and the cuckoo bee *Epeoloides coecutiens* are present on the near continent. This last species lays its eggs in the brood cells of *M. europaea* and *M. fulvipes*.

On the wing
Sussex records run from mid June until the end of August.

Feeding and foraging
While Yellow Loosestrife is the principal source of pollen and floral oils, adult bees need to visit other plant species too as Yellow Loosestrife does not produce nectar. Among the other species visited are agrimony, Alder Buckthorn, bird's-foot-trefoil, bramble, Creeping Thistle, Gypsywort, Great Willowherb, Hogweed, knapweed, Rough Hawkbit, sow-thistle, stitchwort, Water Mint and Water-pepper. Gypsywort is the only other plant species cited as an additional source of pollen.

Macropis europaea ♂s roosting on Yellow Loosestrife [APh]

The Bees of Sussex

Melitta – Blunthorn Bees

The 'Blunthorn Bees'—so named because of their truncated antennae—are ground-nesting solitary bees. The shape of the antennae can be a great help for identifying this genus. The *Melitta* resemble *Andrena* and *Colletes* but, among other differences, have an expanded final tarsal segment just before the claws at the end of each leg. Most specialise in collecting pollen from small groups of closely related plants or even from just a single species. There are four species of *Melitta* in Britain, with three found in Sussex. The fourth species, *Melitta dimidiata*, is restricted to Salisbury Plain in Wiltshire.

Melitta haemorrhoidalis Gold-tailed Melitta
FIRST RECORD 1968, Lewes (Kenneth Guichard) TOTAL RECORDS 64

Geography and history
Melitta haemorrhoidalis is largely restricted to chalk and sandy grasslands in southern England, although it is also known from as far north as Scotland. In Sussex, almost all the records are from flower-rich grasslands on the South Downs, and these are the likeliest places to find this bee—it is widely distributed here although it is rarely abundant. Of sandy grasslands, *M. haemorrhoidalis* was previously known from Sullington Warren and near Pett but has not been recorded at either of these places since 1997. Kingley Vale, Fairmile Bottom and Blackcap are good places to look.

Out in the field and under the microscope
It is worth investigating any places that support bellflowers, particularly areas of flower-rich chalk grassland but also places with sandy soils. David Porter has also noted that on one occasion he found females on Nettle-leaved Bellflower that was growing on the edge of arable fields near Ringmer.

In the Tunbridge Wells area, Ian Beavis reports that males spend a lot of time patrolling around Common and Musk Mallow plants, including in gardens. Mike Edwards has observed the same behaviour in west Sussex. In gardens, females might be seen collecting pollen from ornamental bellflowers. In all these areas, it is worth looking inside any 'bell' to see if a female is foraging inside or if a male has taken up residence for the night or during poor weather.

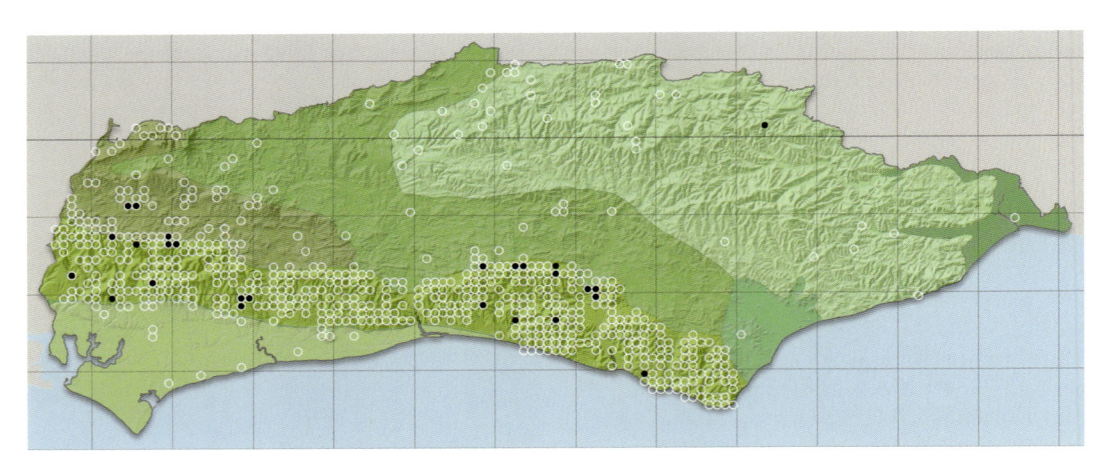

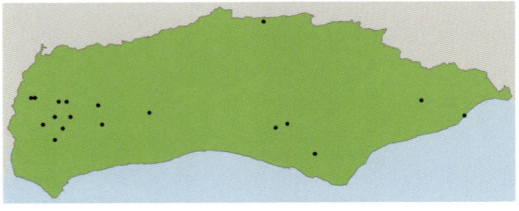

Large map: distribution 2000–2023, with white circles showing distribution of plants in the Bellflower family
Small map: distribution 1844–1999

Melitta haemorrhoidalis ♀ on Clustered Bellflower [PC]

Melitta haemorrhoidalis ♂ on Harebell [SF]

This bee is relatively straightforward to identify because of its association with bellflowers and because of the distinctive orange-haired tip to the abdomen in both males and females, although the colour does fade with age. Otherwise, both sexes are a dull brown overall. Males lack the white hair bands found on *M. leporina* and *M. tricincta*.

Behaviour and interactions
Males fly around bellflowers searching for females, and, once they have mated, the females are thought to establish their nests in small, loose aggregations. As they gather pollen, the females moisten it with nectar, forming a sticky lump attached to their hind legs. One European study showed that a little over 16 Harebell plants were required to provision a single brood cell.

Brood cells are targeted by *Nomada flavopicta* on the continent, and this is likely to be the case in Britain too.

On the wing
Sussex records run from the beginning of July until the end of August.

Feeding and foraging
While adult bees have been recorded visiting plants such as Great Willowherb, Hemp-agrimony, Jacob's-ladder and mallow, pollen is collected from plants in the Bellflower family such as Harebell, Clustered Bellflower and Nettle-leaved Bellflower. Round-headed Rampion, a locally frequent plant on the South Downs, is also a potential source of pollen.

Harebell [NS]

Melitta haemorrhoidalis ♀ collecting pollen from Harebell [PB]

Melitta leporina Clover Melitta
FIRST RECORD 1900, Eastbourne (Charles Nurse) TOTAL RECORDS 193

Geography and history
Melita leporina is widely distributed in Sussex but very local. It is found on grasslands with clovers (especially Red and White Clover) and other members of the Pea family. It even takes advantage of areas of species-poor grassland or improved pasture, as well as gardens and parks, if the right foodplants are present and if relaxed mowing regimes are in place.

Out in the field and under the microscope
It is worth checking any patch of clover. Here, females might be seen foraging for pollen or nectar, while males might be seen flying fast and low as they seek out females. Large groups of males (30 in one count) can sometimes be found close together on a plant stem or sheltering within a cluster of flower heads, especially at night or during poor weather.

Like other species in the genus, *M. leporina* superficially resembles an *Andrena*. However, the combination of truncated antennae and the expanded final tarsal segments help to distinguish *M. leporina* from all other species of bee except for the other species in this genus.

Of these, it most closely resembles *M. tricincta* and the two are difficult to separate. However, differences in flight season and the choice of foodplants are a very good guide. Other differences are subtle and include wider hair bands on the second, third and fourth tergites in *M. leporina* females. Males lack a band of short black hairs found on the second tergite in *M. tricincta*.

Behaviour and interactions
Males emerge from their brood cells first, followed soon after by the females. After mating, the females excavate nests in the ground (a wide range of soil types are exploited) and create brood cells that are lined with a wax-like hydrophobic coating and sealed with soil. In one report, between 12 and 15 brood cells were created within one nest. Sometimes there will be a handful of nests in an area of grassland, but in other cases there will be far more—in one example, around 200 nest burrows were found across an area of 10 m². Nests are sometimes established in lawns.

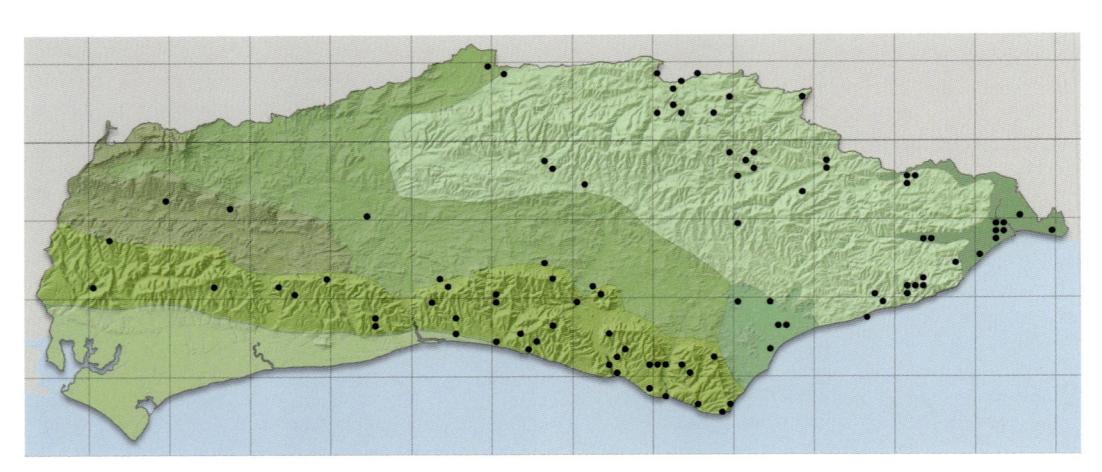

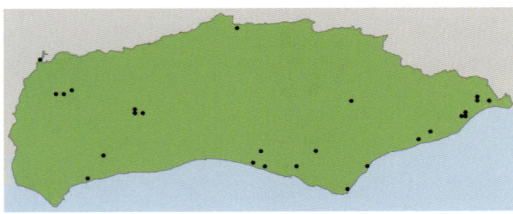

Large map: distribution 2000–2023
Small map: distribution 1844–1999

White Clover, Rye Harbour Nature Reserve [PG]

One study showed females travelling up to 60 m from the nest site to forage, completing around 16 to 18 flights a day gathering pollen. This is then moistened with nectar, prior to being carried back to a brood cell. Once a cell is fully provisioned, an egg is laid on top of the pollen mass and the cell is sealed with soil.

Brood cells are targeted by *Nomada flavopicta*.

On the wing
Sussex records run from the end of May until early August.

Feeding and foraging
Larvae need pollen from clovers and vetches to complete their development, and these plants provide most of the pollen collected (more than 91 per cent in one report). Most is collected from Red Clover and White Clover but also from restharrow, melilot, medick and bird's-foot-trefoil.

The adult bees have also been observed visiting a range of other plants, including bramble, dandelion, Musk Thistle and ragwort.

Melitta leporina ♀ [SF]

Melitta leporina ♂ [PC]

Melitta tricincta Red Bartsia Bee
FIRST RECORD 1951, Bosham Hoe (unknown) **TOTAL RECORDS** 189

Geography and history
Most Sussex records for *Melitta tricincta* are from the chalk grasslands of the South Downs, with just a small number of additional records from places such as Thorney Island, Stedham Common, Knepp, Haywards Heath and the High Weald close to Hastings. Within the South Downs, locations such as Seaford Head, Castle Hill, Blackcap and Malling Down have all generated recent records.

Although the key requirement is the presence of its sole pollen source, Red Bartsia, the distribution of *M. tricincta* does not match the distribution of this plant—Red Bartsia is locally very common over a much wider area of the county.

Red Bartsia growing along a track, Castle Hill [JP]
BELOW Red Bartsia [PC]

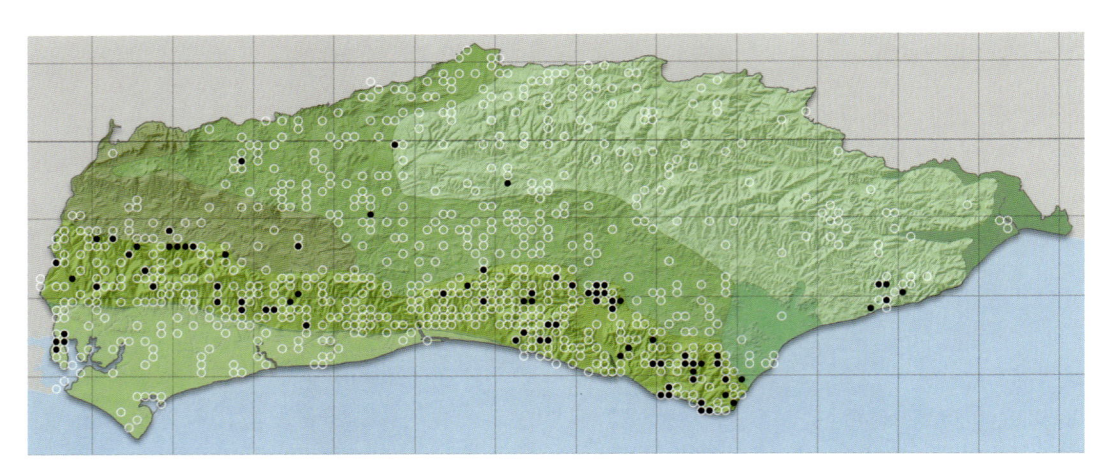

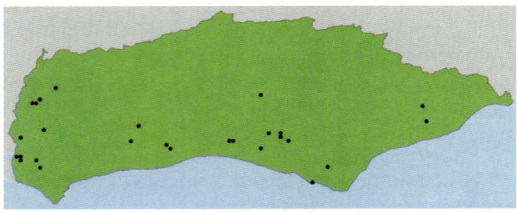

Large map: distribution 2000–2023, with white circles showing distribution of Red Bartsia
Small map: distribution 1844–1999

The Bees of Sussex

Melitta tricincta ♀ [PB]

Melitta tricincta ♂ [SF]

Out in the field and under the microscope
Any patches of flowering Red Bartsia are worth checking during the flight period, including locations away from the South Downs. Red Bartsia is a plant of rough grassland, track-sides and disturbed ground. Here, males might be seen flying rapidly just above the Red Bartsia plants or females might be seen foraging.

M. tricincta often flies with the very similar-looking *M. leporina*, a clover and vetch specialist—the choice of foodplants should help to identify the species, as does the later flight period of *M. tricincta*. This is most apparent in the males. Among other differences, female *M. tricincta* have narrower and whiter hair bands on the hind margins of the tergites, while males have a band of short black hairs running across the second tergite. These hairs are absent in *M. leporina*.

Behaviour and interactions
Little is known about the nesting habits of this bee, aside from the fact that it excavates its shallow nest burrow in the ground close to Red Bartsia plants. One study from the continent describes a nest at the base of a tussock of grass with a one centimetre-high mound of soil at the entrance. Males tend to be seen more often than females and fly fast and low close to patches of forage plants. Females will also move rapidly from plant to plant searching for nectar and pollen.

Emergence is synchronised with Red Bartsia coming into flower, with research showing that *M. tricincta* is a very efficient pollinator of this plant. As well as being the sole food source for the larvae, Red Bartsia is an important food source for the adults. Males frequently stop for nectar in between their fast, low-flying flights. A handful of other plants may also occasionally be visited for nectar.

It is likely that brood cells are targeted by *Nomada flavopicta*, although this has not been confirmed.

On the wing
Sussex records run from late June until early September.

Feeding and foraging
There are few records of this bee visiting plants other than Red Bartsia, which one study showed contributed 97 per cent of the pollen collected. The only other plant species cited in British literature are restharrow and Water Mint. While pollen is collected predominantly from Red Bartsia in Britain, on the continent *M. tricincta* also forages from related species such as Yellow Bartsia.

Melitta tricincta nectaring on Red Bartsia [PC]

The Bees of Sussex

Anthophora – Flower Bees

These are fast-flying, long-tongued solitary bees that superficially resemble bumblebees. They nest either in the ground, in deadwood, or in cliff faces and walls, often in large aggregations. Eggs are laid on a semi-liquid pollen 'loaf'. Males emerge first—in the case of *Anthophora plumipes* two to three weeks ahead of the females. In all species, the pollen is transported dry on the hind legs. *Coelioxys* and *Melecta* parasitise this genus. There are five species in Britain, all of which occur in Sussex.

Anthophora bimaculata Green-eyed Flower Bee
FIRST RECORD 1879, Hastings (Edward Saunders) TOTAL RECORDS 380

Geography and history
Anthophora bimaculata is restricted to southern England, with the most northerly record being from Norfolk. In Sussex, *A. bimaculata* is strongly associated with the sandy soils of the Lower Greensand and the High Weald south of Tunbridge Wells. It is also frequently found along the coast between Peacehaven and Beachy Head, and between Pevensey Bay and the dunes at Camber Sands.

There is also a handful of records from Pagham Harbour and the eastern South Downs away from the coast, including from a garden in Lewes. It can be abundant in locations with sandy deposits, such as along cliff edges and in sandpits.

Out in the field and under the microscope
This is the smallest species in the genus, and the males in particular have very obviously olive-green eyes. Unusually, both males and females have yellow-marked faces, although these markings are less extensive and not as bright in the females. The combination of its smaller size, the olive-green eyes and the pattern of the yellow facial markings makes this perhaps the easiest species in the genus to identify.

A. bimaculata can be found in good numbers and its presence is often first revealed by the shrill whine it makes in flight. They have a fast, darting flight, interspersed with periods of hovering accompanied by the very distinctive whine. It is often found feeding on plants such as Viper's-bugloss.

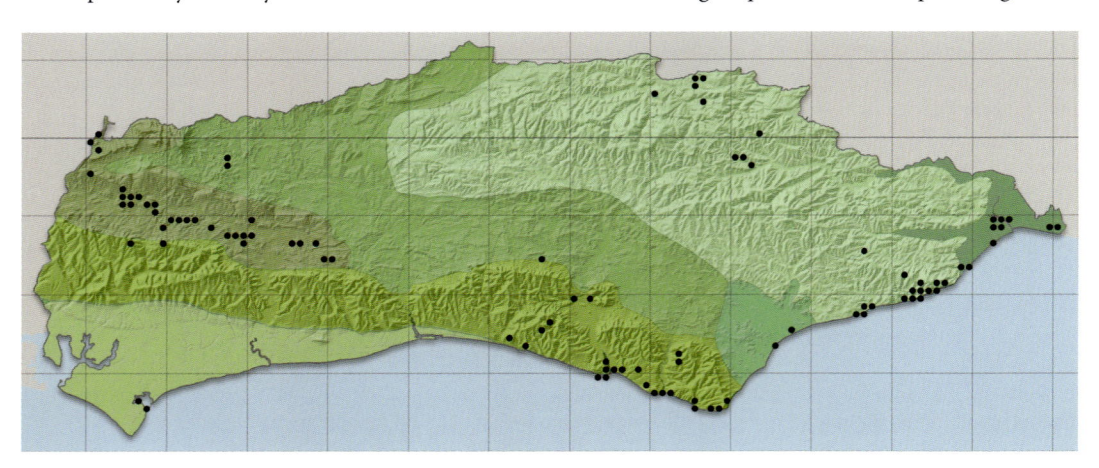

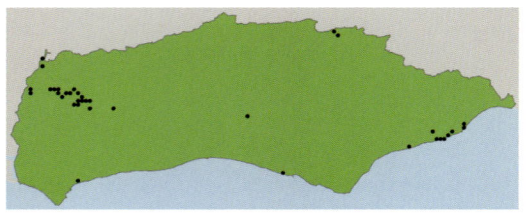

Large map: distribution 2000–2023
Small map: distribution 1844–1999

Anthophora bimaculata ♀ [SF]

Anthophora bimaculata ♂ [PB]

Behaviour and interactions
A. bimaculata excavates its nest burrows in locations with exposed sandy soil. The surface can be level, sloping or with a vertical face. Burrow entrances are about the thickness of a pencil, and, in optimal locations, significant numbers are excavated near each other, creating large aggregations.

Coelioxys rufescens lays its eggs within the brood cells of this bee, its larvae emerging first to consume the food mass. *C. elongata* may also attack this species.

On the wing
Sussex records run from late May until the beginning of September.

Feeding and foraging
This bee collects pollen and nectar from a wide range of plant species including bramble, plants in the Daisy family such as knapweed and ragwort, sea-lavender, Sheep's-bit, Viper's-bugloss, willowherb and dead-nettle.

Anthophora bimaculata ♀ nectaring on **Viper's-bugloss** [BY]

The Bees of Sussex

Anthophora furcata Fork-tailed Flower Bee
FIRST RECORD 1879, Guestling (Edward Saunders) **TOTAL RECORDS** 165

Geography and history
This bee is widely distributed in Britain with records from as far north as southern Scotland.

In Sussex it is widely distributed across much of the county. It takes advantage of a wide range of habitats including chalk grassland, heathland, reedbed and woodland. It is also known from gardens. Most records have been made since the 1970s, and it is rarely numerous.

Out in the field and under the microscope
Anthophora furcata has a brownish appearance that is suggestive of the bumblebee *Bombus pascuorum* and, because of this and its rather drab appearance, it is often overlooked. This means that any bee visiting plants in the Deadnettle family—especially Black Horehound, any of the woundworts and Wood Sage—should be examined carefully.

Females have reddish hairs at the tip of the abdomen and lack the pronounced hair bands found in *A. bimaculata* and *A. quadrimaculata*. Males have a characteristic, yellow-marked face. Within the yellow marking are two small black spots adjacent to the upper margin of the clypeus and two larger black markings on the uppermost corners of the labrum. Unlike other species of *Anthophora*, both sexes have three teeth at the tips of the mandibles.

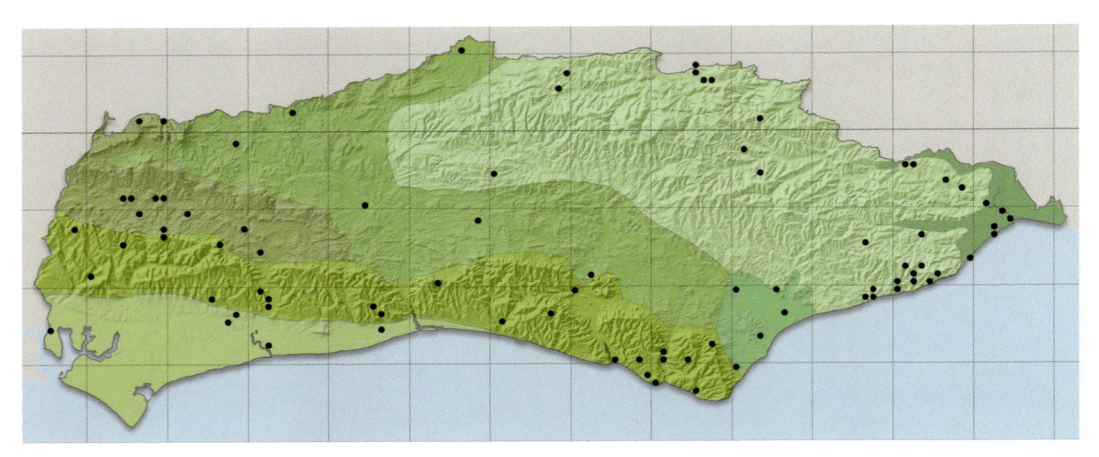

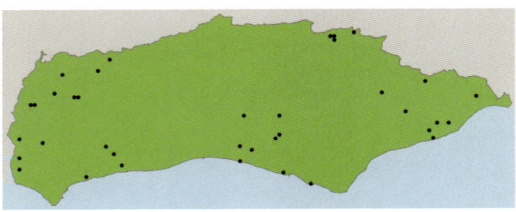

Black Horehound [SF]

Large map: distribution 2000–2023
Small map: distribution 1844–1999

32 *The Bees of Sussex*

Anthophora furcata ♀ [SF]

Anthophora furcata ♂ [PB]

Behaviour and interactions
This is the only species in the genus to excavate its nests in wood. These are created in rotting deadwood, the females using trees, fence posts, driftwood and even pieces of Rhododendron.

Using their mandibles, they excavate a nest tunnel. Some of the sawdust generated is chewed into a pulp and is used to line the walls of the brood cells, possibly mixed with a salivary solution to create an impermeable lining. This is applied to the walls of the brood cell with the aid of the pygidium.

The females are unusual in that they have stout bristles on the face with which they brush pollen from the anthers within the narrow flower tubes of their foodplants. At the same time, they can vibrate the thorax to help loosen the pollen.

Two species, *Coelioxys quadridentata* and *C. rufescens*, are known to lay their eggs within the nests of this bee.

On the wing
Sussex records run from late May until late August.

Feeding and foraging
This is the only species in the genus which collects pollen from a narrow range of plants, with females specialising in foraging from plants in the Dead-nettle family. Species regularly visited include Black Horehound, woundwort and Wood Sage, but also plants such as Bugle and Self-heal.

There are also records of females visiting a wide range of other plants, including bramble, Cross-leaved Heath, Creeping Thistle and Deadly Nightshade.

Anthophora furcata ♀ foraging [SF]

Anthophora plumipes Hairy-footed Flower Bee
FIRST RECORD 1882, Hastings (Edward Saunders) TOTAL RECORDS 925

Geography and history
Anthophora plumipes is a very widely distributed bee that can be found just about anywhere in Sussex. It is most frequently seen in gardens but can also be abundant in coastal locations with soft-rock cliffs, such as Glyne Gap and Seaford Head, or with eroding banked shorelines, as at Chichester Harbour.

Out in the field and under the microscope
This is one of the very first bees to appear each spring and should be looked for on warm days, even from mid February onwards. It resembles a bumblebee but has very distinctive colouring. The females are all black apart from the yellow pollen-collecting hairs on the hind legs, while males are a bright brown when they are freshly emerged, fading to a greyish white as they age.

A. plumipes has a very characteristic flight, with both males and females flying rapidly close to the ground, their flights interspersed with periods of hovering and darting. Males and females can also be seen at nest sites. These are often sited in cavities in walls.

One location where it is important to take particular care with identification is Seaford Head. Here, there is a long-established colony of *A. plumipes* that has an overlapping flight period with the very similar-looking *A. retusa* at the latter's only surviving Sussex location.

Both sexes of *A. plumipes* have dark eyes, while these are an olive-green in *A. retusa*. Other subtle differences include the colour of the innermost spine on the hind tibia in the females (yellowish orange in *A. retusa* and black in *A. plumipes*) and, in the males, the position and length of hairs on the tarsi

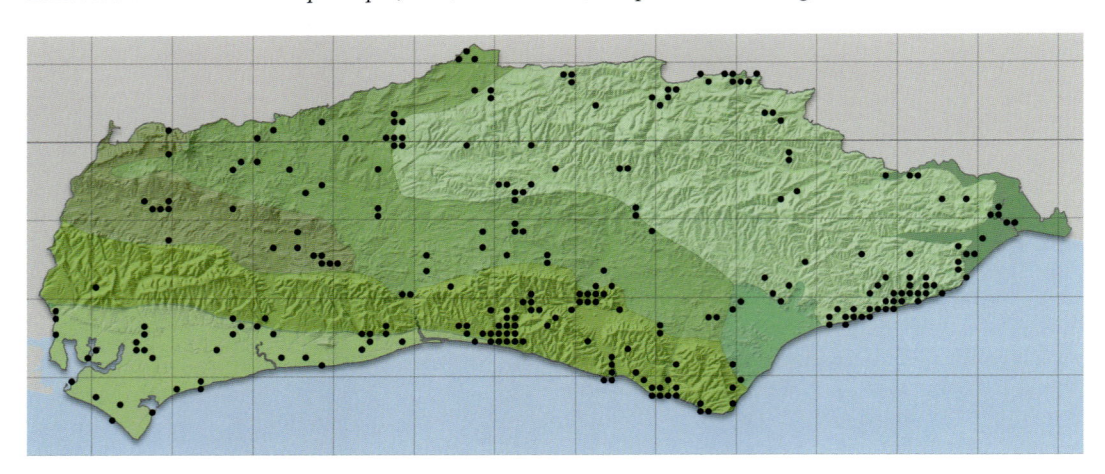

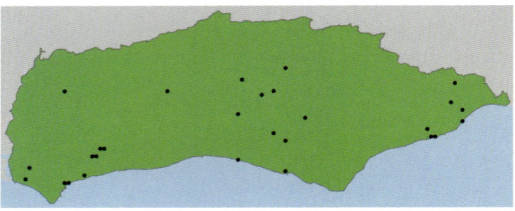

Nest site, Glyne Gap [AP]

Large map: distribution 2000–2023
Small map: distribution 1844–1999

on the middle legs. Very long hairs are found on all tarsal segments here in *A. plumipes*, while even the longest hairs in *A. retusa* are much shorter and are confined to the basitarsi.

Behaviour and interactions

Males establish circuits that they patrol rapidly, searching for females. If a male locates a female that has already mated, she may well move to a new location to forage.

Most nests are established in vertical surfaces, suitable locations including soft-rock cliff faces, vertical banks, the sides of sandpits, and walls that have been built with friable material. This is a gregarious bee that can establish large nesting aggregations.

A. plumipes females have bristles on part of their tongues with which they scrape pollen from narrow flower tubes before transporting it back to the nest on their hind legs. They are able to transport large quantities of pollen and nectar, with 50 per cent of the female's body weight cited in one study. Like other early-spring emerging bees, the larvae complete their development by late summer and overwinter in their brood cells as adults.

The cuckoo bee *Melecta albifrons* is a parasite of this species and is often seen close to nest sites, usually in small numbers.

Anthophora plumipes ♀ [PC]

Anthophora plumipes ♂ [PC]

On the wing

Sussex records run from early February until July.

Feeding and foraging

A. plumipes collects pollen from a very wide range of plants, including members of the Borage, Pea, Dead-nettle, Primrose and Rose families.

Comfrey and lungwort are regularly visited in gardens.

Russian Comfrey [JP]

Anthophora plumipes ♀ foraging on Lungwort [PC]

Anthophora quadrimaculata Four-banded Flower Bee
FIRST RECORD 1905, Brighton (William Unwin) **TOTAL RECORDS** 63

Geography and history
In Sussex, *Anthophora quadrimaculata* is predominantly a garden species, although it has been recorded from a handful of chalk grassland sites as well. Elsewhere in the country, it is also known from coastal grasslands and heathlands. Recent Sussex records are thinly scattered across the county from places such as Bexhill, Brighton, Great Dixter, Portslade, Lewes and Midhurst. Chalk grassland locations such as Seaford Head and Southerham have also provided recent records.

Nationally, *A. quadrimaculata* occurs as far north as Norfolk and as far west as Cornwall, but most records are clustered around London, South Essex and North Kent. It may be expanding its range.

Out in the field and under the microscope
This bee is superficially similar in appearance to the smaller *A. bimaculata* which also has a banded abdomen. The eyes are less brightly coloured in *A. quadrimaculata*, and are more blue-grey than green. Female *A. quadrimaculata* are plumper and, unlike *A. bimaculata*, have a completely dark face. Males have two large black markings on the clypeus which are absent in *A. bimaculata*.

Given its close relationship with plants in the Dead-nettle family, any bee foraging from garden plants such as Rosemary, lavender and other dead-nettles should be checked closely. Out on the downs, patches of Black Horehound particularly merit a look. It does sometimes fly with another species in the same genus, *A. furcata*, so care should be taken with any identifications.

Behaviour and interactions
A. quadrimaculata nests in small aggregations in south-facing sandy banks and cliffs as well as walls that are constructed with a loose, friable material. In flight, this bee makes a shrill whine and can occur in numbers around foodplants. Males are often seen patrolling rapidly along a circuit around gardens, with flights interspersed with periods of hovering and feeding.

While *Coelioxys quadridentata* is suspected of parasitising this species, this has not been confirmed in Britain. This species of *Coelioxys* has, however, been observed flying close to *A. quadrimaculata* nest

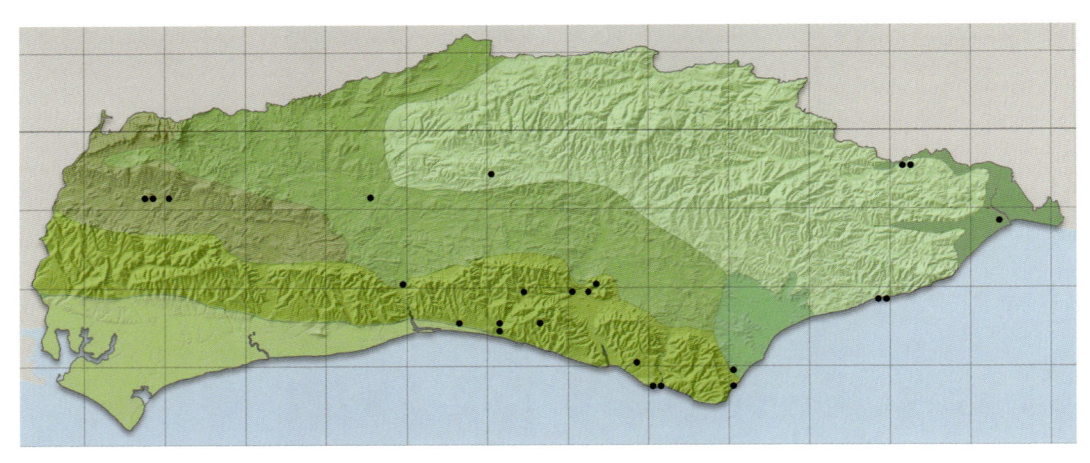

Large map: distribution 2000–2023
Small map: distribution 1844–1999

Anthophora quadrimaculata ♀ **nectaring on sage** [PC]

Anthophora quadrimaculata ♂ [PC]

sites in Surrey and in Northampton and may well lay its eggs within provisioned brood cells. On the continent, the cuckoo bee *C. rufescens* is also reported as occasionally targeting this species.

On the wing
Sussex records run from early May until mid August.

Feeding and foraging
Pollen is collected from a range of plants, but especially from those in the Dead-nettle family. Black Horehound, for example, is a particularly important source of pollen on the downs, while Rosemary, cat-mint and lavender are regularly used in gardens. Mike Edwards has also reported seeing a female collecting pollen from Viper's-bugloss, a member of the Borage family.

There are also records of this bee visiting plants from other families. These plants include Purple Toadflax, euphorbia, restharrow and Red Valerian.

Anthophora quadrimaculata ♂ **visiting garden thyme** [PC]

Anthophora quadrimaculata is often seen flying in and around garden plants such as cat-mint [WP]

The Bees of Sussex

Anthophora retusa Potter Flower Bee
FIRST RECORD 1878, Hastings (Edward Saunders) **TOTAL RECORDS** 93

Geography and history
This bee was formerly widespread across southern England but is now known only from cliffs near Sandown on the Isle of Wight, the MOD ranges around Purbeck in Dorset, and Seaford Head in Sussex. It was last seen at a fourth location near Farnborough in Hampshire in the 1990s.

All modern Sussex records are centred on the cliffs at Seaford Head where evidence of a possible colony was first discovered in the Cuckmere Valley in 1967. This is when Alfred Jones swept a male feeding on Horseshoe Vetch. At the time, this record was not shared, and the colony only came to wider prominence when Martin Jenner came across it in 1995.

Since that time, virtually all the records have been from a 2.5 km stretch of cliffs at this one location. There have also, however, been sporadic records from close to the mouth of the Cuckmere River and from High and Over nearby. This last location is where a new nest site was briefly established in 2011.

Historically, it was first reported in the county from Hastings in 1878 and 1896 and then subsequently from Worthing, Brighton, Broadwater Forest, Ditchling and, finally, Albourne, where it was last recorded in 1954.

Seaford Head, main foraging area with Ground-ivy [SF]

Out in the field and under the microscope
Anthophora retusa generally forages within some 75 m of the cliff edge at Seaford Head, especially at the lower end of Hope Gap. The males are often a grey blur, flying extremely rapidly between patches of Ground-ivy to feed or search for females. The females also move quickly, gathering pollen and nectar

Large map: distribution 2000–2023
Small map: distribution 1844–1999

38 *The Bees of Sussex*

before flying rapidly back to the cliffs to build or provision their brood cells. The steps onto the beach at Hope Gap offer good opportunities to observe nesting behaviour.

This species is on the wing slightly later than the very similar-looking *A. plumipes*, although the flight seasons of the two do overlap. The pale blue-green eyes of *A. retusa* help to distinguish this species from *A. plumipes*. However, females can be definitively separated by the colour of the hind tibial spurs— yellowish orange in *A. retusa* and black in *A. plumipes*—while male *A. retusa* lack the exceptionally long hairs found on all of the tarsal segments on the middle legs of *A. plumipes* (long hairs are restricted to the basitarsi in *A. retusa*). The males are slightly smaller with a duller colouring than *A. plumipes* males.

Behaviour and interactions

Nests are established in the sandy deposits that lie above the chalk at Seaford Head, with as many as 40 nest holes counted in an area of 20 m² on one occasion. Cells are lined with a fine mud concretion, possibly a mix of nectar and soil that, once dry, resembles eggshell. The nest provisions are a combination of pollen and nectar and are very liquid.

Melecta luctuosa (extinct in Britain) is known to parasitise this species. At Seaford Head, *M. albifrons* is also suspected of this behaviour, a relationship that was also noted at a sandpit near Ditchling by Alfred Brazenor in 1942.

Anthophora retusa ♂ nectaring on Ground-ivy [PB]

Nest site, Seaford Head [NS]

Anthophora retusa ♀ [SF]

Stinking Iris [NS]

On the wing
Sussex records run from early April until mid June.

Feeding and foraging
Four pollen samples collected at Seaford Head by Mike Edwards and Martin Jenner were found to contain 60 to 90 per cent maple pollen. Females were also observed here gathering pollen from Stinking Iris. Members of the Pea family have otherwise been described as important sources of pollen.

A. retusa has also been reported as visiting a wide range of other plant species, including dandelion, Honeysuckle, Hound's-tongue, Viper's-bugloss and Yellow Horned-poppy. Ground-ivy is a particularly important source of nectar.

Apis – Honey Bees

These are small to large bees that live in complex, long-lived colonies that may number a few thousand to some 60,000 workers. The genus is principally tropical. The colonies have two female castes (queens and workers). There is one species in Britain, *Apis mellifera*, which occurs in Sussex.

Apis mellifera Honey Bee
FIRST RECORD 1893, Bognor Regis (Henry Guermonprez) TOTAL RECORDS 1,822

Geography and history
Apis mellifera—one of the few British species to have a long-established common name—is believed to be a very early introduction to Britain, and its presence on a site is generally because a hive has been sited nearby. *A. mellifera* is abundant throughout Sussex and is often the most frequently encountered species present at any given location.

Out in the field and under the microscope
This bee can be found anywhere in the county where plants are flowering, even on cool days. As they approach a flower, workers dangle their hind legs, a characteristic that is shared with some species of hoverfly but not with other species of bee.

A worker's abdomen appears to be almost hairless and has varying degrees of black and orange banding. This is one of the few species with hairs on the surfaces of the eyes. Workers can be seen at any time of year, including on mild days in winter.

Males (otherwise known as drones) are broader than the workers, and they have longer wings and eyes that meet on top of the head, but they are otherwise very similar.

Apis mellifera worker in flight, with its hind legs dangling characteristically [PC]

Apis mellifera queen with identification tag [PB]

Apis mellifera ♂ (drone) [PC]

Apis mellifera nesting in a tree [PG]

Apis mellifera swarm on a tree [PC]

Behaviour and interactions

A. mellifera is a short-tongued species that lives in highly evolved social groups, each composed of a queen, workers and drones, and each numbering many thousands of individual bees. Wild-living colonies nest in cavities in trees but also take advantage of roof voids and unused chimneys.

It most commonly nests, however, in hives maintained by beekeepers. Nests are made of combs of cells that are made from wax secreted by the females. Food for the larvae is provided progressively by the workers, with the cells only closed once the larvae have finished feeding.

The large number of food resources gathered and stored by colonies—plus the abundance of adult bees, pupae, larvae and eggs—mean that *A. mellifera* is targeted by many species that feed in and around nests. Species of wasp such as *Philanthus triangulum* and *Vespa crabro* (or Hornet), for example, prey on adult *A. mellifera*, while the wax moths feed on the honeycomb.

Philanthus triangulum with *Apis mellifera* prey [PC]

On the wing

A colony is potentially active at any time of year, with the main period of activity running from April until September.

Feeding and foraging

A. mellifera workers can forage up to a couple of miles from the colony, collecting pollen and nectar from an enormous range of plant species. Queens and drones do not visit flowers to feed, instead obtaining their nutrition from the pollen and nectar collected by the workers.

Apis mellifera worker visiting willow catkins [PC]

The Bees of Sussex

Bombus – Bumblebees

This is a genus of social bees, of which 27 species have been recorded in Britain. Of these, two are extinct and a third (*Bombus pomorum*) is thought to have been a temporary colonist in the nineteenth century. A total of 22 species have been confirmed in Sussex, with at least one of the three species that make up the aggregate referred to as *B. lucorum* agg. also present. This gives at least 23 bumblebee species for the county (and potentially 25), four of which are now extinct. Six of the British species are cuckoos. These are incapable of transporting pollen on their hindlegs, and females instead take over the nests of other bumblebee species, subduing or killing the resident queen and then using the resident workers to forage and tend their own brood. It is not uncommon for some species of 'non-cuckoo' bumblebee to adopt a similar strategy.

Bombus barbutellus Barbut's Cuckoo Bee
FIRST RECORD 1905, Brighton (William Unwin) TOTAL RECORDS 54

Geography and history
Historically, this bee was known from across Sussex, but it is now very scarce. There have been just five records in Sussex since 2000, with three of these from Fairmile Bottom over a span of four days in 2007. These are the last records from Sussex. This species may be in decline in the south-east, a view that is supported by the lack of many modern records from Kent—there have been just three since 1988.

This bumblebee has been recorded from across the county and could potentially be found just about anywhere—its main host *Bombus hortorum* is widely distributed and occurs in gardens, woodlands, flower-rich grasslands and hedgerows.

Out in the field and under the microscope
Despite the abundance of its main host, *B. barbutellus* has rarely been numerous. As with *B. hortorum*, the basic colour pattern for males is a white tail, two yellow bands across the thorax, and one yellow band across the abdomen. Males have a pair of rounded swellings at the tip of sternite six. These almost meet in the middle.

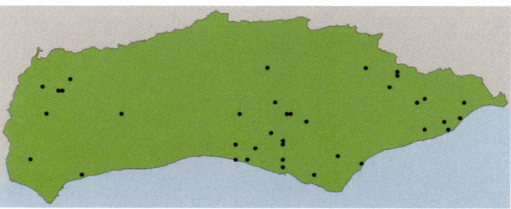

Large map: distribution 2000–2023, with orange circles showing distribution of *Bombus hortorum*
Small map: distribution 1844–1999

Bombus barbutellus ♀ [SF]

Bombus barbutellus ♂ [PB]

Females are most readily identified by examining sternite six, which has a characteristic U-shaped swelling. There is also a fine ridge running lengthwise along tergite six. In addition, the wings are darkened and the abdomen is sparsely haired. This bumblebee is most likely to be seen feeding on plants such as bramble and thistle.

Behaviour and interactions

This cuckoo bumblebee takes over the nests of *B. hortorum* and possibly of *B. ruderatus*. Females emerge from hibernation in April or May and seek out nests of the host species after a short period of feeding. New females and males appear from late May onwards.

One nest that was investigated in the early twentieth century contained 16 *B. barbutellus* daughters and two males, as well as several *B. hortorum* larvae (indicating that the host species had still been able to reproduce). Forty-nine *B. hortorum* workers were also present.

On the wing

B. barbutellus is on the wing from mid April until September.

Feeding and foraging

Adult bees feed on nectar and pollen from a range of plants, including bramble, Bugle, knapweed, members of the Pea Family, scabious, thistle, Viper's-bugloss and White Dead-nettle.

Bombus hortorum queen, the principal host [SF]

Viper's-bugloss [PC]

Bombus bohemicus Gypsy Cuckoo Bee
FIRST RECORD 1921, Hastings (Butterflield) TOTAL RECORDS 34

Geography and history
Populations of this bumblebee are concentrated in northern and western Britain where it can be abundant. It was formerly distributed throughout Britain, but it has become very scarce in south-east and central England—it appears to be retreating north and west. It is also becoming scarce across mainland Europe. These declines are linked to climate change.

In line with this trend, the population in Sussex has declined and it may now be lost from the county. It has previously been recorded in small numbers from a range of different habitats, including woodland, heathland, chalk grassland and coastal dunes. Records are concentrated in Ashdown Forest and other parts of the High Weald, the Wealden Greensand, and the western section of the South Downs. There are also two records from Rye Harbour Nature Reserve, where it was last recorded in the county in 2010 by S. R. Miles.

Out in the field and under the microscope
Although *Bombus bohemicus* may be extinct in Sussex, any bee with yellow patches above a white tail should be examined carefully. Females are most likely to be confused with the widespread *B. vestalis*, not least as fresh specimens of each species share these features. The yellow patches can fade which means that this species can also resemble its hosts, although darkened wings and a sparse covering of hairs on the abdomen help to identify this cuckoo.

As well as yellow patches, *B. bohemicus* females have a broad yellow collar behind the head and can have yellow bands at the back of the thorax and the front of the abdomen, although these may be weak or even absent. The tail is white.

Males generally have yellow bands at the front and back of the thorax, and at the front of the abdomen. Antennal segment three is about the same length as segment five in this species, a character that helps to separate *B. bohemicus* from species such as *B. vestalis*, *B. barbutellus* and *B. campestris*. In these, antennal segment three is obviously shorter than segment five. The structure of the genital capsule will confirm an identification.

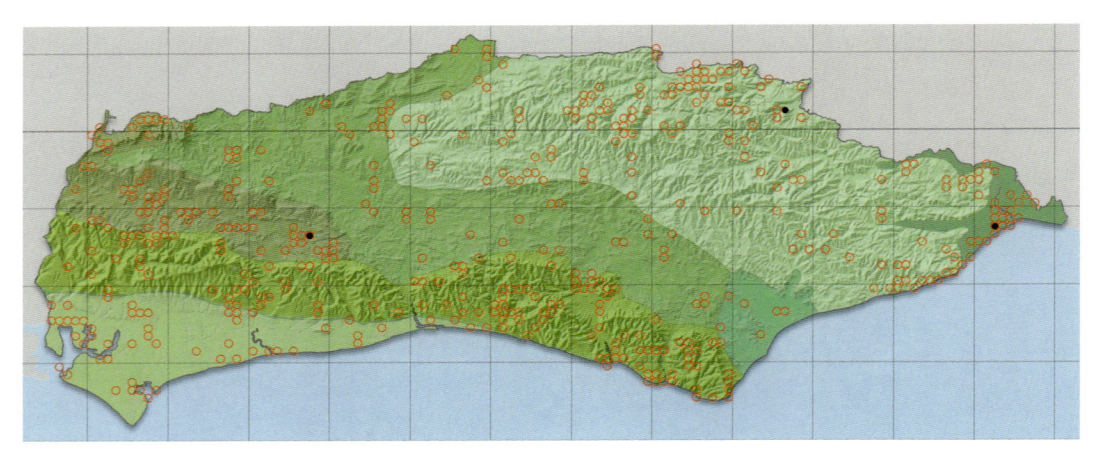

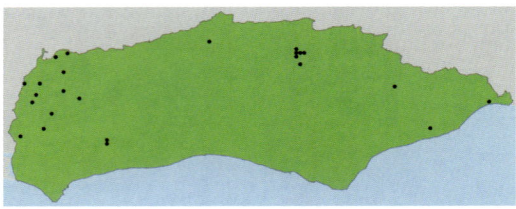

Large map: distribution 2000–2023, with orange circles showing distribution of *Bombus lucorum* agg.
Small map: distribution 1844–1999

Bombus bohemicus ♀ [SF]

Bombus bohemicus ♂ [PB]

Bombus lucorum agg. queen, the host [SF]

Behaviour and interactions
B. bohemicus takes over nests established by queens in the *B. lucorum* aggregate. This is composed of the species *B. cryptarum*, *B. lucorum* and *B. magnus*. These are three very similar and closely related species that cannot be distinguished in the field. The precise relationships between these three species and the cuckoo are not fully understood.

Overwintered females emerge from hibernation in April and are believed to enter nests of the host species early in the colony cycle. These nests are sited in old rodent burrows. In a couple of studies, the host queens did not respond aggressively to the cuckoo and were able to continue to reproduce. The new generation of males and females emerge in June and may still be active in early autumn, resting and feeding on flowers.

On the wing
B. bohemicus is active from mid April until the end of September.

Feeding and foraging
Adult bees visit a wide range of flowers to feed on pollen and nectar, including Bilberry, bramble, dandelion, Heather, knapweed, Common Ragwort, White Clover and willow.

Bombus campestris Field Cuckoo Bee
FIRST RECORD 1905, Brighton (William Unwin) **TOTAL RECORDS** 173

Geography and history
As recently as 1975, this was regarded as a widely distributed and relatively common bumblebee. However, it is now a scarce species and is never as abundant as its main host *Bombus pascuorum*. Although it has declined in abundance since the 1970s, there can be considerable variation in numbers from place to place and from year to year.

Modern records for this bee are thinly distributed across Sussex, and it should be looked for wherever one of its host species is found, particularly *B. pascuorum*. This latter species occurs throughout Sussex, occupying a very wide range of habitats such as chalk grassland, meadows and pasture, woodland and heathland, as well as gardens and parks.

Out in the field and under the microscope
B. campestris is a very variable species. Typical females have a densely haired thorax and sparsely haired and often shiny abdomen that is strongly downcurved. They have two yellowish bands on the sides of the abdomen towards the tip, a white tail, and yellowish hairs at the front and back of the thorax.

Males are particularly variable and can be either pale or entirely black. They always have tufts of long black hairs towards the tip on the underside of the abdomen. The genital capsule is very distinctive.

This species is most likely to be encountered feeding on plants such as bramble, thistle and knapweed.

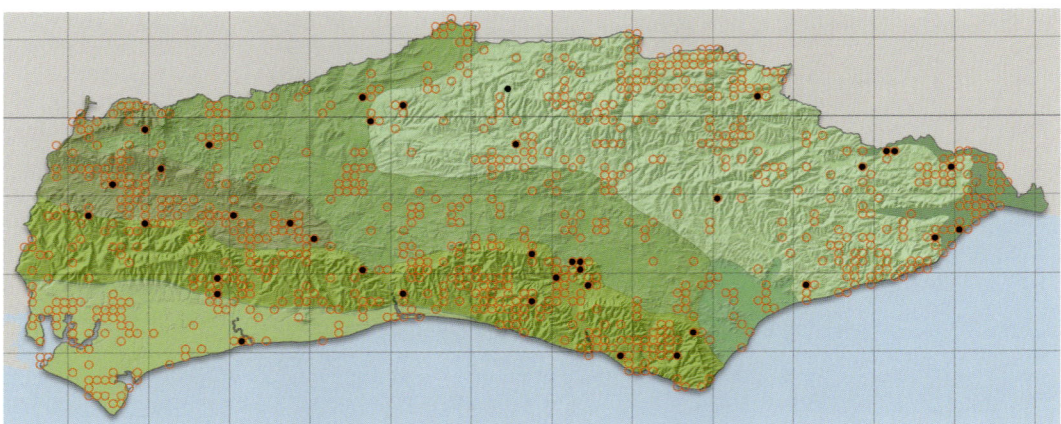

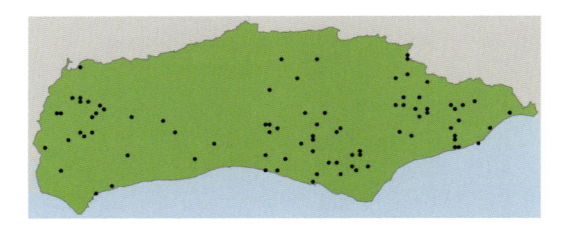

An all-black ♂ *Bombus campestris* [SF]

Large map: distribution 2000–2023, with orange circles showing distribution of *Bombus pascuorum*
Small map: distribution 1844–1999

Bombus campestris ♀ [SF]

Bombus campestris ♂ [PG]

Bombus pascuorum queen, the principal host [SF]

Behaviour and interactions
B. campestris takes over the nests of *B. humilis*, *B. pascuorum* and *B. ruderarius* queens. On the continent, it has been reported that it also takes over the nests of *B. pratorum*. Other hosts are possible.

Females emerge from hibernation in spring, seeking out the nests of one of the host species. Once a nest has been located, they use chemicals to mimic the 'smell' of the resident queen and attempt to establish themselves within the nest where they and the host queen may co-habit. It has been suggested that the host queen does not take part in a defence of the nest—this is left to the workers.

If a nest is successfully invaded, the cuckoo may feed on the host's eggs and eject the host's larvae, while both host and cuckoo may each successfully produce offspring. The cuckoo's offspring emerge from the nest from the middle of May onwards. They can often be seen resting on flowers or feeding during July, August, September and early October. This is a very long flight period and suggests that this species may produce two generations in some years.

On the wing
Sussex records run from mid April until early October.

Feeding and foraging
To feed on pollen and nectar, adult bees visit a wide range of flowers, including bramble, Common Knapweed, Creeping and Spear Thistle, dandelion, Devil's-bit Scabious, Germander Speedwell, Ground-ivy, Heather, Red Clover, Wild Thyme and Wood Sage.

Bombus hortorum Garden Bumblebee
FIRST RECORD 1888, Bognor Regis (Henry Guermonprez) **TOTAL RECORDS** 1,440

Geography and history
Bombus hortorum is one of seven species of bumblebee which is still widespread in Sussex. It can be found just about anywhere in the county, including gardens, parks, woodlands, flower-rich grasslands, hedgerows, flowery field edges on farmland and along riverbanks.

Out in the field and under the microscope
This bumblebee is best looked for as it forages, particularly from plants with long flower tubes such as comfrey, Foxglove and Viper's-bugloss—workers can often be seen systematically working their way up a flowering stem, moving from flower to flower. The tongue is often extended to its full length as a bee approaches a flower to take up nectar.

With a white tail and three yellow bands, *B. hortorum* is very similar to the closely related and much scarcer *B. ruderatus* and also *B. jonellus*. It also resembles the extinct *B. subterraneus*.

B. hortorum can be separated from *B. jonellus* by the much shorter face and shorter tongue of the latter, while distinguishing between *B. hortorum* and *B. ruderatus* is more challenging. In common with *B. ruderatus*, *B. hortorum* has a very long face and a strikingly long tongue which is almost the length of its

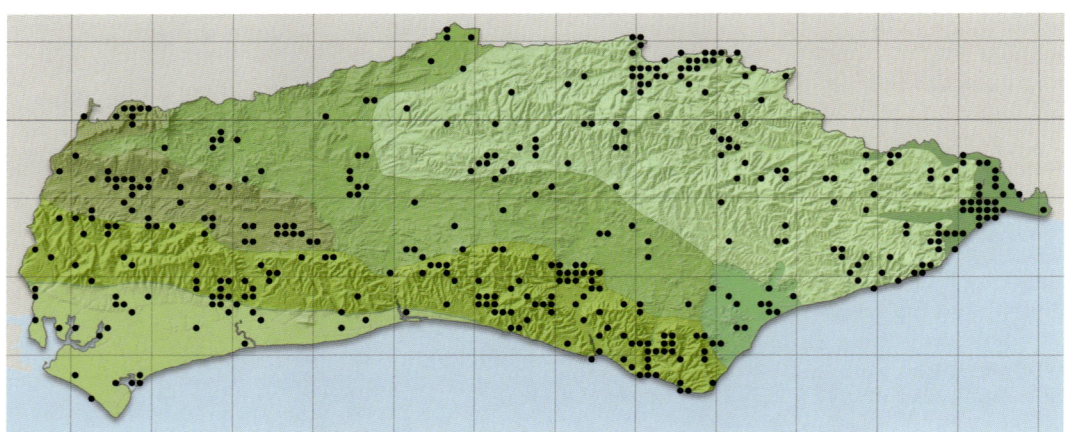

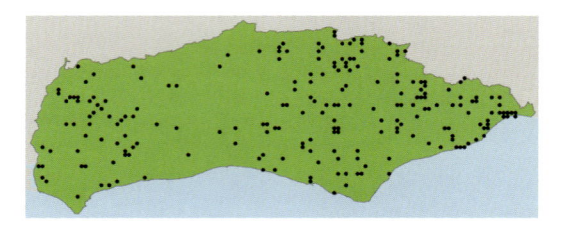

Bombus hortorum queen [SF]

Large map: distribution 2000–2023
Small map: distribution 1844–1999

48 *The Bees of Sussex*

Bombus hortorum worker [SF]

Bombus hortorum ♂ [SF]

body. A key distinguishing feature in females is the extent of yellow on the abdomen which extends onto the second tergite in *B. hortorum* but is restricted to the first in *B. ruderatus*.

For males, a good feature to check is the colour of the hairs on the underside of the mandibles. These are black in *B. hortorum* and golden in *B. ruderatus*. Another more subtle difference is the length of hairs at the top of the front tibiae, which are longer than the width of the tibiae at this point in *B. hortorum* and shorter in *B. ruderatus*.

Behaviour and interactions

This species nests just above or below ground, in old mouse or vole burrows, among plant roots, and in leaf litter. Frederick Sladen, writing in 1912, even describes finding a nest 20 feet above the ground in a sparrow's nest, on top of an addled egg.

The number of workers in a nest ranges between 30 and 130, with Alfred Brazenor noting that he had collected 128 from a nest on Lewes Road, Brighton, in July 1895. Queens generally complete their colony cycle in late June or early July. In some years, daughter-queens start a new colony cycle during July or August.

B. barbutellus parasitises *B. hortorum* nests, the females forcing their way into young nests before usurping the resident queen.

On the wing

Sussex records run from late February until late October.

Feeding and foraging

Pollen is collected from a wide range of plant species, particularly members of the Pea family and especially Red Clover. It will also collect pollen from the Bindweed, Borage and Dead-nettle families, as well as Foxglove, a member of the Speedwell family.

Bombus hortorum worker visiting sage, a member of the Dead-nettle family [PC]

Bombus humilis **Brown-banded Carder Bee**
FIRST RECORD 1876, Brighton (A. Brazenor) TOTAL RECORDS 339

Geography and history
This long-tongued bumblebee underwent a significant population decline during the twentieth century, which coincided with the destruction of large areas of flower-rich grassland. It does, however, appear able to recover to some extent if there is a succession of warm, dry years. This contrasts with its close relative *Bombus muscorum* which favours cooler, wetter summers.

The distribution of *B. humilis* in Sussex has become very restricted—it was formerly found across the county. Today, this bee is largely restricted to the coast between Hastings and Camber and chalk grasslands on the South Downs between Truleigh Hill and Eastbourne. Places such as Rye Harbour Nature Reserve, Cow Gap, Castle Hill and downland within Brighton have all generated good numbers of records in recent years.

Out in the field and under the microscope
B. humilis is best searched for by checking any brown bumblebee foraging on plants such as Red Clover and Common Bird's-foot-trefoil. It is a tricky one to identify, and care

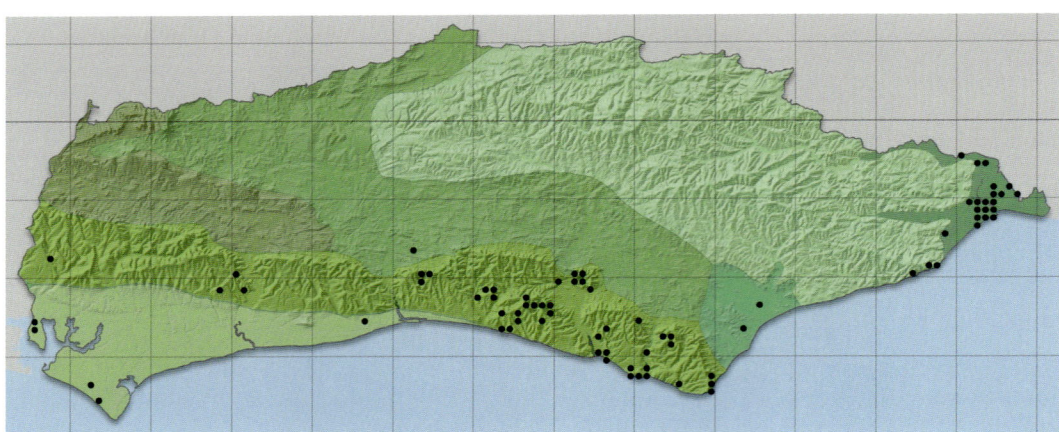

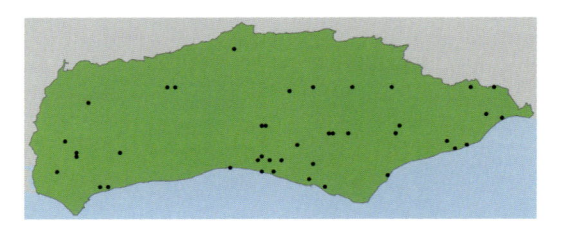

Bombus humilis queen [PG]

Large map: distribution 2000–2023
Small map: distribution 1844–1999

should be taken with identification as it resembles both *B. pascuorum* and the closely related *B. muscorum*.

Fresh specimens of both sexes often have a dark brown band on tergite two on the otherwise pale brown abdomen, as well as noticeably pale sides to the thorax in contrast to the brown hairs covering its upper surface. Under magnification, this bumblebee usually has at least a few black hairs on the thorax just above the wing bases (unlike *B. muscorum* which has none). Like *B. muscorum*, it has no black hairs on the abdomen (in contrast to most *B. pascuorum*). Males can be conclusively identified by examining the genital capsule.

Bombus humilis worker [SF]

Bombus humilis ♂ [SF]

Behaviour and interactions

To locate an unmated queen, males adopt patrolling behaviour that involves passing over a nest containing the new generation of queens. After mating, these queens generally establish their own nests at ground level, at the base of tall grasses where the sun can warm the grass stems. Newly established nests are occasionally usurped either by other *B. humilis* queens or by *B. muscorum* queens. Nests can be at a density of one to five nests per km^2 if the habitat is favourable.

In August 1949, Alfred Brazenor noted that he had reared 20 adults from a nest in Ditchling that was sited in a manure heap, a very unlikely location for this species.

A single colony generally has fewer than 100 workers and may contain just 40. Daughter-queens will emerge from June onwards, while males are most visible during August and September.

No cuckoos have been reported from Britain, although *B. campestris* is known to invade nests on the continent, and this could well be the case here too.

Red Clover [PG]

On the wing

Sussex records run from late April until late September.

Feeding and foraging

This species is very reliant on plants in the Pea, Dead-nettle, Daisy, Borage and Broomrape families for pollen. Red Clover, Common Bird's-foot-trefoil and vetches are especially important.

Bombus hypnorum **Tree Bumblebee**
FIRST RECORD 2002, Hastings (Jonty Denton) TOTAL RECORDS 1,023

Geography and history
Bombus hypnorum was first recorded in Britain in 2001, in Wiltshire. Since its arrival, it has spread throughout England and Wales, and it had reached the north-east tip of Scotland by 2022. It even reached Ireland in 2014. This species is also continuing to expand its range across much of the continent, and this expansion includes ongoing movement to Britain.

This is a recent arrival in Sussex and was first recorded from Holmhurst St Mary in Hastings in 2002, one year after it was first found in Britain. It rapidly spread and was soon known from all corners of the county. As well as in woodlands, *B. hypnorum* can be found in any village, town or city in Sussex, taking advantage of gardens which replicate the open-canopied woodland that is its natural habitat.

Out in the field and under the microscope
Queens, workers and males are very alike. Typical specimens have a ginger thorax and a black abdomen with a white tail, which makes this one of the most readily identifiable bumblebee species in Britain. There is considerable variation, however, so it is not uncommon to come across examples with varying degrees of black on the thorax.

Workers are often encountered as they forage from garden plants, while groups of males are frequently seen either flying or at rest near the entrance to a nest site, awaiting the emergence of unmated queens.

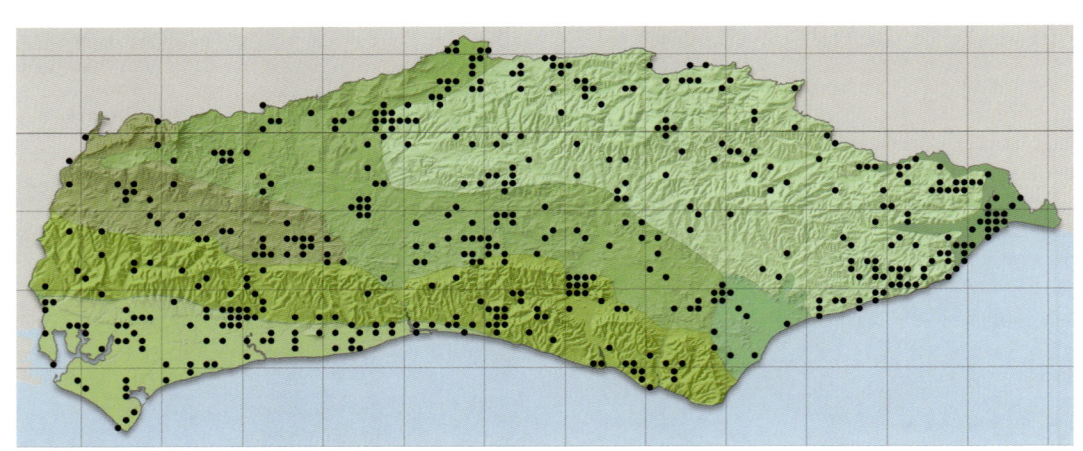

Bombus hypnorum nest [PW]

Large map: distribution 2000–2023
Small map: distribution 1844–1999

Bombus hypnorum queen [SF]

A darkened *Bombus hypnorum* queen [SF]

Bombus hypnorum worker [PC]

Bombus hypnorum ♂ [PB]

Behaviour and interactions
This species is an aerial nester, taking advantage of holes in trees and rocks, as well as bird and dormouse boxes and cavities within buildings. It will also nest in compost bins, and it sometimes uses a nest site more than once.

Nest density varies from year to year, with one study showing that this changed from c.2.5 nests per km² in one year to 0.7 the next. Elsewhere, up to 4 nests per km² has been reported, the highest density for any British bumblebee species. Mature colonies are large and can have between 150 and 400 workers. The males and daughter-queens emerge from their nests from late May or early June. Given that this bumblebee can still be on the wing in the autumn, it is thought likely that it manages a second generation in some years.

On the continent, the very local bumblebee *B. norvegicus* takes over nests of this species. As this species is absent from Britain, it seems that *B. sylvestris* may be taking advantage of this new nesting opportunity here. There is also a record of *B. barbutellus* doing so on the continent.

This is a very irritable species that will vigorously defend its nest.

On the wing
Sussex records run from early February until mid October.

Feeding and foraging
B. hypnorum forages from an enormous variety of plant species. It has also been seen collecting honeydew on the leaves of Beech trees.

Bombus jonellus Heath Bumblebee
FIRST RECORD 1894, Hollington (unknown) TOTAL RECORDS 291

Geography and history
Before 1972, there were just six Sussex records for *Bombus jonellus*, which was almost certainly a reflection of recording effort rather than a sign of scarcity. The first was in 1894 from Hollington (the name of the recorder has not survived), and was followed by isolated records from the Brighton Downs, near Crowborough, Hollington again, Graffham and, finally, Lavant in 1913. Fifty-nine years later, it was re-found on Ashdown Forest by Gerald Dicker.

Today, *B. jonellus* is widely distributed on Ashdown Forest and nearby in the countryside south of Tunbridge Wells but also on the heathlands of the Lower Greensand. These areas probably support the strongest populations in the county.

In other parts of Sussex, heathlands such as Chailey Common and the chalk heath at Lullington Heath also support populations as do areas of open woodland such as Ebernoe Common and Abbot's Wood. It is also known from flower-rich grasslands in the High Weald, from locations along the coast between Hastings and Rye Harbour, and from a scattering of locations on the eastern South Downs.

Out in the field and under the microscope
B. jonellus is most frequently found foraging, and any bumblebee with three yellow bands and a white tail should be checked to see if it is this species. In the spring, males might also be seen zigzagging over open ground, while queens are often seen foraging from willow.

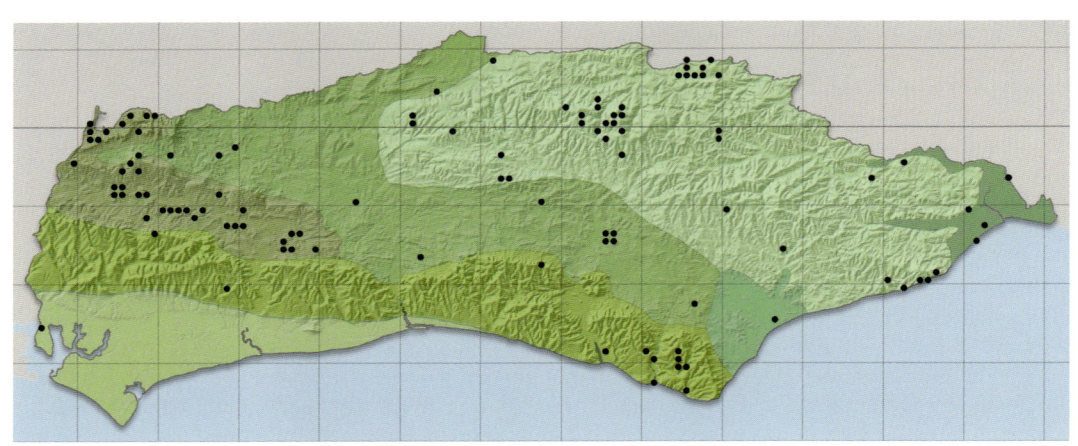

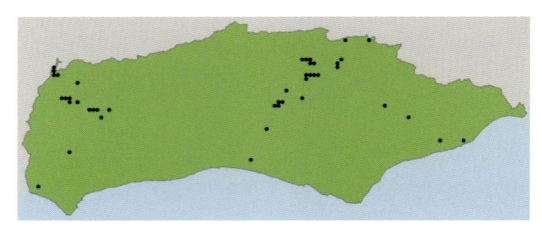

Bombus jonellus queen [SF]

Large map: distribution 2000–2023
Small map: distribution 1844–1999

The Bees of Sussex

Queens, workers and males have a white tail and yellow bands at the front and back of the thorax, and at the front of the abdomen. Superficially, *B. jonellus* resembles two other three-banded species, *B. hortorum* and *B. ruderatus*. It also resembles the extinct *B. subterraneus*. However, *B. hortorum* and *B. ruderatus* are generally (but not always) significantly larger and have a much longer face and tongue—the shape of the face is a good distinguishing feature.

Bombus jonellus worker visiting Heather [NO]

Behaviour and interactions
Nests are established both above and below ground, with the first workers appearing from late April onwards. Mature nests are small and may have as few as 30 workers, although between 100 and 150 has also been reported.

B. jonellus has a short colony cycle and frequently produces two generations a year. Daughter-queens and males from the first generation may appear in June or July and can overlap with the second generation.

B. sylvestris is a cuckoo of this species, and it has been reported that the host and cuckoo can co-exist in the nest and that both may succeed in reproducing.

Bombus jonellus ♂ [SF]

On the wing
Sussex records run from the beginning of February until mid September.

Feeding and foraging
Pollen is collected from a wide range of plants. Members of the Pea family are especially important, and Bell Heather, Cross-leaved Heath and Heather are important sources of pollen on heathlands.

The velvet ant *Mutilla europaea*, is a parasitoid of bumblebee populations, escpecially on heathlands [PB]

The Bees of Sussex **55**

Bombus lapidarius Red-tailed Bumblebee
FIRST RECORD 1878, Brighton (unknown)   TOTAL RECORDS 3,610

Geography and history
Bombus lapidarius is found in all corners of Sussex, occurring in all sorts of open habitats. It is often found in towns and villages.

One record of this bumblebee from 1889, collected by Francis Morice, was initially reported as *B. pomorum*: "On September 9th I secured a female of this extremely rare species at Beachy Head, just under the lighthouse." *B. pomorum* had otherwise only been recorded twice before in Britain, in Kent in 1857 and 1864, so this would have been a notable record, but Francis Morice later confirmed that it was actually an unusual form of *B. lapidarius*.

Out in the field and under the microscope
The enormous queens are frequently seen clasping onto flowers as they feed high up on willow during spring. They can also be seen prospecting for nest sites, flying close to the ground. Workers are most frequently encountered feeding or foraging.

Typical queens and workers have black bodies and red tails, and could only be confused with *B. ruderarius* and *B. rupestris*. Historically, they were also confused with the extinct *B. cullumanus*. *B. lapidarius* queens and workers have a much redder tail than *B. ruderarius*, and have black (as opposed to reddish-orange) pollen-collecting hairs on the hind tibiae. They lack the blackened, smoky wings of *B. rupestris* females. Yellow-collared *B. lapidarius* queens and workers do occasionally occur, a variety that is frequent elsewhere in Europe.

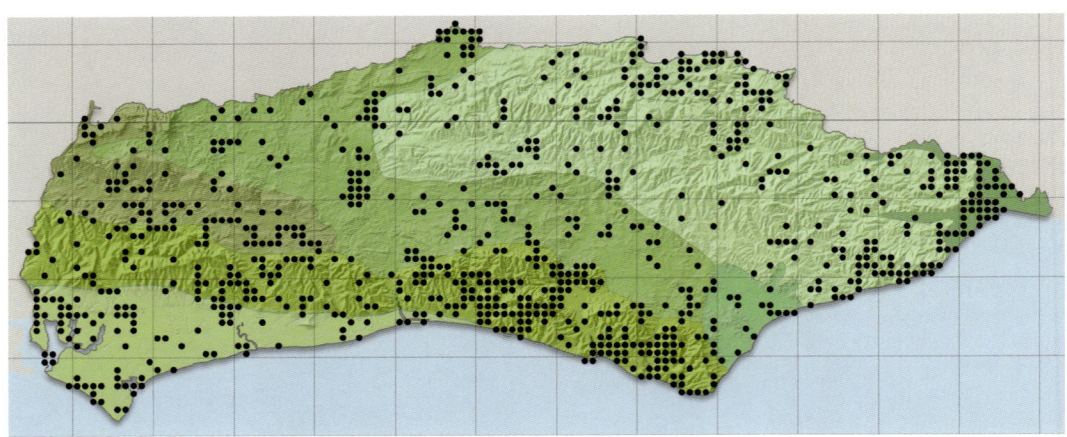

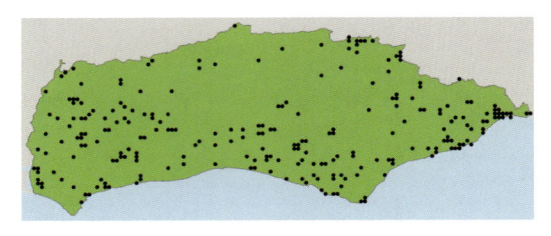

Bombus lapidarius queen [PB]

Large map: distribution 2000–2023
Small map: distribution 1844–1999

The Bees of Sussex

Bombus lapidarius worker [SF]

Bombus lapidarius ♂ [PC]

Mating pair—the queen's sting is extruded during mating [PC]

Males can be seen feeding or coursing above blocks of gorse or scrub. They can be easily identified by the yellow hairs on the head and front of the thorax, as well as the black abdomen and red tail. Fresh specimens generally have a weak yellow band at the back of the thorax and the front of the abdomen.

Behaviour and interactions
B. lapidarius queens hibernate under loose soil, emerging from hibernation later than the other widespread species. On emergence, they seek out suitable nest sites which are usually sited underground, often in abandoned mouse and vole burrows. These can have a long entrance tunnel—in one exceptional example, reported in the early twentieth century, the tunnel was more than seven feet long.

Workers appear in May, with numbers often rising to more than 150. The colony cycle is long in comparison with other species, lasting up to five months. New queens and males generally appear from July.

In common with other species of bumblebee, some queens will attempt to take over a nest that has already been established, generally resulting in the death of one of the queens.

The cuckoo bumblebee *B. rupestris* targets this species, taking over nests and in most cases killing the resident queen.

On the wing
Typically, *B. lapidarius* is active from late April until late October, although it has been recorded as early as late February in Sussex.

Feeding and foraging
This is a very adaptable forager, gathering pollen and nectar from plants with a variety of flower shapes.

Bombus lucorum agg.
FIRST RECORD 1890, Bognor Regis (Henry Guermonprez) **TOTAL RECORDS** 2,447

Geography and history
Since 2004 it has been known that the 'White-tailed Bumblebee' is in fact three very closely related species (it had already been recognised as two in the mid-1970s) and it is therefore treated as an aggregate. The three species in this grouping are *Bombus cryptarum*, *B. magnus* and *B. lucorum*.

Because they cannot currently be distinguished from each other on physical characters, their relative distributions are not well understood. As a result, although the aggregate is widely distributed in the county, it is not known which of the three species is definitely present in Sussex.

B. lucorum is, however, understood to be the most abundant and therefore the likeliest to be here, with *B. cryptarum* thought to potentially be in the county too. *B. magnus* is largely restricted to the north and west of Britain, and is the least likely to be in Sussex.

Out in the field and under the microscope
Only queens, with a single lemon-yellow band on the thorax and a pure white tail, and males, with yellow facial hair in combination with a white tail, can be identified. Even then, this will only confirm that they belong to the aggregate.

Workers from this aggregate cannot be identified with confidence. This is because they

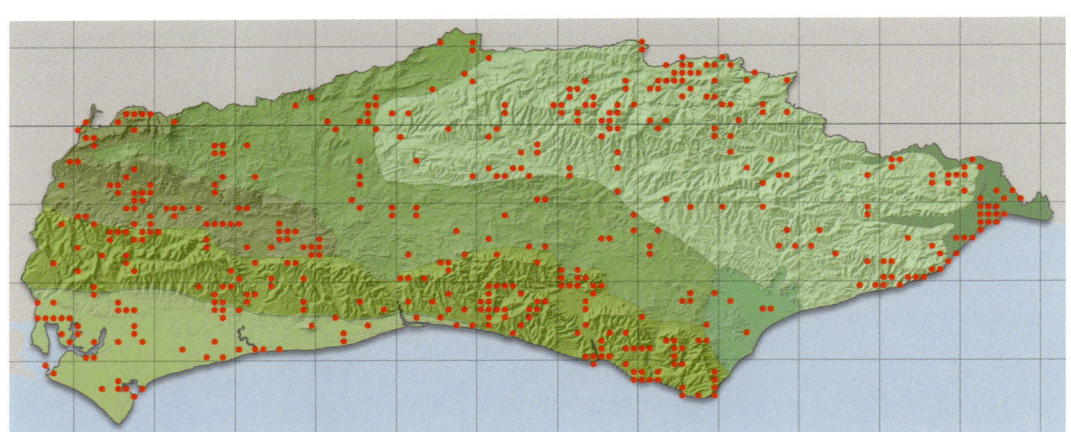

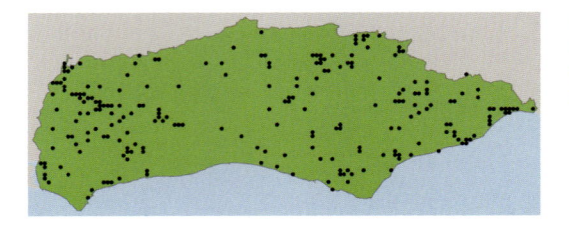

Bombus lucorum agg. queen [PB]

Large map: distribution 2000–2023, with red dots showing distribution of *Bombus lucorum* agg.
Small map: distribution 1844–1999, showing distribution of *Bombus lucorum* agg.

can be confused with some forms of *B. terrestris* workers which are very variable. For example, while many *B. terrestris* workers have a buff-coloured tail, others have a tail that is also pure white and cannot be distinguished from the species in the aggregate.

Finding *B. lucorum* agg. therefore means searching for newly emerged queens in early spring or for males later in the summer.

Bombus lucorum agg. ♂ visiting Wild Marjoram [PC]

Behaviour and interactions
Queens are among the earliest bumblebees to emerge each spring, establishing their nests below ground, often in abandoned small mammal burrows. Shredded grasses, mosses and leaf litter are used as nesting material.

B. bohemicus takes over the nests of the species in this aggregate but has greatly declined in Sussex and may now be extinct.

In common with other species of bumblebee, *B. lucorum* agg. nests are frequently attacked by the Bee Moth *Aphomia sociella*. Eggs are laid in the nest and the larvae feed on detritus and wax comb before starting to feed on the bumblebee larvae

Bee Moth (♀ left, ♂ right) larvae feed on bumblebee food stores and larvae [both PC]

Ground-ivy [PC]

On the wing
Sussex records run from mid February until the end of November.

Feeding and foraging
A wide range of plants is visited for pollen and nectar, including bramble, Ground-ivy, knapweed, plants in the Pea and Dead-nettle families, thistle, Viper's-bugloss and willow.

The Bees of Sussex

Bombus muscorum Moss Carder Bee
FIRST RECORD 1896, Tunbridge Wells area (Henry Guermonprez) **TOTAL RECORDS** 160

Geography and history
Bombus muscorum used to be widely distributed across Britain but has declined significantly, particularly at inland locations. In southern areas, it has largely retreated to the coast.

It was not, however, thought to be abundant even at the beginning of the twentieth century. Much of the decline has been driven by the destruction of large areas of flower-rich grassland. It remains widely distributed in parts of Scotland, particularly on islands and coastal areas.

In Sussex, it is an elusive bee. In common with national trends, it has disappeared from most inland areas of the county. Landscapes where it can still be found include the coast between Pett and Camber, the Pevensey Levels, the eastern South Downs, vegetated shingle near the mouth of the Cuckmere River, and grasslands upstream from Newhaven. It has also been known from Chichester Harbour, but has not been recorded there since 2011.

Pevensey Levels [SF]

Out in the field and under the microscope
During late spring and summer, the known locations are worth checking for foraging bees by focusing a search on clovers and vetches that are in flower. Any brown bumblebee encountered is worth a close look.

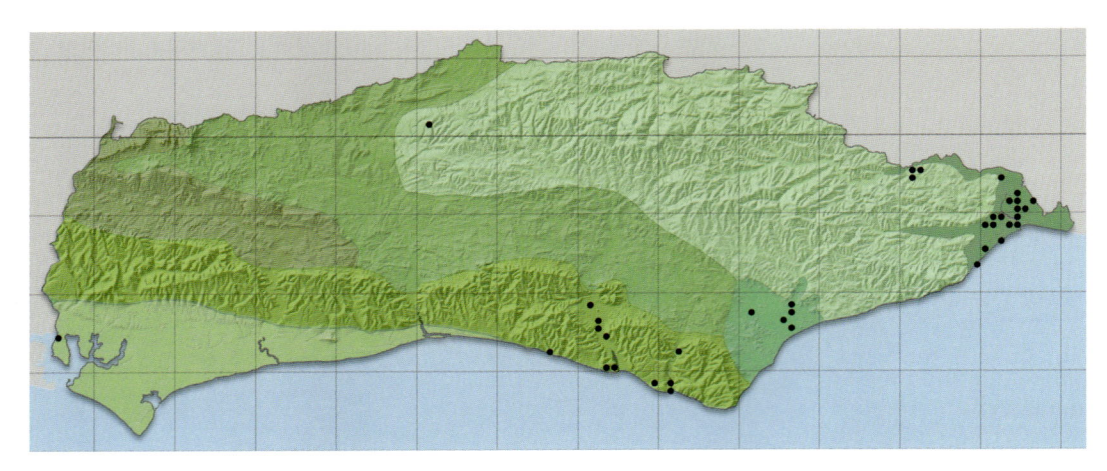

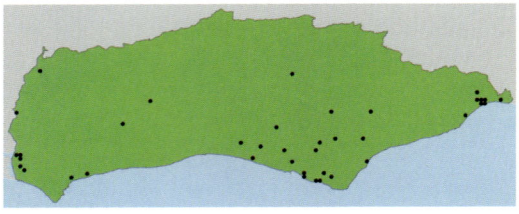

Large map: distribution 2000–2023
Small map: distribution 1844–1999

The Bees of Sussex

Queens, workers and males broadly have the same distinguishing features. One key character to check is the absence of any black hairs on the thorax or on the first five abdominal tergites. This separates *B. muscorum* from Sussex populations of *B. pascuorum* (which always have at least some black hairs on the abdomen) and from *B. humilis* (which should have at least some black hairs on the thorax, especially close to the wing bases).

Behaviour and interactions
Queens emerge from hibernation much later than most other species of bumblebee, generally appearing during May. This is a surface-nesting bumblebee that uses dead grasses and moss to build its nest. The top of the nest is often exposed to the sun. Colonies are small and number between 40 and 120 workers. Males generally start to appear from late June onwards, with daughter-queens active during August and September.

B. campestris takes over the nests of this species on the continent, and this may also be the case in Britain.

Bombus muscorum queen [SF]

Bombus muscorum worker [SF]

Bombus muscorum ♂ [SF]

On the wing
Sussex records run from late April until late September.

Feeding and foraging
B. muscorum collects pollen from a wide range of plants, but with a preference for plants in the Pea family, particularly clovers, vetches, everlasting-peas and vetchlings. Pollen is also collected from the Dead-nettle, Speedwell, Rose and Daisy families.

Bombus pascuorum Common Carder Bee

FIRST RECORD 1894, Bevendean (A. Brazenor) **TOTAL RECORDS** 4,182

Geography and history
Bombus pascuorum occurs throughout Sussex and can be especially abundant in gardens and parks. It is quite a distinctive bumblebee, and as a result it is the most frequently recorded bee in Sussex, accounting for almost seven per cent of all bee records.

Out in the field and under the microscope
B. pascuorum is most often seen foraging. Most predominantly brown bumblebees encountered are likely to be this species, especially if they are found in built-up areas. Queens, workers and males are broadly similar. There are three species that they could be confused with, particularly *B. humilis* and *B. muscorum* but also *B. hypnorum*. Typical specimens can be distinguished from the first two by the presence of black hairs on the abdomen, and from the usual form of *B. hypnorum* by the lack of white on the tail.

B. pascuorum is a very variable bumblebee, both in size and colouring. For example, some individuals have large numbers of black hairs on the thorax and abdomen and can appear very dark, while others have very few black hairs.

Behaviour and interactions
B. pascuorum nests either on the surface or just below the ground, generally in areas of tussocky

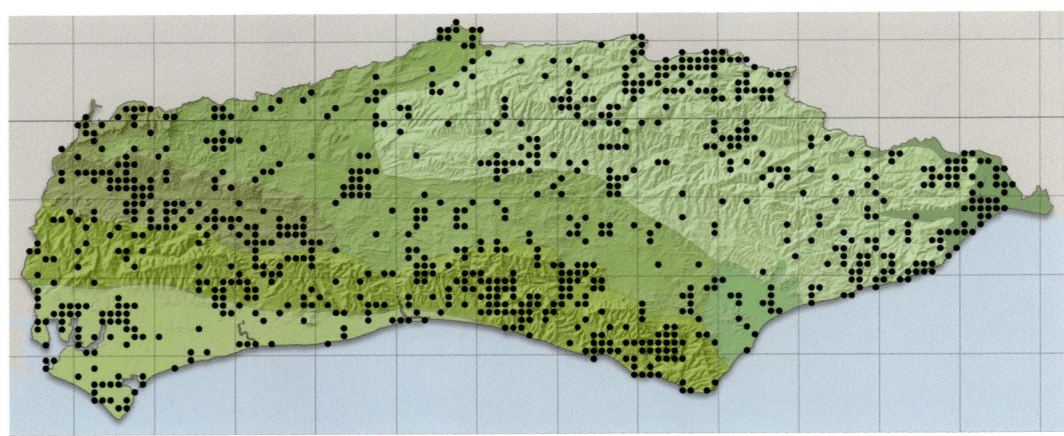

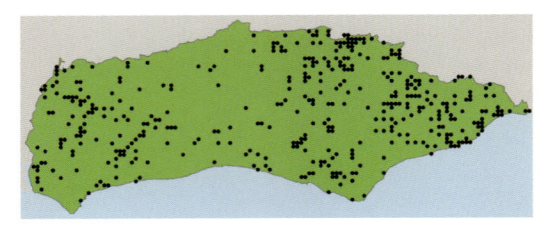

Bombus pascuorum **queen** [NO]

Large map: distribution 2000–2023
Small map: distribution 1844–1999

The Bees of Sussex

Bombus pascuorum worker [WP]

Bombus pascuorum ♂ [SF]

or unmown grassland. The nest is constructed of grasses and mosses, with perhaps as many as 50 per cent failing in any one year. The number of workers can be as low as 30 to 40 per colony, or as high as 150. Pollen-gathering trips might last some 45 minutes. One continental study showed that just over a fifth of the workers can live for 49 days or more.

Males emerge from the nest from May onwards, with daughter-queens from June onwards. This is one of the last bumblebees to still be active each autumn.

The wide variation in the size of this bee is thought to be because they compete for food as larvae in the nest.

Nests are taken over by the cuckoo *B. campestris*.

On the wing
Sussex records run from early March until mid November.

Feeding and foraging
This bumblebee has a strong preference for plants that produce flowers with a deep corolla, including the Borage family, Foxglove, knapweed, members of the Pea family such as Common Bird's-foot-trefoil and Red Clover, scabious, thistle and White Dead-nettle. Unusually, it is a pollinator of both a plant (Greater Knapweed) and its parasite (Knapweed Broomrape).

Sicus ferrugineus, a conopid fly, is a parasitoid of bumblebees. The fly deposits an egg in the bee's abdomen whilst in flight. Its larva develops inside the bee and pupates when the bee is finally killed. [PC]

Knapweed Broomrape [PC]

The Bees of Sussex

Bombus pratorum Early Bumblebee
FIRST RECORD 1882, Hastings (Edward Saunders) **TOTAL RECORDS** 1,769

Geography and history
Bombus pratorum is a very widespread species and is found in all parts of Britain except for the Western Isles, Orkney and Shetland. It is very widely distributed across Europe too and is found from the shores of the Mediterranean to the Arctic. It is one of the commonest species across Britain.

Although it is essentially a woodland species, it is well adapted to gardens and parks and can be found just about anywhere in Sussex.

Out in the field and under the microscope
This small, agile bumblebee can fly in low temperatures and, apart from winter-active *B. terrestris*, is often the first on the wing each spring. The yellow-banded and red-tailed queens and small workers are quite distinctive. These have a yellow band at the front of the thorax, a yellow band on tergite two, and a reddish-orange tail. The yellow band on tergite two is often missing in workers.

Males are very distinctive and have a reddish tail in combination with yellow hairs on the face, on the front of the thorax, and on the front of the abdomen.

The queens are occasionally seen searching for nest sites but, as with the workers and males, are most readily seen while foraging, often hanging beneath small flowers.

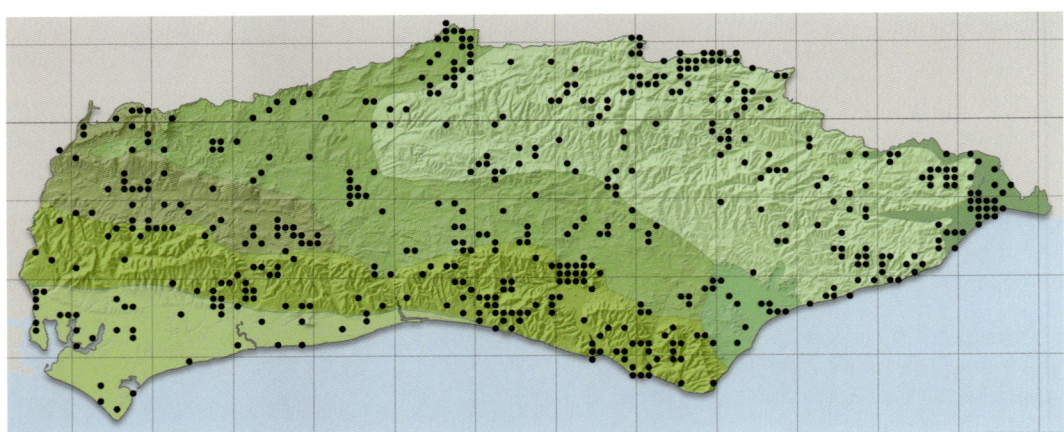

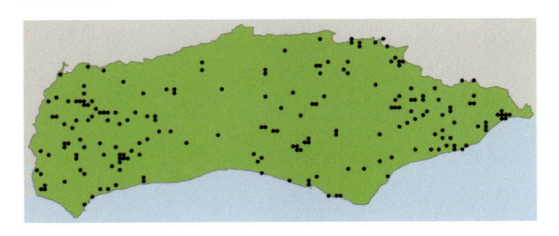

Bombus pratorum queen [SF]

Large map: distribution 2000–2023
Small map: distribution 1844–1999

The Bees of Sussex

Bombus pratorum worker [NO]

Bombus pratorum ♂ [PC]

Behaviour and interactions

B. pratorum was given the common name 'Early Bumblebee' because it is the first bumblebee species to complete its colony cycle each year. The first workers are often on the wing as early as late March, with daughter-queens and males emerging during May and June. Some of these new queens then go on to establish a colony that same summer, although the majority wait until the following spring, passing the rest of the summer and winter below ground.

Courtship behaviour involves males patrolling along a circuit and scent-marking the tips of branches. Queens nest in a variety of locations, including abandoned mouse and vole nests, cavities within trees, abandoned bird nests, and occasionally on the ground. Colonies usually have 40 to 100 workers, although as many as 200 has been reported. In one study, a high proportion of the workers and young queens present in the colonies studied had been born in other nests—the workers foraged and cared for the brood in their adoptive nest as normal.

Nests are taken over by the cuckoo *B. sylvestris*.

On the wing

This species is bivoltine in Sussex. Here, records run from the beginning of February until late October.

Feeding and foraging

This is a short-tongued bumblebee that forages from a wide variety of plants, including those with both short and long flower tubes. Among those visited are apple, bramble, comfrey, geranium, Green Alkanet, Ground-ivy, Hound's-tongue, White and Red Dead-nettle, Red Bartsia, Sycamore, Field Maple, Viper's-bugloss and willow. They also take advantage of holes cut in the base of a flower tube by species such as *B. terrestris*. This is so that they can take nectar that might otherwise be inaccessible. It is an important pollinator of soft-fruit.

A *Bombus pratorum* ♂ visiting Field Maple flowers [PC]

Bombus ruderarius Red-shanked Carder Bee
FIRST RECORD 1878, Brighton (A. Brazenor) TOTAL RECORDS 175

Geography and history
Bombus ruderarius was formerly widely distributed across England, parts of Wales and the Inner Hebrides, and in the early part of the twentieth century was described by Frederick Sladen in his book The Humble Bee as "a widely-distributed bee in Great Britain and Ireland, and common in many places".

During the twentieth century, however, it underwent a rapid and severe decline which coincided with the destruction of large areas of flower-rich grassland. This is now a very local bee, although it does have strong populations in locations with extensive species-rich grassland such as Salisbury Plain in Wiltshire and the Scottish islands of Coll and Tiree. It is now very scarce in Sussex and declining further.

Beachy Head: *Bombus ruderarius* is now only recorded intermittently on the eastern South Downs. [SF]

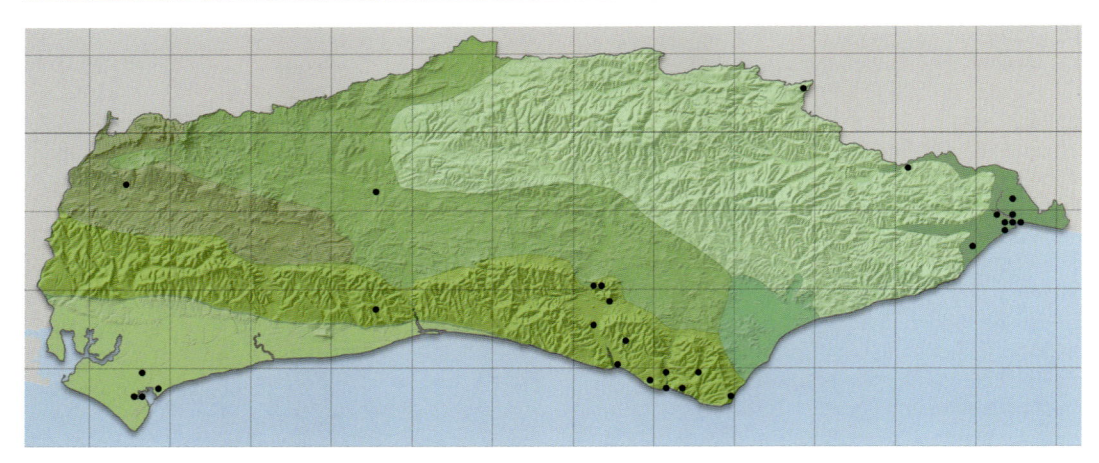

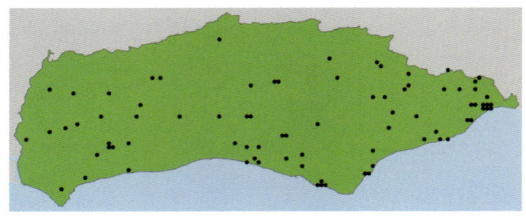

Large map: distribution 2000–2023
Small map: distribution 1844–1999

The Bees of Sussex

Historically, *B. ruderarius* was found right across Sussex, but the small number of modern records are now mostly from the South Downs between Lewes and Beachy Head, and from the coast between Pett and Camber Sands. There is also a handful of outlying records from scattered locations such as Great Dixter.

Bombus ruderarius queen showing the red hairs that fringe the pollen basket [SF]

Out in the field and under the microscope

This bumblebee closely resembles *B. lapidarius* and for this reason may be under-recorded—any black bumblebee with a red tail should be examined closely. Like *B. lapidarius*, *B. ruderarius* queens and workers have an all-black body except for the tail, which is a reddish colour, although this is a more orangey-red than the deep red tail of *B. lapidarius*.

Other key distinguishing features are its smaller and rounder shape and the red hairs fringing the pollen baskets on the hind legs. These are black on *B. lapidarius* and are just about visible to the naked eye. Another clue is that a feeding or foraging *B. ruderarius* worker is likely to fly off very rapidly when approached, whereas *B. lapidarius* is more likely to remain on the plant.

Males are more straightforward and, as well as an orangey-red tail, have black hairs on the face and very weak hair bands on the front of the thorax and on the abdomen.

Bombus ruderarius worker [SF]

Bombus ruderarius ♂ [PB]

Behaviour and interactions

B. ruderarius emerges from hibernation later than many of the more common species, with the queens appearing in April and early May. These establish their nests with pieces of grass and moss at, or just under, the surface in among tall vegetation.

Colonies are small and at their peak may have between 50 and 100 workers. Males and daughter-queens emerge from July onwards.

B. campestris is known to target this species on the continent, but this is not confirmed in Britain.

On the wing

Sussex records run from mid March until mid August. Elsewhere, it is reported to be active until September.

Feeding and foraging

This species visits a range of plants, including Black Horehound, bramble, Ground-ivy, Hound's-tongue and plants in the Pea family, as well as Viper's-bugloss and White Dead-nettle. The Pea family is an especially important source of pollen.

Bombus ruderatus Large Garden Bumblebee
(also known as the Ruderal Bumblebee)
FIRST RECORD 1877, Brighton Downs (A. Brazenor) TOTAL RECORDS 143

Geography and history
This bumblebee species was very widespread during the early twentieth century—it used to be a regular visitor to gardens—but is now very scarce. Most modern records are from Romney Marsh, with additional recent records from scattered locations such as Ebernoe Common, Great Dixter, Lewes, Midhurst and Robertsbridge. It is most closely associated with sites that hold extensive areas of flower-rich grassland, especially those supporting Red Clover.

More recently, there is evidence that this bumblebee has made a small recovery, particularly in areas where there has been targeted conservation work. A good example of this is the work of the Short-haired Bumblebee Reintroduction Project, centred on Romney Marsh. Here, the population of *Bombus ruderatus* is thought to have increased eightfold during the first ten years of the project, which ran between 2009 and 2022.

Bombus ruderatus queen [SF]

Out in the field and under the microscope
This bumblebee is most likely to be found foraging, but it is a tricky species to identify as it closely resembles *B. hortorum* (as well as the extinct *B. subterraneus*).

As with *B. hortorum*, typical queens and workers have two yellow bands on the thorax and one on the front of the abdomen. This last band is very weak and indistinct in *B. ruderatus* and is restricted to tergite one, whereas in *B. hortorum* there are at least some yellow hairs on tergite two.

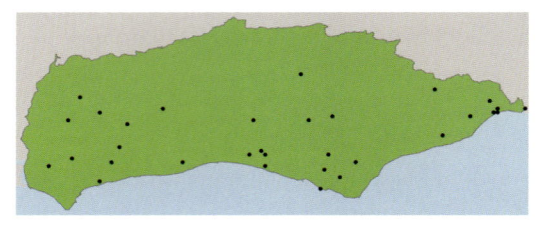

Large map: distribution 2000–2023
Small map: distribution 1844–1999

The Bees of Sussex

Bombus ruderatus worker [SF]

Bombus ruderatus worker, black form [SF]

Bombus ruderatus ♂ [SF]

Red Clover, Rye Harbour Nature Reserve [PG]

    B. ruderatus males also resemble *B. hortorum* but have reddish-gold—as opposed to black—hairs on the mandibles. Overall, males, queens and workers tend to have a neater covering of hairs than this other species.

    B. ruderatus is quite variable in its colouring, with all-black individuals not uncommon. This can be a good clue and any all-black bumblebee should be checked, especially if it is flying with bumblebees that have three yellow bands and a white tail.

Behaviour and interactions

This bumblebee nests underground, often using disused small mammal burrows. Workers appear from the middle of May, with mature nests holding as many as 150 workers. Males and daughter-queens are reported to be active from the middle of June or early July.

    Although this has not been confirmed, it is thought likely that *B. barbutellus* takes over the nests of this species as well as those of *B. hortorum*.

On the wing

Sussex records run from late March until mid August, although elsewhere it is reported to be active until mid October.

Feeding and foraging

B. ruderatus visits a range of plants, including bramble, cat-mint, comfrey, Common Mallow, Field Beans, Foxglove, Ground-ivy, Marsh Woundwort, members of the Pea Family such as Red and White Clover, plus Red and White Dead-nettle and Viper's-bugloss.

Bombus rupestris Red-tailed Cuckoo Bee
(also known as the Hill Cuckoo Bee)
FIRST RECORD 1894, Brighton Downs (A. Brazenor) TOTAL RECORDS 239

Geography and history
Bombus rupestris used to be widely distributed across England and Wales, its range extending as far north as Northumberland. From the 1940s onwards, it then underwent a significant decline. It has now re-established itself in many of its former locations and has become abundant in some areas. It remains scarce in south-west England and Wales and had reached Scotland in 2023.

In Sussex, this bumblebee is most often found on the South Downs and along the coast between Pett and Camber Sands. However, it does use a wide range of habitats so could be found elsewhere in Sussex.

Out in the field and under the microscope
To find *B. rupestris*, it is worth checking every bumblebee with a red tail. The females are large and very distinctive: the combination of their size, a large head, a red tail and very smoky wings means that this is one of the easier bumblebees to identify. The females also lack the smooth, shiny surface on the hind tibiae found in the otherwise similar species *B. lapidarius* and *B. ruderarius*.

Like the females, males are black with a red tail, but they are trickier to identify, not least as they are quite variable and can resemble *B. ruderarius* males. Identification should be confirmed by examining the genital capsule. Yellow-collared individuals of both sexes are known.

Generally, *B. rupestris* males and females are found at rest or feeding, but males are also often found patrolling.

Behaviour and interactions
B. rupestris is the last cuckoo bumblebee to emerge each year. It is a cuckoo of *B. lapidarius*, and it seems that queens of the host species are generally stung and killed by the parasite. More than one *B. rupestris* female might also attempt to occupy the same nest with only one of them surviving.

Males adopt mate-seeking behaviour, typically flying in a line across areas of open grassland. Unusually, more than 50 widely dispersed males were observed in 2022 flying within 1 m of the

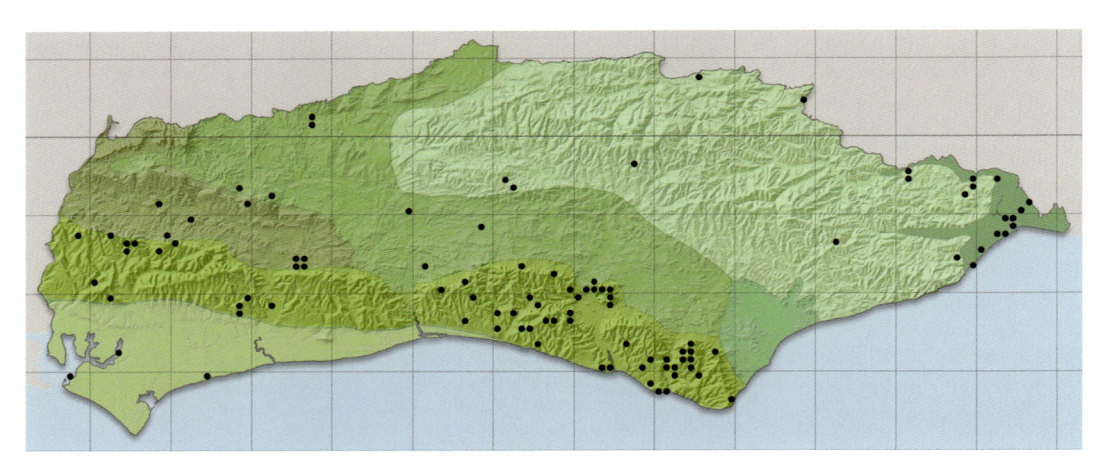

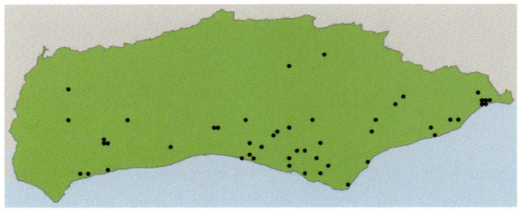

Large map: distribution 2000–2023
Small map: distribution 1844–1999

Bombus rupestris ♀, with mites [PG]

Bombus rupestris ♂ [PG]

Bombus lapidarius queen, the host [SF]

ground across Swanborough Hill, an extensive area of chalk grassland.

Males and females emerge from the nest in July or August, with mated females generally entering hibernation before the end of August.

On the continent, *B. sylvarum* and *B. pascuorum* nests are also targeted.

On the wing
While Sussex records run from late March until mid September, this is more typically a late-emerging species, with queens generally active from late May.

Feeding and foraging
This bumblebee feeds from a wide range of plant species, including Devil's-bit Scabious, Field Scabious, Ground-ivy, knapweed, mallow, Marsh Thistle, Ox-eye Daisy, Self-heal, Small Scabious, Spear Thistle, Viper's-bugloss and willow, as well as plants in the Pea family such as White Clover and Sainfoin.

Bombus soroeensis Broken-belted Bumblebee
FIRST RECORD 1876, Brighton Downs (Frederick Smith) TOTAL RECORDS 27

Geography and history
Bombus soroeensis used to be found throughout Britain but its range is now very much reduced, and it is only doing well in Scotland. There, it has expanded its range, colonising Orkney in 2011. It seems to require very extensive areas of flower-rich grassland, and its decline in other areas of Britain coincided with the destruction of much of this habitat. As well as its Scottish strongholds, its principal populations are now restricted to the west coast of Wales and to Salisbury Plain in Wiltshire. Elsewhere, it is possibly overlooked.

The earliest reference to *B. soroeensis* in Sussex is by Frederick Smith, writing in 1876. He describes how "a few specimens have been found on the Downs, near Brighton". Two years later, in August 1878, two specimens of *B. soroeensis* were collected near Ovingdean. These survive in the Booth Museum which also holds a further 58 specimens collected by Alfred Brazenor from the same area between 1943 and 1948. The final records for the county were mapped as part of the Bumblebee Distribution Map Scheme and span the period 1961 to 1976. These are from the eastern South Downs, from near Lewes and from near Hailsham.

While these represent the last records for the county, *B. soroeensis* was known from the Lydd Ranges between 1998 and 2014, and may still be present here. While much of this site is in Kent, it does extend into Sussex. Although none of the

Bombus soroeensis queen [SF]

Large map: distribution 2000–2023, including records from Lydd Ranges in Kent
Small map: distribution 1844–1999

Bombus soroeensis worker [SF]

Bombus soroeensis ♂ [SF]

records are from the Sussex side of the county boundary, the habitat is of comparable quality throughout the ranges, and there is every reason to presume that *B. soroeensis* was present in Sussex at this time. As well as Lydd Ranges, *B. soroeensis* may well be present at other nearby coastal sites such as Rye Harbour Nature Reserve.

Out in the field and under the microscope
This is a small bumblebee. As with *B. lucorum* agg. queens and workers and white-tailed *B. terrestris* workers, *B. soroeensis* queens and workers have a yellow band on the front of the thorax, a yellow band on tergite two and a white tail. Unlike these other species, however, the yellow band on tergite two usually extends onto tergite one at the sides.

Males have black hairs on the face, a yellow hair band at the front of the thorax, yellow bands on tergites one and two, and a tail that can be white or buff.

This bumblebee is most frequently found foraging or feeding. A number of Alfred Brazenor's male specimens in the Booth Museum are accompanied with the observation "On Greater Knapweed".

Behaviour and interactions
Queens emerge from hibernation later than most other bumblebee species and establish their nests during June and July, usually below ground in abandoned small mammal nests. The number of workers produced averages 100, but can be between 80 and 150. This species can forage at particularly low temperatures, with 1°C reported from Scotland.

No species of cuckoo bumblebee targeting *B. soroeensis* is known from Britain..

On the wing
Sussex records run from mid July until the beginning of October.

Feeding and foraging
In southern Britain, this bee forages from plants in the Pea family such as Common Bird's-foot-trefoil, Meadow Vetchling, melilot, Red Clover, Sainfoin and White Clover, plus plants such as bramble, Clustered Bellflower, Devil's-bit Scabious, Field Scabious, knapweed, thistle, Red Bartsia, Viper's-bugloss and White Dead-nettle.

Bombus sylvestris Forest Cuckoo Bee
FIRST RECORD 1905, Hastings (Edward Saunders)　TOTAL RECORDS 294

Geography and history
This cuckoo bumblebee could be found just about anywhere its various hosts occur, but especially in places where its main host *Bombus pratorum* is found. It occurs throughout most of Sussex and is often seen in gardens, parks and woodlands. Most records are from the South Downs, High Weald and the Lower Greensand heaths.

Out in the field and under the microscope
Males are often seen feeding, but are sometimes found patrolling around a short circuit. For example, four were observed in woodland near Isfield in 2014, flying within one metre of the ground along a short circuit.

Females are much more elusive, but might be seen feeding or as they search for the nest of a potential host in the spring

This is a small bumblebee. The females have a yellow band at the front of the thorax, a faint yellow band at the front of the abdomen, and a white tail. They also have a characteristically curved tip to the abdomen.

Typical males have a broad yellow band at the front of the thorax, white hairs on tergites three and four, followed by black hairs on tergites five and six that grade into pale orange hairs at the tip of the abdomen. Males have particularly long hairs on the hind tibiae.

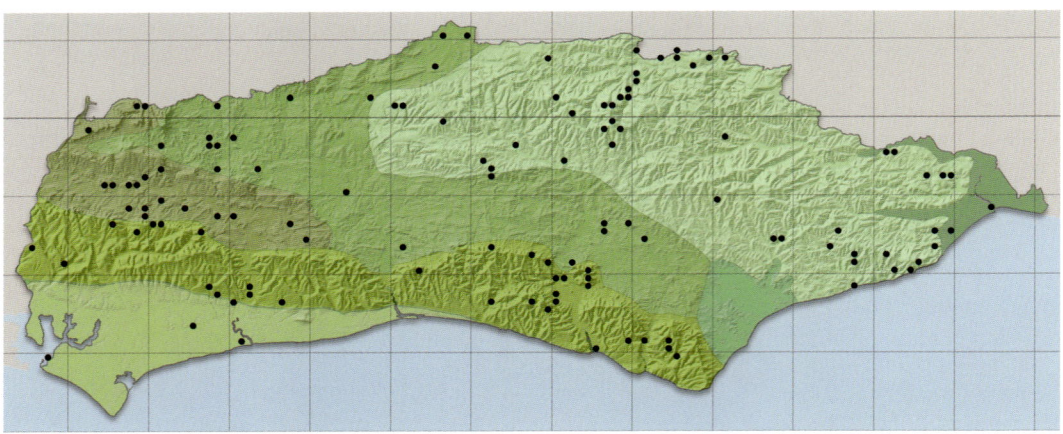

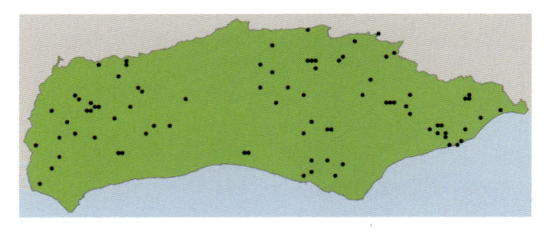

Bombus sylvestris ♀ [CKL]

Large map: distribution 2000–2023
Small map: distribution 1844–1999

74　*The Bees of Sussex*

Behaviour and interactions

This is usually the first of the cuckoo bumblebees to emerge from hibernation each year. Its principal host is *B. pratorum*, although it is also believed to take over the nests of *B. jonellus*, *B. hypnorum* and *B. monticola* (absent in Sussex). These are all members of a group of closely related bumblebees.

The typical pattern of behaviour is for mated females to emerge from winter hibernation in April and shortly afterwards to invade the nest of a host queen. It seems that *B. sylvestris* does not exhibit aggressive behaviour towards the host and that both host and parasite may co-exist and successfully produce offspring. The new generation of males and females generally emerges during May.

Bombus sylvestris ♂ [PB]

Unlike other species of cuckoo bumblebee, *B. sylvestris* can produce a second generation in southern counties including Sussex, mirroring the behaviour of its principal host.

Bombus pratorum queen, the principal host [PB]

On the wing
Sussex records run from early March until early October.

Feeding and foraging
This bumblebee feeds from a very wide range of plant species. Females have been recorded on Bugle, buttercup, Crab Apple, dandelion, Horse-chestnut, Mahonia, mallow, sedum, White Dead-nettle and willow, while males have been recorded feeding on Common Bird's-foot-trefoil, bramble, Hound's-tongue, knapweed, thistle, thyme, Viper's-bugloss and White Clover.

Bombus terrestris Buff-tailed Bumblebee
FIRST RECORD 1882, Hastings (Edward Saunders) TOTAL RECORDS 3,385

Geography and history
This is a very widespread bumblebee that can be found just about anywhere in the county. It can be very abundant wherever it is found.

Out in the field and under the microscope
Bombus terrestris closely resembles the three species referred to collectively as *B. lucorum* agg., and only males, queens and some workers are distinguishable from these other species. Males, for example, have black facial hair (as opposed to yellow in these other species), while the queens always have a buff-coloured tail.

Workers, on the other hand, are quite variable, and only those with a buff-coloured tail or with a faint brown band above a white tail can be identified as *B. terrestris*. Many individuals, however, have an all-white tail and are very similar in appearance to *B. lucorum* agg. workers. For this reason, workers with a tail that is completely white cannot be identified to species except through DNA analysis.

The very large queens can readily be found in March and April as they search through undergrowth, prospecting for nest sites. They are also frequent visitors to willow trees. Workers are usually seen as they forage, often vibrating

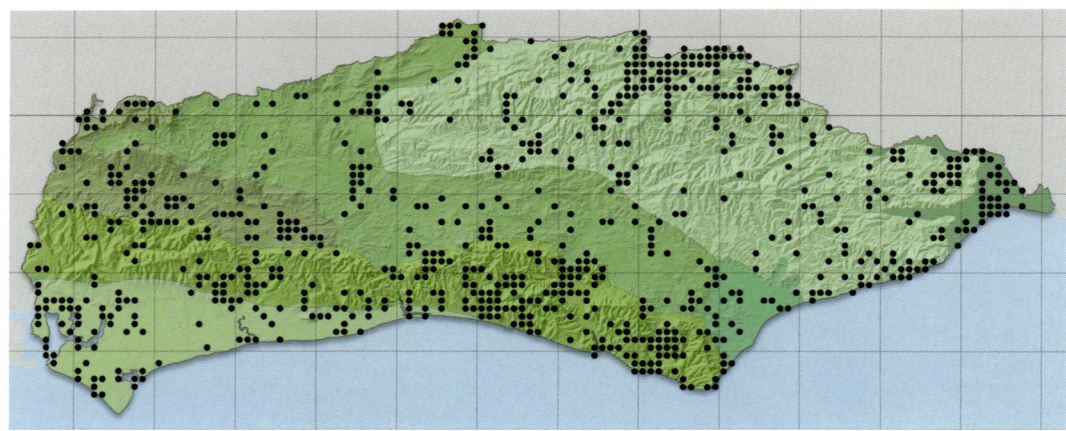

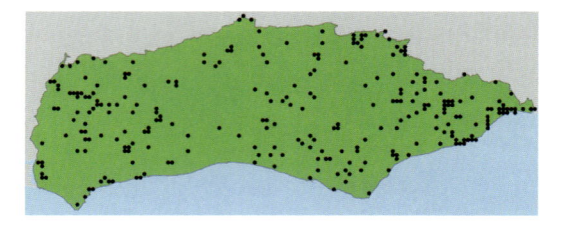

Bombus terrestris queen [SF]

Large map: distribution 2000–2023
Small map: distribution 1844–1999

76 *The Bees of Sussex*

Bombus terrestris worker with a faint brown band of hairs above a white tail [SF]

Bombus terrestris ♂ [SF]

Bombus terrestris daughter-queens emerging from nest [PG]

their wing muscles to shake pollen free from plants such as poppies and tomatoes. Workers from winter-active colonies are frequent visitors to winter-flowering plants such as Mahonia.

Behaviour and interactions

Typically, mated queens emerge in late winter or early spring. After some initial feeding, they search for a suitable nest site which is generally in an abandoned nest created by a small mammal. Workers are active from the middle of May, and numbers can exceed 300 at a colony's peak. Males and daughter-queens normally appear between July and October.

There has, however, been a recent shift away from this pattern. Some colonies are now being established in the autumn and remain active during the winter months, with varying degrees of success. This means workers, daughter-queens and males can be seen throughout the year.

B. vestalis is a widespread cuckoo of this species. There is also evidence that *B. terrestris* will behave as a cuckoo, taking over the nests of other *B. terrestris* queens.

On the wing

B. terrestris is potentially active at any time of the year, although most Sussex records run from March or April until late October.

Feeding and foraging

This bumblebee is a very versatile forager and feeds from an extremely wide range of plant species. Pollen is collected from plants in the Cabbage, Carrot, Dead-nettle, Pea, Plantain, Poppy, Purple-loosestrife, Rose and Teasel families. It is also known to bite holes in the base of flowers that have a deep flower tube, gaining access to nectar that it would not otherwise be able to reach.

Bombus vestalis Vestal Cuckoo Bee
(also known as the Southern Cuckoo Bee)
FIRST RECORD 1894, Brighton (A. Brazenor) TOTAL RECORDS 939

Geography and history
This is a widely distributed species that for many years was restricted to England and Wales. It appears, however, to be expanding its range, and it was reported from south-east Scotland for the first time in 2009 and had re-established itself in Ireland in 2014 (the last previous record had been in 1926). It is very common in southern England.

Like its host *Bombus terrestris*, this bumblebee is found throughout Sussex. It is the most frequently recorded cuckoo bumblebee in the county and is often abundant where it is found.

Out in the field and under the microscope
B. vestalis occurs wherever its host is found. Females can be identified by their large size, smoky wings, and largely black colouring that is combined with a yellow collar, plus lemon-yellow flashes just above a white tail. At the tip of the abdomen, on the underside, there is also a distinctive V-shaped swelling.

Although they are smaller and slimmer than the females, males are still large. Like the females, they have a broad yellow band at the front of the thorax and, in fresh specimens, yellow flashes above a white tail. Another good character to check is the length of the third antennal segment which should clearly be shorter than the fifth, a character shared with *B. barbutellus* and *B. campestris*.

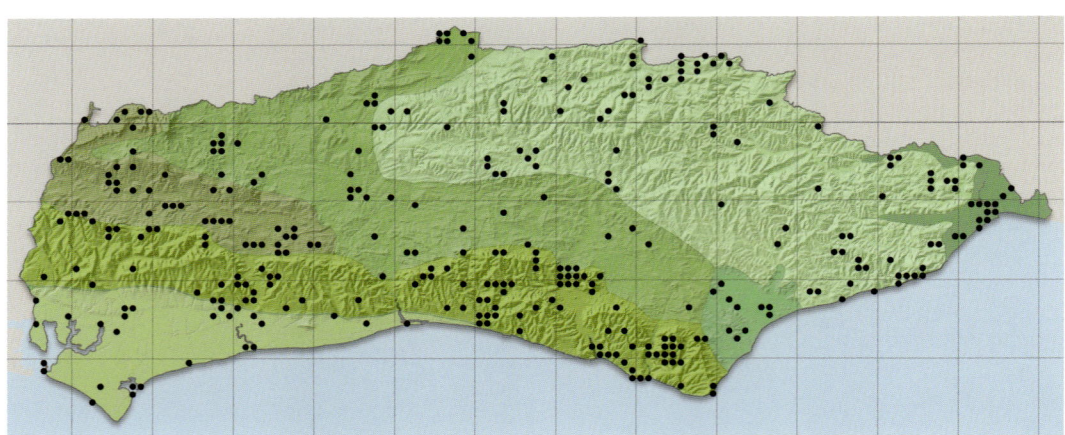

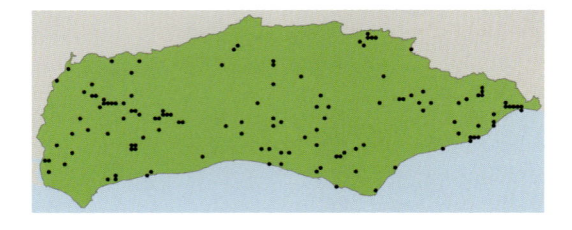

Bombus vestalis ♀ [SF]

Large map: distribution 2000–2023
Small map: distribution 1844–1999

The Bees of Sussex

Bombus vestalis ♂s on Musk Thistle [PC]

The very large females have a slow ungainly flight and are most often seen feeding or resting on flowers. The males are often seen at rest but also gather in large numbers to feed on a single patch of a suitable foodplant. Gardens are among the best places to find this species.

Behaviour and interactions
Females emerge from hibernation in early spring and, after a period of feeding, seek out nests of the host species, *B. terrestris*. To successfully invade a nest, the cuckoo may need to defend itself from attacks by workers and often kills the host queen a few days after invading the nest. The cuckoo will also destroy the host's eggs and larvae. Some of the surviving workers will manage to lay eggs and, as they are unmated, will only produce male offspring.

The new generation of males and females emerge towards the end of May or in early June.

Bombus terrestris **queen, the host** [SF]

On the wing
Sussex records run from mid March until the end of August.

Feeding and foraging
This bumblebee feeds from a wide range of plants, including Blackthorn, Bluebell, Common and Tufted Vetch, clover, Crab Apple, dandelion, Devil's-bit and Field Scabious, Green Alkanet, Ground-ivy, knapweed, Red and White Dead-nettle, Teasel, thistle, Viper's-bugloss and willow. Males can often be seen feeding on cat-mint in gardens.

Ceratina – Carpenter Bees

This is a large genus of solitary bees found worldwide. They excavate their linear nests within pithy plant stems, each nest starting where there is a break in the stem. Adults emerge in the summer and pass the winter within excavated plant stems, often in small groups. Pollen is transported on the hind legs and within the crop. Just one species, *Ceratina cyanea*, is known from Britain and is found in Sussex.

Ceratina cyanea Little Blue Carpenter Bee

FIRST RECORD 1911, Midhurst (J. G. Dalgleish) TOTAL RECORDS 144

Geography and history

Ceratina cyanea is restricted to southern England, especially to counties in the south-east. In Sussex, this bee is usually found in warm, scrubby places on heathlands, on the downs and in open-canopied woodlands in the west of the county. Levin Down, for example, is a good location, as are the downs between Devil's Dyke and Lullington Heath. Key requirements are warmth and the presence of pithy-stemmed plants such as bramble, Buddleia and rose.

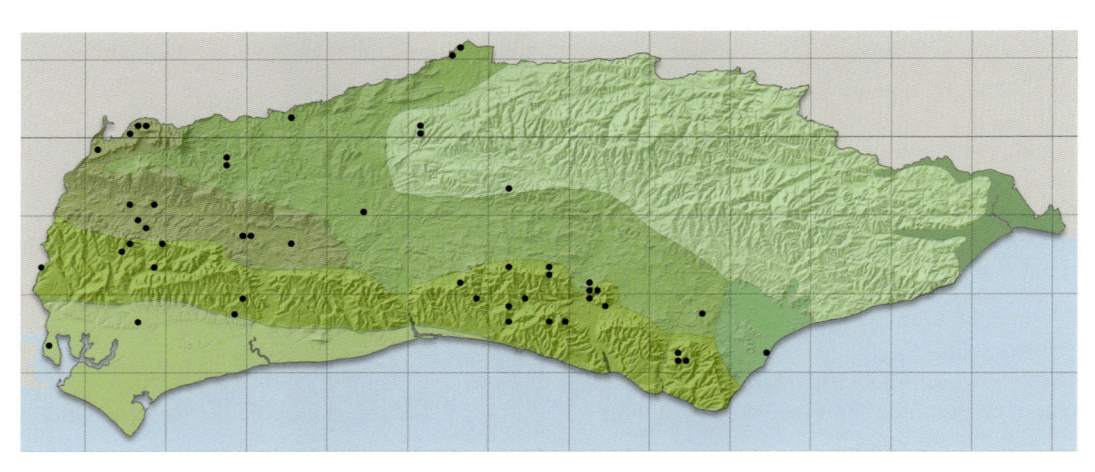

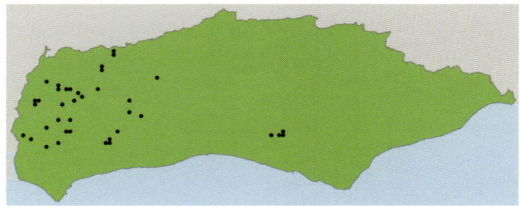

Levin Down, with its abundant scrub, supports a strong population of *Ceratina cyanea* [SR/SWT]

Large map: distribution 2000–2023
Small map: distribution 1844–1999

The Bees of Sussex

Ceratina cyanea ♀ [SF]

Ceratina cyanea ♂ [CKL]

Ceratina cyanea ♂ [PB]

Developing larvae in bramble stem [PW]

Out in the field and under the microscope

This is a very distinctive bee that is unlikely to be confused with any other species—it is small, has an elongated body that widens towards the tip, a long tongue and dark, metallic-blue colouring. Males have cream-coloured markings on the clypeus and labrum.

Despite its distinctiveness, it is very elusive and not easy to find—searching along the edges of bramble brakes and patches of scrub on south-facing slopes may be successful. It might also be found feeding on flower heads.

An alternative approach is to gather bramble stems in late winter. Breaking these open will sometimes lead to the discovery of overwintering *C. cyanea* adult bees.

Behaviour and interactions

Individual bees can live up to 18 months. After mating in May, females search for the broken end of a dead stem of a plant such as bramble or thistle and excavate a tunnel within the pith. These stems might be on or close to the ground. Two females may even use opposite ends of the same stem.

Within the stem, a series of brood cells is then provisioned with a small pollen mass. An egg is laid on each lump of pollen and the individual cells separated from each other by loose particles of chewed pith. A female may remain in the nest to guard the brood, sometimes staying with her adult offspring until the following year.

New adult bees emerge in August and September and then overwinter in an excavated plant stem.

On the wing

Most Sussex records run from late April until the end of September, although this bee can be found at other times of the year too.

Feeding and foraging

C. cyanea forages from a wide range of plants, including those in the Bellflower, Daisy, Dead-nettle, Pea and Pink families. Confirmed sources of pollen include buttercup, Common Bird's-foot-trefoil, cinquefoil and Yellow-rattle, as well as species in the Daisy family.

Epeolus – Variegated Cuckoo Bees

This is a genus of cuckoo bees with black and white markings, all of which lay their eggs within the brood cells of various species of *Colletes*. Females make a cut through the cellophane-like lining of the brood cell wall using the tip of the abdomen. An egg is then laid in the cell wall. This genus is very distinctive, although separating the species is not always straightforward. All species have white patches on the abdomen that are made up of short felt-like hairs. Two species are known from Britain, *Epeolus cruciger* and *E. variegatus*. Both of these occur in Sussex.

Epeolus cruciger Red-thighed Epeolus
FIRST RECORD 1935, Camber Sands (Owain Richards) TOTAL RECORDS 183

Geography and history
Epeolus cruciger has a very distinctive distribution in Sussex with two main centres of population, one on the Lower Greensand heathlands and the other in the High Weald between Ashdown Forest and the area south of Tunbridge Wells. There are also records from the coast around Rye, plus isolated locations such as Chailey Common and East Head. Iping Common, Graffham Common and Rye Harbour Nature Reserve are all reliable places to find this bee.

Heathlands such as Iping Common are reliable places to search for *Epeolus cruciger*. [ME]

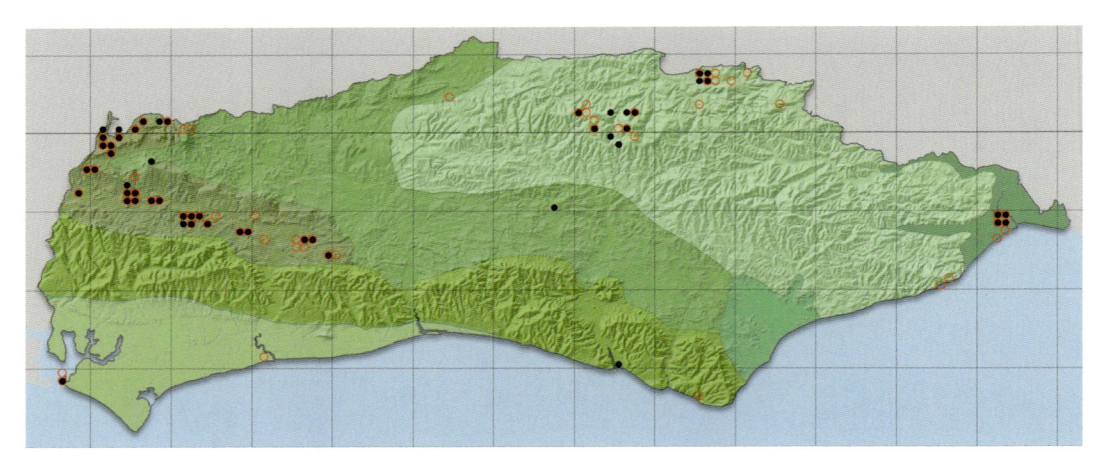

Large map: distribution 2000–2023, with orange circles showing combined distribution of *Colletes marginatus* and *C. succinctus*
Small map: distribution 1844–1999

82　*The Bees of Sussex*

Out in the field and under the microscope
E. cruciger can be particularly abundant on heathland sites close to nests of its main host *Colletes succinctus*. Here, female *E. cruciger* might be seen flying slowly as they search for suitable nests.

E. cruciger closely resembles the other species in the genus, *E. variegatus*, but females of the two species can be separated by checking characters such as the shape of the underside of the tip of the abdomen—straight in this species, gently curved in *E. variegatus*—and the colour of the labrum and femora, which are usually red in *E. cruciger* and dark in *E. variegatus*. Male *E. cruciger* have a red plate at the tip of the abdomen, but this is black in *E. variegatus*.

Epeolus cruciger ♀ [SF]

Epeolus cruciger ♂ [SF]

Behaviour and interactions
E. cruciger targets the brood cells of *C. succinctus*, *C. marginatus* and possibly *C. fodiens*. In some parts of the continent, this species also parasitises *C. hederae*, and recent observations from Midhurst support the view that this relationship may also have become established in Britain.

Colletes succinctus ♀, the principal host [PC]

Colletes marginatus ♀, a secondary host [SF]

The first stage larvae have large mandibles that are used to destroy the eggs or larvae of the host before consuming the provisions that have been stored in the brood cell.

Ongoing work may show that *E. cruciger* encompasses other closely related species. The occurrence of more than one possible host species and differences in flight season are two of the clues that this may be the case.

On the wing
Sussex records run from early June until mid September.

Feeding and foraging
This bee has been recorded feeding from plants such as clover, Common Ragwort, hawkbit, Heather, mint, Sheep's-bit and Wild Thyme.

Epeolus variegatus Black-thighed Epeolus
FIRST RECORD 1873, Littlehampton (Edward Saunders) TOTAL RECORDS 115

Geography and history
The distribution of *Epeolus variegatus* across Britain is not dissimilar to *E. cruciger*, and it can be found as far north as southern Scotland and as far west as Cornwall and the west coast of Wales.

In Sussex, it is especially frequent on the heathlands of the Lower Greensand. It might also be found in several other locations, including coastal habitats such as cliff-top grassland, sand dunes and saltmarsh, plus isolated sandpits, quarries, and open heathy woodland. Stedham Common, the dunes at Littlehampton, Hargate Forest and Rye Harbour Nature Reserve have all generated recent records.

Out in the field and under the microscope
This bee can be hard to find, and the best approach is to search in the vicinity of nests of its various possible hosts, including low, vertical banks on the edge of saltmarsh for *Colletes halophilus* and vertical sandy banks for *C. daviesanus*. Here, females may be prospecting for suitable brood cells in which to lay their eggs.

In cooler temperatures, it is also worth searching the tops of grass stems close to the nest site. This is where this small dark bee can sometimes be seen, using its mandibles to clasp onto the stem near the top of the stalk. The bees can be perpendicular to the stem and look as though they have been glued head on.

Like its close relative *E. cruciger*, this is a tricky bee to identify to species. The two are very similar in appearance and also have an overlapping distribution. There is also the potential for other similar-looking species in the genus to colonise Britain from the continent.

Epeolus variegatus ♀ at *Colletes halophilus* nest [PG]

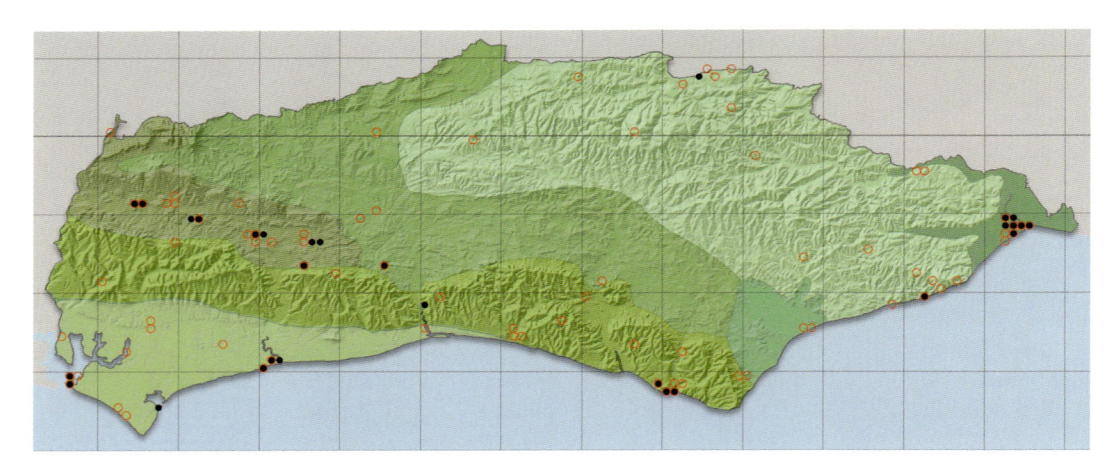

Large map: distribution 2000–2023, with orange circles showing combined distribution of *Colletes daviesanus*, *C. fodiens*, *C. halophilus* and *C. similis*
Small map: distribution 1844–1999

Epeolus variegatus ♀ roosting at the top of a grass stem in cool weather [CB]

Epeolus variegatus ♂ [SF]

Like *E. cruciger*, *E. variegatus* is variable in its colouring. Typical females have a black labrum and black femora, while the underside of the tip of the abdomen is slightly concave. Males have a black plate at the tip of the abdomen.

Behaviour and interactions

The lifecycle of this species is very similar to that of *E. cruciger*, with females seeking out the nests of the host species and laying an egg between the inner and outer linings of an unsealed cell. On hatching, the larvae use their sickle-shaped mandibles to destroy the egg or larva of the host before consuming the provisions that have been collected.

This bee is a cuckoo of several species of *Colletes*, although many of the relationships need confirmation. In 1948, Alfred Brazenor reared three specimens that were "bred from cells formed by *C. daviesanus*". These are in the Booth Museum. Other hosts listed in the literature include *C. fodiens*, *C. halophilus* and *C. similis*. It seems likely that it is very fluid in its choice of host. In 2023, several *E. variegatus* were observed investigating nests of enormous *C. hederae* colonies, one near Plumpton and the other at Seven Sisters

The occurrence of more than one possible host species and differences in flight season suggest that *E. variegatus* may yet prove to be more than one species.

Colletes daviesanus ♀, a host [PC]

Colletes halophilus ♀, a host [SF]

On the wing
Sussex records run from late June until mid October.

Feeding and foraging
As well as visiting members of the Daisy family such as Common Ragwort and Common Fleabane, this species has been recorded feeding from Bog Pimpernel and Creeping Buttercup, as well as plants in the Borage, Carrot, Dead-nettle, Pea and Rose families.

Eucera – Long-horned Bees

Males in this genus of ground-nesting solitary bees have spectacularly long antennae and a protruding clypeus. Pollen is moistened, probably with nectar, and transported on the hind legs. This genus is parasitised by species of *Nomada*. Two species, *Eucera longicornis* and *E. nigrescens* (now extinct), have been known from Britain, including from Sussex.

Eucera longicornis Long-horned Bee
FIRST RECORD 1890, Storrington (Fordham) TOTAL RECORDS 249

Geography and history
This bee is much declined but may be expanding its range in some areas of Britain, including in Sussex. Most records are from areas of flower-rich grassland in the Low and High Weald.

An area of grassland adjoining Gatwick Airport and the garden and meadows at Great Dixter support significant populations, while grasslands close to Ebernoe Common have also been a regular source of records. Other recent locations include Batemans, Nymans and Rye Harbour Nature Reserve.

Nest site (left) and a ♀ within a nest burrow (below) at Gatwick Airport Biodiversity Area [both RB]

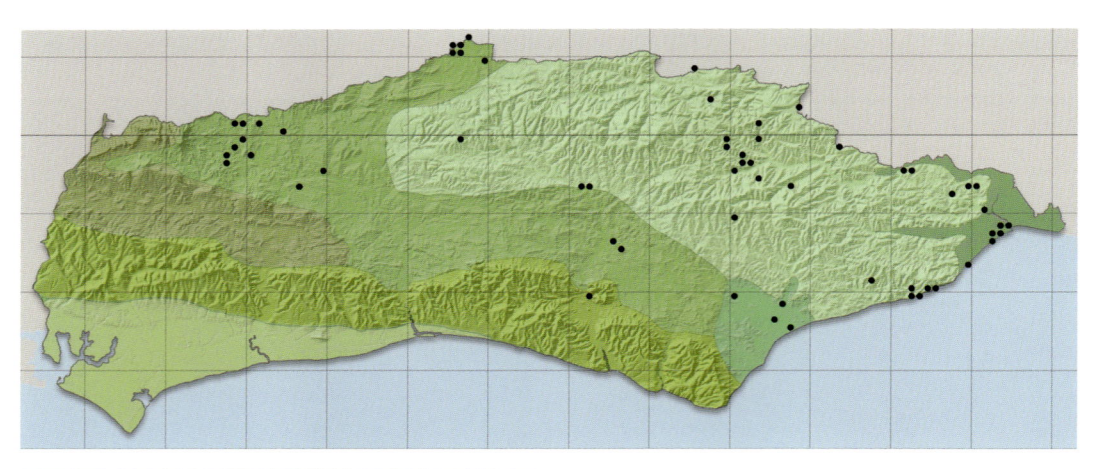

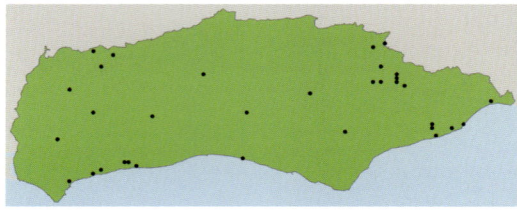

Large map: distribution 2000–2023
Small map: distribution 1844–1999

Out in the field and under the microscope

Eucera longicornis is usually found in low numbers, although in some locations, especially those close to a nest site, it can be numerous. In flower-rich grasslands, the silvery-grey males are often seen flying rapidly at or just below the height of flowering plants and grasses. Both males and females are frequently found feeding or foraging on plants in the Pea family, and especially on patches of Meadow Vetchling. These are always worth checking.

Males are characterised by their long antennae, while females superficially resemble *Anthophora*. Among other differences, however, they have two—as opposed to three—sub-marginal cells.

E. longicornis very closely resembles the extinct *E. nigrescens*, but female *E. longicornis* have a short tooth at the tip of the front basitarsi which is absent in *E. nigrescens*. In males, the hind basitarsi widen towards the tip and are clearly down-curved, while they are parallel-sided in *E. nigrescens*.

Eucera longicornis ♀ [SF]

Eucera longicornis ♂ on bird's-foot-trefoil, Gatwick Airport [RB]

Behaviour and interactions

Males emerge from the nest first and patrol patches of foodplants, seeking out females. They fly rapidly just above the sward, with their antennae held erect. Once mated, the females typically excavate a nest in bare or sparsely vegetated ground. Occasionally, two females may share the same nest. Cells are coated in a liquid that ensures the mixture of pollen and nectar does not seep into the soil.

Patches of Meadow Vetchling are worth checking for *Eucera longicornis* [NS]

Nest sites are not easy to find. However, there are recent records of nesting aggregations sited on three low and sparsely vegetated banks of clay soil close to Gatwick Airport. In 2018, one of these banks was alive with returning females and patrolling males, all within 20 m of the nest site.

Nomada sexfasciata, extinct in Sussex, is a cuckoo of this species.

On the wing

Sussex records run from the beginning of May until the beginning of August.

Feeding and foraging

E. longicornis females collect pollen from plants in the Pea family such as Bush Vetch, Greater Bird's-foot-trefoil, and especially Meadow Vetchling. Ian Beavis has observed males feeding on Red Clover and Bugle, while Andy Phillips has observed this bee feeding on Two-flowered Everlasting Pea in the gardens at Great Dixter.

Melecta – Mourning Bees

The British species in this genus are robust black bees with white hair patches on the sides of the abdomen. They lay their eggs within brood cells belonging to different species of *Anthophora* and other closely related bees. Unusually, the *Melecta* target brood cells that have already been sealed. Two species have been known from Britain, including Sussex—*Melecta albifrons* and *M. luctuosa*. This last species is extinct.

Melecta albifrons Common Mourning Bee
FIRST RECORD 1894, Polegate (A. Brazenor) TOTAL RECORDS 112

Geography and history
Melecta albifrons is found throughout Sussex, its range closely matching the distribution of its principal host *Anthophora plumipes*. Some coastal locations with soft-rock cliffs such as Galley Hill and Glyne Gap or those with low eroding banks such as Chichester Harbour can support good populations. It is also frequently found in gardens.

Out in the field and under the microscope
This bee is rarely as numerous as its host and is usually found in ones or twos, often as it feeds in gardens. It can appear to be quite a lethargic bee and is often seen at rest on a flower or leaf.

At *A. plumipes* nest sites, it is often very active and very abundant, and these are generally the most reliable locations to search for

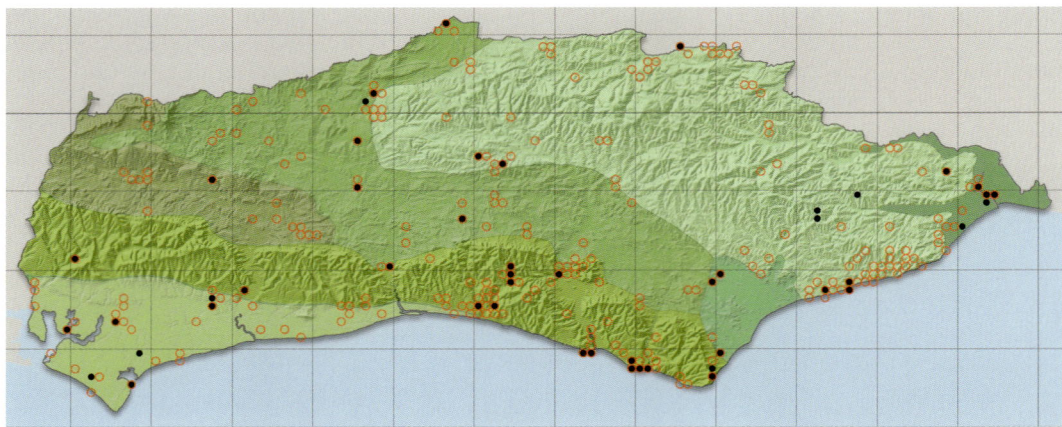

The cliffs here at Galley Hill and nearby at Glyne Gap are reliable places to find *Melecta albifrons*. [PG]

Large map: distribution 2000–2023, with orange circles showing distribution of *Anthophora plumipes*
Small map: distribution 1844–1999

The Bees of Sussex

Melecta albifrons ♀ [NO]

Melecta albifrons ♂ [AP]

The black form of *Melecta albifrons* closely resembles its principal host (see right) [PC]

Anthophora plumipes ♀, the principal host [PC]

it. In these places, females can be seen close to nest entrances or at rest on a cliff face, while the males fly rapidly close to the surface, searching for females.

This species is nearly always completely black apart from white hair patches on tergites one, two, three and four (and could only ever have been confused with the extinct *M. luctuosa*).

All-black forms of *M. albifrons* are not uncommon and could be confused with females of either of the host species, *A. plumipes* and *A. retusa*.

Behaviour and interactions

This spring bee lays its eggs within *A. plumipes* and *A. retusa* brood cells. During mating, the male clasps the female in such a way that she is unable to fly off. Mated females then look for a provisioned, closed cell of the host species and, on finding one, break into the cell and lay an egg before resealing the cell opening with moistened soil. On hatching, the larvae use their large mandibles to destroy the host larvae or eggs before consuming the pollen and nectar provisions.

This species overwinters as an adult bee within the brood cell. Female *A. plumipes* have been seen dragging *M. albifrons* out of cells.

On the wing

Sussex records run from late March until early June.

Feeding and foraging

M. albifrons has been recorded visiting apple and Blackthorn, as well as plants in the Cabbage, Mint, Borage and Dead-nettle families.

Nomada – Nomad Bees

This is a large genus of slender bees which are sometimes mistaken for wasps—many have yellow and black markings, while others are red and black. They lay their eggs within the brood cells of a variety of different species from several different bee genera, but principally *Andrena*. Eggs are laid in unsealed cells, usually in a pocket that the female *Nomada* excavates into the cell wall. The number of species recognised in Britain is likely to rise in the near future. This is because work is ongoing to determine whether some species should be separated into two or more distinct species. In addition, three species have colonised Britain in the last five years, and others are likely to follow. Currently 34 species are currently recorded from Britain, 28 of which have been found in Sussex.

Nomada alboguttata Large Bear-clawed Nomad Bee
FIRST RECORD 2017, Rye Harbour Nature Reserve (Chris Bentley) TOTAL RECORDS 7

Geography and history
This bee is a new arrival in Britain. It was first recorded in 2016 in Kent and was found in Sussex the following year at Rye Harbour Nature Reserve. In 2018, 2020 and 2023 it was recorded again at Rye Harbour Nature Reserve. In 2021 a male and female were found near Lewes, where it was again recorded in 2022 and 2023.

As its host *Andrena barbilabris* can be such an abundant bee, it seems likely that *Nomada alboguttata* will continue to expand its range. It should therefore be looked for in any locations known to support populations of *A. barbilabris*. This last bee is found in sandy

Sand patch at Rye Harbour Nature Reserve, where *Nomada alboguttata* was first recorded in Sussex in 2017 [PG]

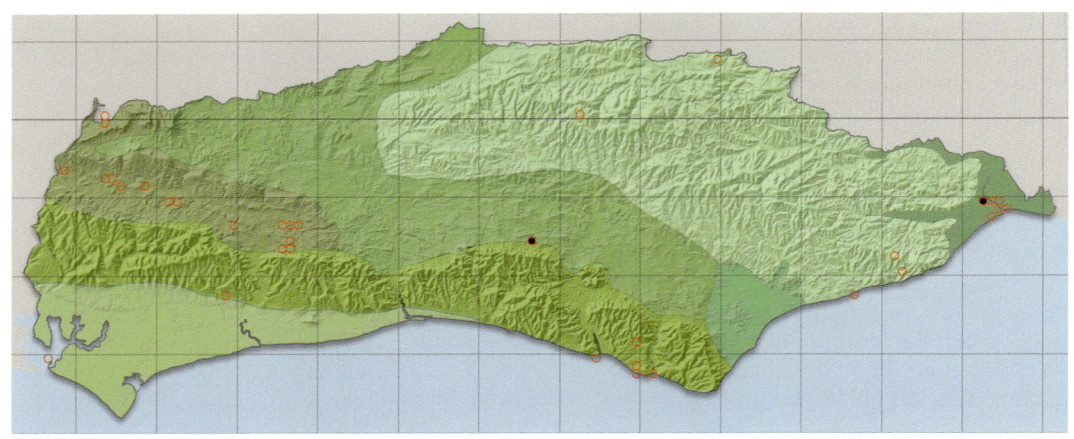

Large map: distribution 2000–2023, with orange circles showing distribution of *Andrena barbilabris*
Small map: distribution 1844–1999

90   *The Bees of Sussex*

Nomada alboguttata ♀ [SF]

Nomada alboguttata ♂ [CB]

places, including the heathlands of the Lower Greensand, sandpits across Sussex, and sandy coastal habitats.

Out in the field and under the microscope
N. alboguttata is a reddish-brown and black bee with cream-coloured markings on the abdomen. It is very similar in appearance to the closely related *N. baccata*. These two species are unique among British *Nomada* in that both males and females have very distinctive spines on the hind tibiae which resemble a bear's claws. *N. alboguttata* flies earlier in the year than *N. baccata* and is a larger insect.

A. barbilabris nesting aggregations are the best places to check for *N. alboguttata*, and females might be found searching here for suitable nests.

These two *Nomada* attack different hosts in Britain—*N. baccata* targets *A. argentata*—which means that identification of these is a first indication as to which of the *Nomada* is likely to be present.

Behaviour and interactions
N. alboguttata lays its eggs within *A. barbilabris* brood cells. On the continent, other forms of *N. alboguttata* are known, each with a different host and slight variations in flight period and colouration. It is thought these could be hidden species or physical adaptations to the different hosts.

Andrena barbilabris ♀, the host [PC]

On the wing
Sussex records run from early April until early May.

Feeding and foraging
The only sources of information on this are from European literature, with Alder Buckthorn, dandelion, hawkweed and willow all listed as being visited.

Nomada argentata Silver-sided Nomad Bee
FIRST RECORD 1846, Arundel (S. Stevens) TOTAL RECORDS 10

Geography and history
Nationally, *Nomada argentata* has become much scarcer, matching the decline of its host *Andrena marginata*. In Kent, for example, it was last recorded in 1918. In Sussex, it is restricted to chalk grassland, while elsewhere in the country it is also known from acid grassland sites.

N. argentata was first recorded in Sussex in 1846 near Arundel and again 98 years later in 1944, and then in 1946 and 1948. These records are from the Brighton Downs near Ovingdean. On his visit during August 1944, Alfred Brazenor collected 13 specimens here on three different dates, an indication that *N. argentata* was reasonably abundant at this time.

N. argentata then went unrecorded on the Brighton Downs until it was recorded by George Else near Falmer in 1976 and then at Castle Hill in 1977 and 1997.

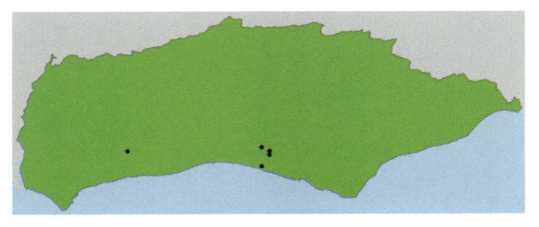

Large map: distribution 2000–2023, with orange circles showing distribution of *Andrena marginata*
Small map: distribution 1844–1999

92 The Bees of Sussex

Until very recently, the 1997 record was the last for the county. In 2020, however, it was found in a completely new part of Sussex, on the Devil's Jumps by Mike Edwards. This is an area that had been extensively surveyed over the years and is where its host had recently been re-recorded after a gap of 31 years. Castle Hill and the Devil's Jumps are the only locations for its host, *A. marginata*, in Sussex.

Out in the field and under the microscope

N. argentata is an extremely elusive bee and is best searched for by checking Field Scabious and Small Scabious flower heads on scabious-rich chalk grassland sites that support populations of the host.

This is a small dark bee with a reddish-brown and black abdomen and thorax, and is one of the few *Nomada* where the males and females lack yellow markings on the abdomen. Females are most like *N. integra* and *N. facilis* (not known from Sussex). Among the characters to look for to separate *N. argentata* from these other two species are the dense patches of silvery-white hairs on the sides of the propodeum and, in most cases, the presence of reddish marks on the sides of the thorax.

Males are variable and can be very dark, but, like the females, they have dense patches of silvery-white hairs on the sides of the propodeum.

Behaviour and interactions

N. argentata takes over the nests of its host, *A. marginata*. It is one of the last species of *Nomada* to appear each year. Males emerge about a week before the females. There have been observations of females following the host to its nest site, while a cluster of bees can sometimes be seen on a flower head.

Andrena marginata ♀, the host [PC]

On the wing

Sussex records run from mid July until mid August. In the literature, it is reported that this species is on the wing until mid September.

Feeding and foraging

In Sussex, *N. argentata* feeds principally from Field Scabious and Small Scabious. It has also been seen on Creeping Thistle and Red Bartsia. Elsewhere in the country, it is also known from acid grassland sites where it mainly visits Devil's-bit Scabious. In these places, *N. argentata* and its host fly slightly later in the year.

Field Scabious [PC]

Small Scabious [PC]

Nomada baccata **Small Bear-clawed Nomad Bee**
FIRST RECORD 1910, Midhurst (Henry Guermonprez) TOTAL RECORDS 24

Geography and history
This species is strongly associated with heathland, all but one of the Sussex records being from the heaths on the Lower Greensand. The one outlying record is from Rewell Wood which is where *Nomada baccata* was recorded in 1982. Ambersham Common and Iping Common have provided most of the modern records, although none since 2007. The last Sussex record is from Weavers Down in 2008.

Nationally, *N. baccata* has a very restricted distribution and has mainly been recorded from Dorset, Hampshire and Surrey, as well as from west Sussex.

Iping Common is one of just three locations where *Nomada baccata* has been recorded since 2000 [SWT]

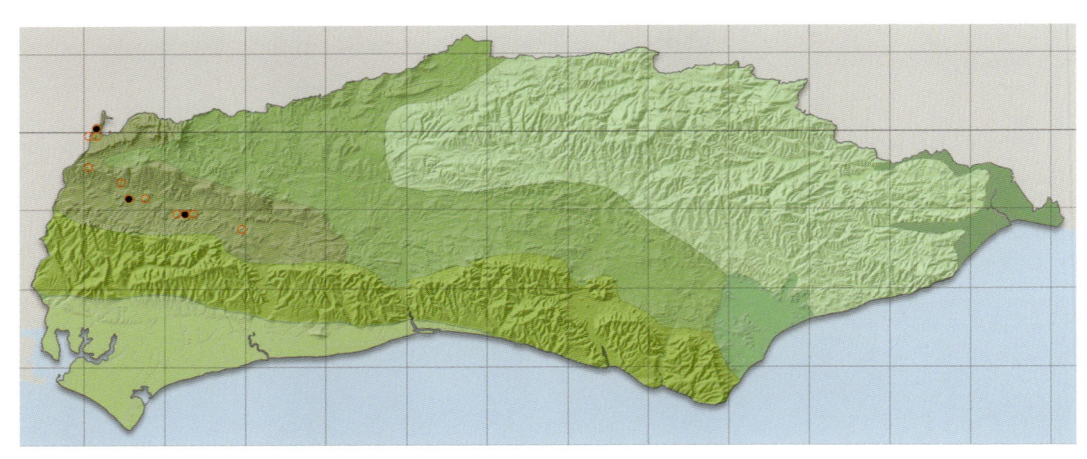

**Large map: distribution 2000–2023, with orange circles showing distribution of *Andrena argentata*
Small map: distribution 1844–1999**

94 *The Bees of Sussex*

Out in the field and under the microscope

This reddish-brown *Nomada* is similar in appearance to *N. alboguttata* but much smaller. The females are extensively marked with red on the head and thorax and have creamy white markings on the tergites. The males resemble the females but are white-marked on the antennal scapes, on the lower clypeus and on the mandibles.

Like *N. alboguttata*, this species has distinctive spines on the hind tibiae that resemble a bear's claws—this is the best character to use to separate this bee from all *Nomada* except *N. alboguttata*. Differences in size and host will help to distinguish between these two species.

N. baccata is found on heathlands close to nesting aggregations of its host species *Andrena argentata*. These tend to be sited on areas of loose, very hot and dry sand, where males of the *Nomada* can be seen patrolling and females can be seen testing the ground for suitable nest burrows of the host.

Nomada baccata ♀ [ME]

Nomada baccata ♂ [JE]

Behaviour and interactions

The males emerge first, a few days ahead of the females. They fly fast and low above an area of bare sand, waiting for the females to emerge. As they do so, the males respond immediately, grabbing hold of a female and attempting to mate. Two or three males may try to mate with the same female, forming a frenetic ball of activity that rolls around on the ground.

Andrena argentata ♀, the host [PB]

Once mated, the females start a search across the site of an *A. argentata* aggregation, looking for a suitable nest within which to lay an egg, sometimes following a female host back to her burrow.

N. baccata can be very abundant in some years, and as many as 50,000 have been reported by David Baldock on a visit to Hankley Common in Surrey.

On the wing

Sussex records run from late June until late August.

Feeding and foraging

N. baccata feeds on plants such as bramble, Common Ragwort, Creeping Thistle, heather and thyme.

Nomada conjungens Fringeless Nomad Bee
FIRST RECORD 1970, Laughton (Alfred Jones) **TOTAL RECORDS** 8

Geography and history
Nationally, there are indications that the population of *Nomada conjungens* is increasing, a view that is supported by the number of recent Sussex records.

It was unknown from the county until a specimen was collected in 1970 near Laughton but was not seen again until 2005 when a female was collected on Mount Caburn. Since then, it has been found on a further eight occasions. These records are from six very widely separated places: the Heyshott escarpment, Mount Caburn (where it was recorded for a second time in 2008), Mayfield, Fishbourne, Rye Harbour Nature Reserve and Lewes Cemetery (where it was most recently recorded in 2022).

Mount Caburn (in the foreground), where *Nomada conjungens* was recorded again in Sussex after a gap of 35 years. Since then, it has been recorded with increasing frequency. [SF]

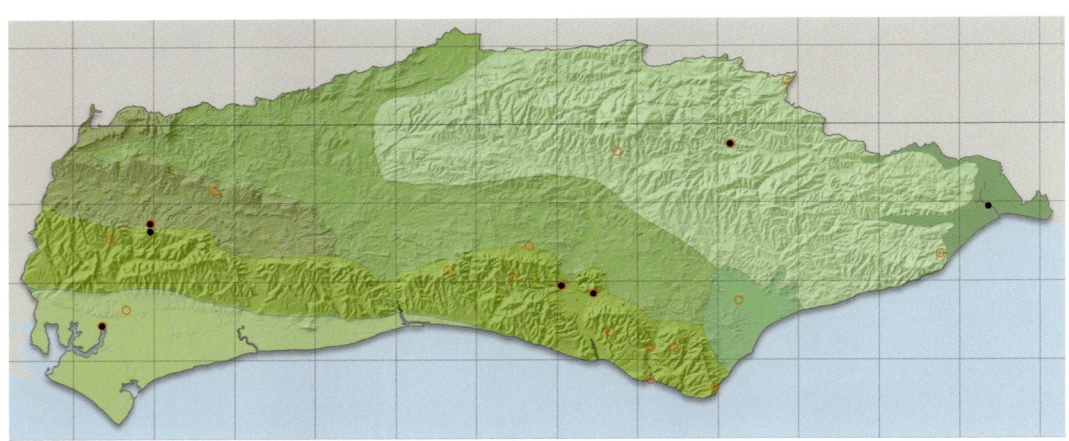

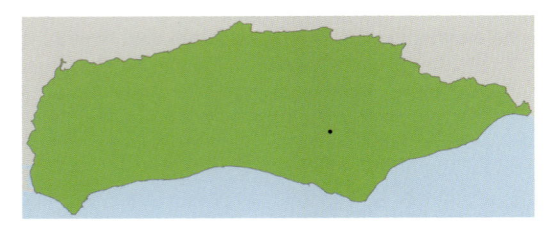

Large map: distribution 2000–2023, with orange circles showing distribution of *Andrena proxima*
Small map: distribution 1844–1999

The Bees of Sussex

The host of *N. conjungens* was long thought to be a single species, *Andrena proxima*. However, in 2019 this last species was recognised as two distinct species, *A. ampla* (not known from Sussex) and *A. proxima*. Following this separation, investigations are underway to determine whether *N. conjungens* is also two species.

Out in the field and under the microscope
This is a reddish-brown *Nomada* which closely resembles *N. flavoguttata*, and, as in this species, females have a dark labrum and red markings on the top of the thorax. However, female *N. conjungens* lack the tufts of silvery-white hairs found on the sides of the propodeum in this other species.

In males the thorax is completely black, while the underside of the front femur is fringed with very short hairs. The labrum and lower clypeus are yellow, while the extensive black and reddish-brown markings on the abdomen give males a smoky appearance.

Nomada conjungens ♀ [SF]

Nomada conjungens ♂ [TF]

Its host, *A. proxima*, is more abundant than this *Nomada*, but it is still not found very often. As *A. proxima* and *N. conjungens* are usually found together, this *Nomada* is best searched for by checking sites that support plants in the Carrot family, and also by looking for both host and parasite. *A. proxima* specialises in collecting pollen from plants in the Carrot family such as Cow Parsley and Hemlock Water-dropwort.

Andrena proxima ♀, the host [SF]

Behaviour and interactions
In 2022 a group of approximately ten males was observed in Lewes Cemetery flying rapidly around each other in a tight swirling mass within a metre of the ground. This was within a small area of rank grassland in dappled sunlight, close to male and female *A. proxima* feeding nearby.

On the wing
Sussex records run from late February until late October.

Feeding and foraging
There are records of this species visiting plants in the Carrot family such as Alexanders, Cow Parsley, and Hemlock Water-dropwort, as well as visiting spurge.

Nomada fabriciana Fabricius' Nomad Bee
FIRST RECORD 1879, Hastings (Edward Saunders) **TOTAL RECORDS** 394

Geography and history
This bee can be found just about anywhere in Sussex—it is very widely distributed throughout the county. It uses many different habitats, which reflects the range of places used by its host species. These include heathland, chalk grassland and woodland, and it is also found in gardens.

Out in the field and under the microscope
Nomada fabriciana is most likely to be found flying slowly close to a nest site. In common with other species of *Nomada*, it flies just above the surface of the ground searching for a suitable nest. It might also be found feeding, head down in a flower such as dandelion, Bugle or Pignut.

In typical females, the combination of a reddish-brown abdomen with four small yellow spots and a black labrum makes this one of the easier *Nomada* to identify. In the field the pale tips to the

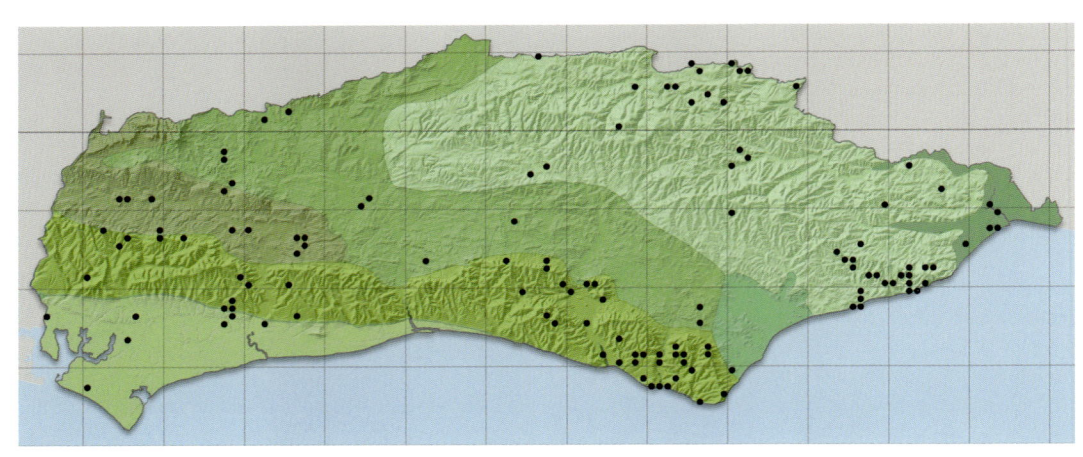

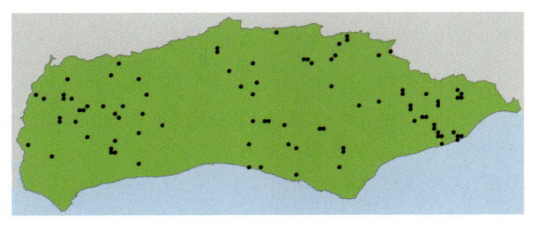

Nomada fabriciana ♀ [SF]

Nomada fabriciana ♂ [PB]

Large map: distribution 2000–2023
Small map: distribution 1844–1999

98 *The Bees of Sussex*

antennae form a strong contrast to the adjoining antennal segments which are very dark. With the aid of a microscope, it is also possible to see that each mandible has two teeth. This character is shared with just one other species in the genus, *N. ruficornis*. This last species, however, has extensive yellow markings on the abdomen.

Males are very like the females, only smaller and slimmer, and with entirely dark antennae.

Behaviour and interactions

N. fabriciana lays its eggs within the brood cells of several different species of *Andrena*, including *Andrena angustior*, *A. bicolor*, *A. chrysosceles*, *A. flavipes*, *A. nigroaenea* and *A. varians*. Of these, *A. bicolor* is understood to be the principal host.

The size of the host varies widely, which in turn means that adult *N. fabriciana* vary significantly in size—the larger females (possibly from *A. nigroaenea* nests) can be twice the size of smaller specimens.

Andrena bicolor ♀, the principal host [PC]

Andrena angustior ♀, an alternative host [NO]

Andrena chrysosceles ♀, an alternative host [NO]

Andrena nigroaenea ♀, an alternative host [PC]

On the wing

This species has two flight seasons, the first running from early March until the beginning of June and the second running from June until early August.

Feeding and foraging

A wide range of plants are visited, including those in the Daisy, Spurge, Speedwell, Pink, Rose and Willow families.

Nomada ferruginata Yellow-shouldered Nomad Bee
FIRST RECORD 2016, Rye Harbour Nature Reserve (Thomas Wood) TOTAL RECORDS 7

Geography and history
Nomada ferruginata is restricted to southern England. Its numbers and distribution have fluctuated considerably over time. In the latter half of the twentieth century, for example, it was a very scarce and declining species across Britain, since when it is thought to have increased its range.

There have been just seven confirmed records of this bee in Sussex. It was first seen at Rye Harbour Nature Reserve in 2016 and then again here on three separate dates in 2021, and it has also been seen twice near Ifold, close to the Surrey border.

In other parts of the country, it is known from habitats such as coastal shingle (as at Dungeness in Kent, close to Rye Harbour), woodland, heathland and parkland. The key factor is the presence of willow, the main source of pollen for its host *Andrena praecox*.

The host, *Andrena praecox*, is closely tied to willow-rich habitats, and these are worth searching for *Nomada ferruginata* too. [PG]

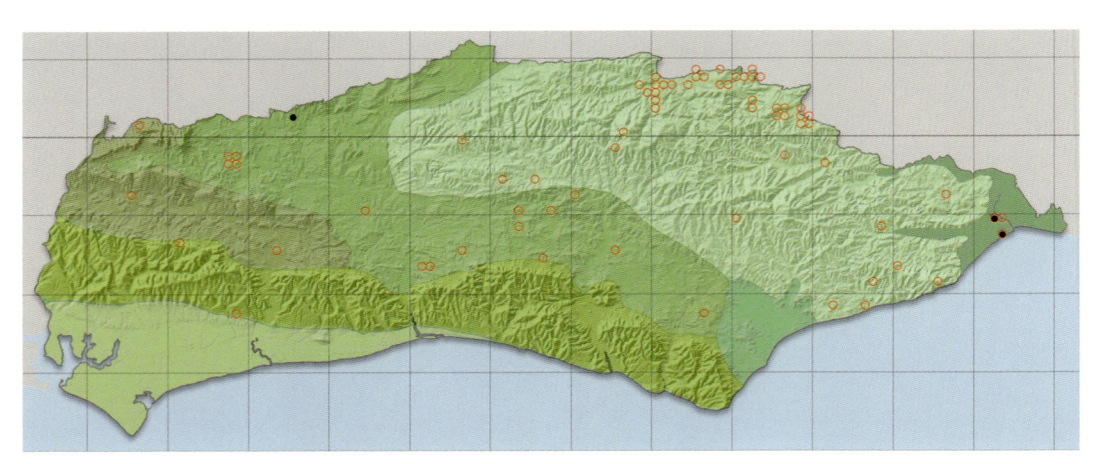

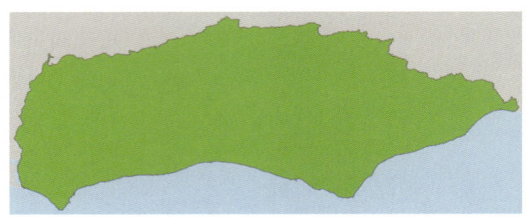

Large map: distribution 2000–2023, with orange circles showing distribution of *Andrena praecox*
Small map: distribution 1844–1999

100 *The Bees of Sussex*

Nomada ferruginata ♀, showing the yellow spot at the front of the thorax [PC]

Nomada ferruginata ♂ [SF]

Out in the field and under the microscope
April is probably the best month to search likely places, even on quite cool days—the first Sussex specimen was found on a windy and overcast day, for example. Both males and females are predominantly reddish brown and black, and superficially resemble *N. fabriciana*. However, they can be distinguished from all other reddish-brown and black *Nomada* by the presence of a clear yellow spot on each side of the front of the thorax.

The key to finding *N. ferruginata* is to search damp places that support willow and where the host is present. Here, this bee might be feeding on willow or on other flowering trees, shrubs and plants nearby.

The first specimen, collected at Rye Harbour Nature Reserve by Thomas Wood, was found on a Daisy flower growing in stony ground within an area of Blackthorn and Goat Willow scrub. The first clue that *N. ferruginata* might be present was the presence of *A. praecox* feeding on a nearby willow tree.

Behaviour and interactions
This *Nomada* targets nests of *A. praecox*, laying its eggs within brood cells that are being provisioned by the host. It may also target *A. varians*.

On the wing
Sussex records run from the end of March until late June.

Feeding and foraging
N. ferruginata has been recorded feeding on Blackcurrant, Blackthorn, dandelion and Lesser Celandine, plus Grey and Goat Willow. While the first Sussex specimen was collected from a Daisy flower, it is not known if this bee was feeding.

Andrena praecox ♀, the host [PB]

Nomada flava Flavous Nomad Bee
FIRST RECORD 1895, Rewell Wood (Henry Guermonprez) **TOTAL RECORDS** 565

Geography and history
This bee is found right across Sussex in a wide range of habitats, as well as on farmland and in gardens. It is most often found in woodland but also occurs on heathland and chalk grassland. It is one of the easier *Nomada* to find and is the most frequently recorded.

Out in the field and under the microscope
This is one of a handful of the bees in this genus that have reddish-brown, black and yellow markings on the abdomen. Identification of *Nomada flava* is challenging, however, especially as the males are very similar in appearance to those of *N. panzeri* and *N. glabella*. Even with a microscope, it is often impossible to distinguish between males of these three species.

Nomada flava ♂ [NO]

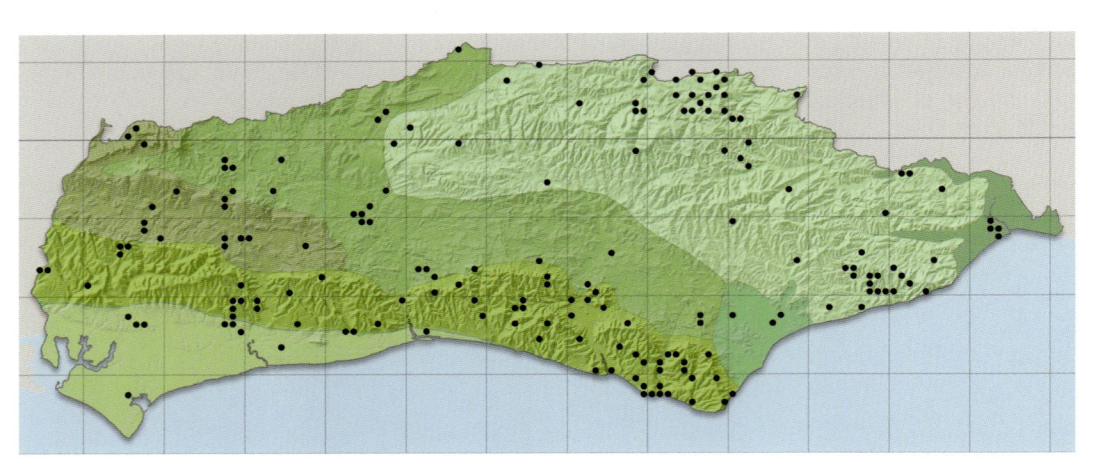

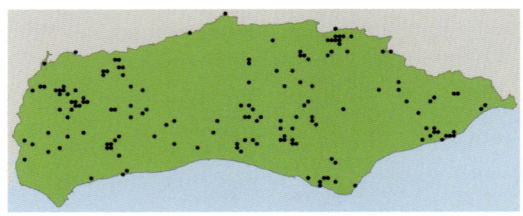

Large map: distribution 2000–2023
Small map: distribution 1844–1999

102 *The Bees of Sussex*

Nomada flava ♀ [PC]

It is possible to identify most females, though they are always tricky. Typical specimens have two black bands and one reddish-brown band on tergite one plus large yellow spots separated by a very narrow reddish-brown band on tergite two. They also have red markings on the scutum and on the sides of the thorax, plus a red-marked scutellum.

Females are often seen flying slowly just above the ground close to path edges, in woodland clearings, or along ditches and banks as they search for the nest sites of their hosts. The males are sometimes seen following the females, or in large numbers chasing back and forth in a patch of sunlight within a woodland. They might also be seen flying in numbers around sunlit trees and shrubs in parks and gardens, and occasionally landing on foliage. Alternatively, this bee might be found feeding on a flower head.

Behaviour and interactions
N. flava lays its eggs in *Andrena scotica* nests but also possibly in those of *A. nigroaenea*. Several other species have also been suggested but are unconfirmed.

On the wing
Sussex records run from the beginning of April until mid June.

Feeding and foraging
This bee visits a very wide range of spring-flowering shrubs and plants to feed on pollen and nectar. These include plants in the Buttercup, Cabbage, Carrot, Daisy, Heather, Pink, Rose and Willow families.

Andrena scotica ♀, the only confirmed host [PC]

Nomada flavoguttata Little Nomad Bee
FIRST RECORD 1884, Hastings (Edward Saunders) TOTAL RECORDS 414

Geography and history
Nomada flavoguttata is found in a wide range of different habitats throughout the county, including chalk grassland, heathland, sandpits, woodland, and meadows and pasture. It is also regularly found in gardens, churchyards and cemeteries. It is very widely distributed across Sussex and is the most frequently recorded of the small *Nomada*.

Out in the field and under the microscope
At most, females measure just 9 mm long. Typical specimens are black and reddish brown overall but with two widely separated small yellow spots on the abdomen. These can be very indistinct. The top and sides of the thorax have red markings, and there are two small dense tufts of silvery-white hairs to the sides of the propodeum.

Males resemble the females (they have a pair of small yellow spots on the abdomen, for example) but are smaller and have a thorax that is usually entirely black. The labrum is also dark.

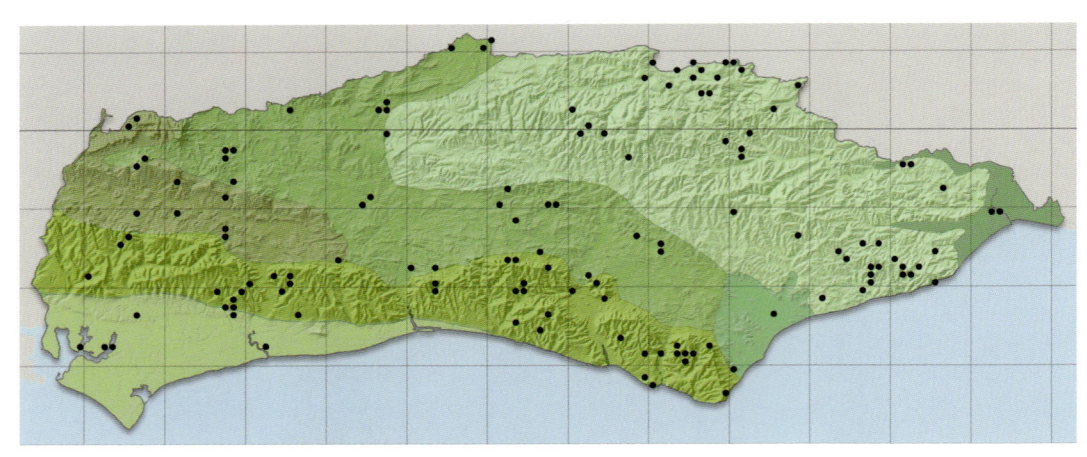

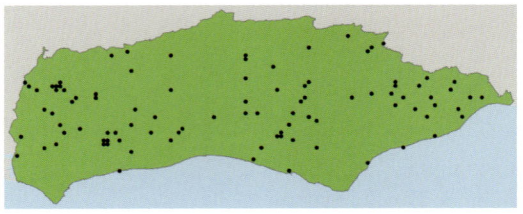

Large map: distribution 2000–2023
Small map: distribution 1844–1999

The small size of this *Nomada* means that this is a very easy bee to overlook. Checking foodplants is probably the most effective way of finding it, although females are also often found searching for a nest site.

Behaviour and interactions

The precise relationships between *N. flavoguttata* and its possible hosts are not completely resolved. While it is known that it lays its eggs in *Andrena falsifica*, *A. minutula* and *A. subopaca* nests, it is also possible that it uses nests of other small black *Andrena*, including *A. alfkenella*, *A. minutuloides* and *A. semilaevis*.

This bee can have two generations a year in Sussex, mirroring the behaviour of *A. minutula* and possibly *A. subopaca*. Further north and at higher altitudes in Britain, it is likely to have just one.

Andrena minutula ♀, a host [SF]

Andrena subopaca ♀, a host [SF]

On the wing

This species has two flight seasons, the first running from late March until late June and the second running from early July until early September.

Feeding and foraging

N. flavoguttata visits a wide range of different plant species in several families, including Agrimony, bramble, Blackthorn, cinquefoil, Daisy, dandelion, Field Pepperwort, forget-me-not, Sea Mayweed, Sheep's-bit, speedwell and Wood Spurge.

Nomada flavoguttata ♀ visiting Germander Speedwell [PC]

The Bees of Sussex **105**

Nomada flavopicta Blunthorn Nomad Bee
FIRST RECORD 1883, Malling Hill (William Unwin) TOTAL RECORDS 122

Geography and history
Most records for *Nomada flavopicta* are from the eastern South Downs, the High Weald and from a handful of locations across the Wealden Greensand. The earliest record is from "Malling Hill" in 1883 and was caught by William Unwin. The specimen is in the Booth Museum.

Its distribution in the county reflects the distributions of its probable hosts, and it is most frequently found on areas of chalk grassland. Its only confirmed host, *Melitta leporina*, collects pollen from widely distributed plants such as Red and White Clover, so *N. flavopicta* can also occur in parks, gardens and in areas of improved pasture.

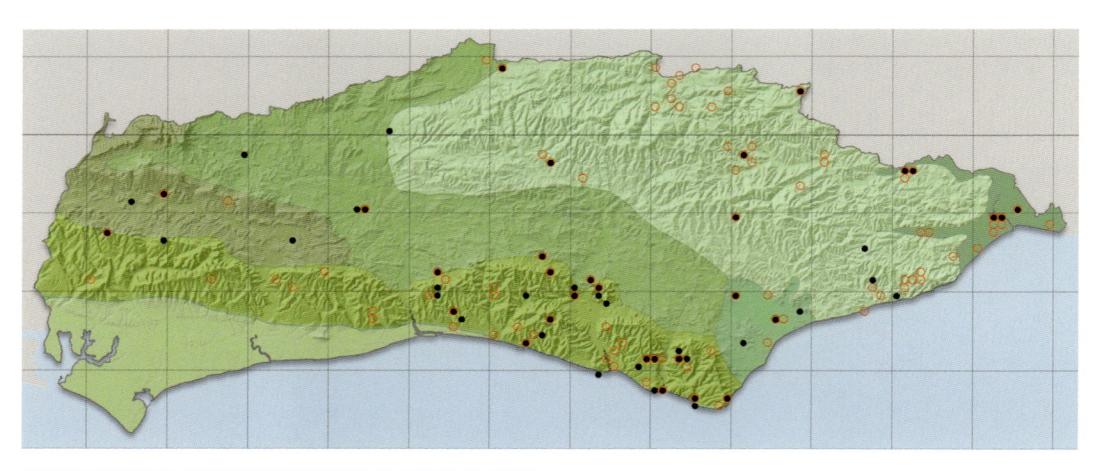

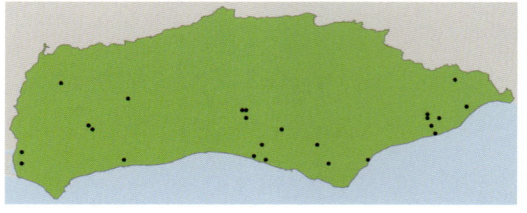

Nomada flavopicta ♀ [PC]

Nomada flavopicta ♂ [SF]

Large map: distribution 2000–2023, with orange circles showing distribution of *Melitta leporina*
Small map: distribution 1844–1999

Out in the field and under the microscope
N. flavopicta generally occurs in places where *M. leporina* is found foraging or feeding on Red and White Clover. It can also be seen close to patches of Red Bartsia where *M. tricincta*, a probable host, is found.

This means that any of the places where these plants grow is worth checking. It has, for example, been found in a garden in Lewes where Red and White Clover grow in the lawn, and where *M. leporina* was also recorded.

This is a very obviously marked *Nomada*. Both males and females have a strongly marked black and yellow abdomen and thorax (with no reddish brown), largely black antennae, and two bright yellow spots on the scutellum. The large yellow spots on the second and third tergites are separated by a broad black stripe.

Behaviour and interactions
N. flavopicta is one of a handful of British *Nomada* that does not attack a species of *Andrena*. Instead, it targets bees in a different genus, the *Melitta*. Of these, only *M. leporina* is confirmed as a host, but it probably targets two other species, *M. haemorrhoidalis* and *M. tricincta*.

On the wing
Sussex records run from late May until the end of August. Elsewhere, it is reported to be active until mid September.

Feeding and foraging
Among the plants visited are bramble, Common Fleabane, Common Ragwort, Creeping Thistle, Cross-leaved Heath, Field Scabious, Sainfoin and Small Scabious.

Melitta leporina ♀, the only confirmed host [PB]

Nomada favopicta ♀ visiting Field Scabious [PC]

Nomada fucata Painted Nomad Bee
FIRST RECORD 1879, Hollington (Edward Saunders) TOTAL RECORDS 458

Geography and history
This has been a very scarce bee in the past, even at times when its host *Andrena flavipes* was quite abundant. For the period 1946 to 1987, there are no Sussex records whereas *A. flavipes* was frequently recorded.

Today, *Nomada fucata* is found throughout Sussex, occurring in the same places as its host. *A. flavipes* is the most frequently found solitary bee in the county and occurs in a very wide range of habitats, accompanied by its cuckoo. Among the habitats recorded are chalk grassland, flower-rich meadows and pasture, heathland, sand dunes, sandpits, soft-rock cliffs and woodland. It also occurs in gardens, and it has been found on the edge of an active landfill site near Small Dole. Although *N. fucata* is an abundant bee, it is still not found as often as its host.

Out in the field and under the microscope
N. fucata is nearly always found close to *A. flavipes* nest sites, although both males and females can also be found feeding on flower heads. In the early 1940s, for example, Alfred Brazenor collected a series of specimens from Field Scabious flowers in the Ovingdean area.

The first indication that *N. fucata* might be present can be the sight of large numbers of *A. flavipes* males flying close to the surface, a sign that a nesting aggregation is present. These are sited in a range of different situations, including lawns, patches of exposed soil, or sand and cliff faces.

This is one of the 'three-coloured' *Nomada*. Its abdomen is predominantly black and yellow but with black and reddish-brown markings on the first tergite. *N. fucata* has a single yellow spot on the scutellum.

N. fucata is one of just four species where the females have two or three small, strongly curved spines on the tip of the hind tibiae. These are so close together that, even with a microscope, it is not easy to see that they are separate. Two of these other three species, *N. bifasciata* and *N. succincta*, are not currently known from Sussex (the former may colonise Sussex in the near future, while the latter is restricted to the Channel Islands). The third, *N. goodeniana*, can be readily distinguished from

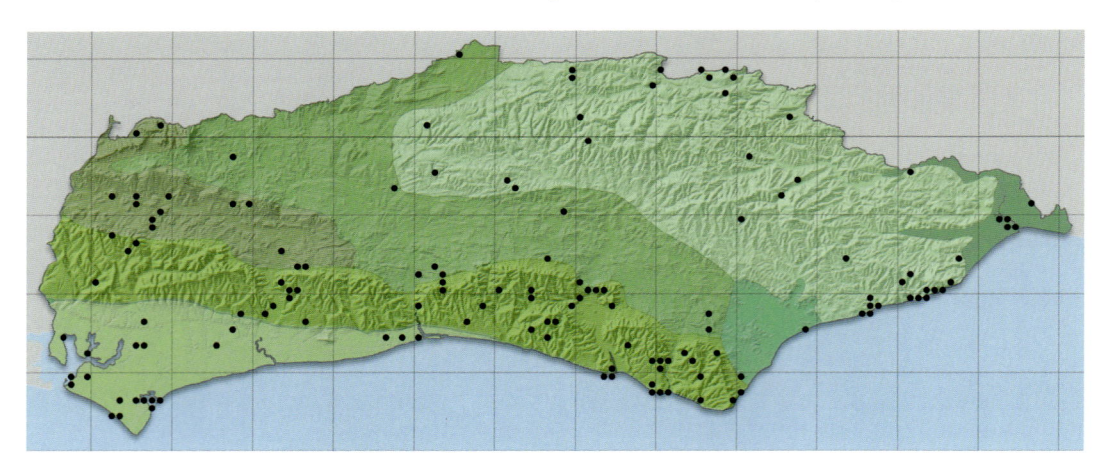

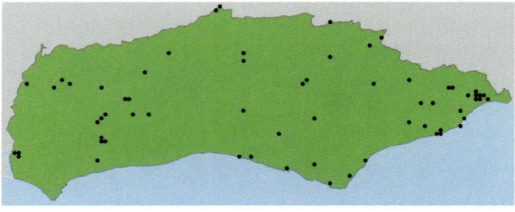

Large map: distribution 2000–2023
Small map: distribution 1844–1999

Nomada fucata ♀ [PC]

Nomada fucata ♂ [PC]

Andrena flavipes ♀, the host [MH]

N. fucata by the presence of two yellow spots on the scutellum and the absence of any reddish-brown on the first tergite.

Behaviour and interactions
This bee lays its eggs within *A. flavipes* brood cells. At some sites, females can occur in large numbers, flying slowly close to the surface or investigating nest burrows on the ground. They are often accompanied by males.

It has two flight periods a year, mirroring the behaviour of its host.

On the wing
Sussex records run from late March until the end of May, and then from June until late October (although the end of August is more typical).

Feeding and foraging
This bee feeds from a wide range of plants from several different families, including Cat's-ear, Common Ragwort, Creeping Buttercup, Greater Chickweed, Cuckooflower, dandelion, fleabane, hawkbit, Lesser Celandine, White Clover and Yarrow.

Nomada fulvicornis Orange-horned Nomad Bee
FIRST RECORD 1905, Hollington (unknown) TOTAL RECORDS 48

Geography and history
Until 2004, *Nomada fulvicornis* was thought to be a single species. It was then recognised through DNA analysis as two separate species, *N. fulvicornis* and *N. subcornuta*. As a result, pre-2004 records for *N. fulvicornis* are now treated as an aggregate of the two. However, in the absence of evidence that neither *N. subcornuta* nor its host *Andrena nigrospina* has occurred in Sussex, the summary below should be read as relating to *N. fulvicornis*.

Since 2000, there has been a concentration of records for *N. fulvicornis* from chalk grasslands on the eastern South Downs, with other records from the heathlands of the Lower Greensand, from Rewell Wood, from Broadwater Warren and from near Battle.

Out in the field and under the microscope
This is a very variable bee with several different forms. The basic colour pattern is black and yellow, sometimes with a little reddish brown on the first two tergites and with varying degrees of yellow. Some forms can be confused with *N. marshamella* and *N. flavopicta*.

It is a tricky bee to find. Ian Beavis, recording in the area south of Tunbridge Wells, suggests that females are usually found either at rest on sunlit foliage or close to the nests of their host species. These are sited in a variety of situations and include grassy slopes, areas of bare sand, and

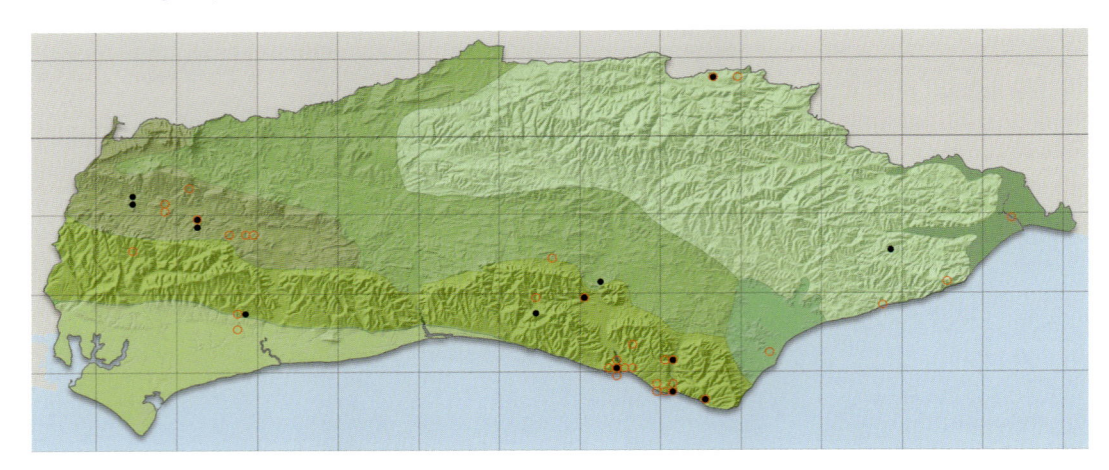

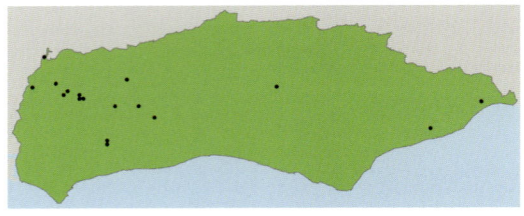

Nomada fulvicornis ♀ [CKL]

Large map: distribution 2000–2023, with orange circles showing distribution of *Andrena bimaculata*, *A. pilipes* and *A. tibialis*
Small map: distribution 1844–1999

Nomada fulvicornis ♂ [SF]

Andrena bimaculata ♀, a host [PC]

Andrena pilipes ♀, a host [SF]

Andrena tibialis ♀, a host [SF]

vertical cliffs on the coast. There is also a report from 2011 of a group of some 15 males that were observed flying in a tight group over an area of sunlit Ivy in woodland at High and Over.

Behaviour and interactions
N. fulvicornis invades the nests of three species of *Andrena*—*A. bimaculata*, *A. pilipes* and *A. tibialis*. It is most frequently found close to *A. bimaculata* nests, and it is possible that those bees attacking *A. pilipes* and *A. tibialis* will prove to be distinct, hidden species.

This *Nomada* has two flight periods, mirroring those of two of its host species, *A. bimaculata* and *A. pilipes*.

On the wing
Sussex records run from mid March until early June, and then from late June until the end of August.

Feeding and foraging
The first generation feeds on plants of the Cabbage family and on Creeping Willow, Daisy, spurge, Thrift and White Willow, while the second has been seen visiting bramble, Creeping Thistle, goldenrod and ragwort.

Nomada glabella Bilberry Nomad Bee
FIRST RECORD 2020, Rake Hanger (Scotty Dodd) TOTAL RECORDS 1

Geography and history
Nomada glabella was recognised as a distinct species in 2022—until then, it had been assumed that it was a form of *N. panzeri*. Like its host *Andrena lapponica*, *N. glabella* is restricted to habitats supporting different species of *Vaccinium*, especially Bilberry and, to a lesser extent, Cowberry. These locations encompass heathland, moorland, and conifer and broadleaf woodland.

Most records are from the west and north of Britain where it appears to be widely distributed, while the solitary record from Sussex is the only one for south-east England—the nearest other records are from Staffordshire, south Wales and Devon. The Sussex specimen was swept by Scotty Dodd in 2020 in Rake Hanger, a Sessile Oak woodland with extensive areas of Heather and Bilberry in the understorey.

Although further investigation of museum and private collections is likely to yield further examples from additional locations, it is thought likely that this species has been lost from some parts of its range in the south-east.

Out in the field and under the microscope
N. glabella is most likely to be found close to *A. lapponica* nest sites. These are sited in sunny locations within areas of Bilberry-rich habitat and encompass footpaths and low unvegetated banks. Scotty Dodd reports that the specimen swept in 2020 "was netted flying low around Bilberry visited by the host *Andrena* on a reasonably open ride".

This bee closely resembles *N. panzeri*, and the differences between them are very subtle. Female *N. glabella* can be separated by the much longer bristles on sternites two, three and four (in *N. panzeri* these are about half as long as those on *N. glabella*), and by the distribution and size of the punctures on the clypeus. In *N. glabella*, these are finer and more evenly distributed all the way to the bottom edge of the clypeus, whereas in *N. panzeri* they are more variable in size and less clearly formed towards the bottom.

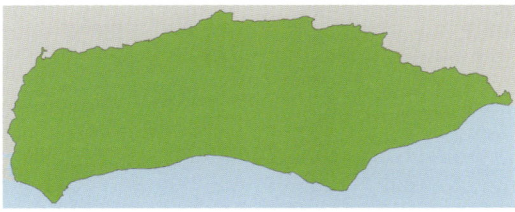

Large map: distribution 2000–2023, with orange circles showing distribution of *Andrena lapponica*
Small map: distribution 1844–1999

112 *The Bees of Sussex*

Nomada glabella ♀ (left) and ♂ (below) [both SF]

Andrena lapponica ♀, the host [SF]

Males are more challenging to identify—the length of the bristles on the sternites is one suggested character.

Behaviour and interactions
Although it is known that this *Nomada* targets the brood cells of *A. lapponica*, there is otherwise little known about its behaviour.

On the wing
The flight season for *N. glabella* overlaps with that of its host, which has been recorded in Sussex between the beginning of April and late June.

Feeding and foraging
There is no information available on which plant species are visited by this species.

The Bees of Sussex 113

Nomada goodeniana Gooden's Nomad Bee
FIRST RECORD 1882, Hastings (Edward Saunders) TOTAL RECORDS 482

Geography and history
Nomada goodeniana is very widely distributed across Sussex and could turn up just about anywhere, including in parks and gardens.

Out in the field and under the microscope
This bright, strongly marked bee is distinctive. It has no red or reddish-brown markings on the thorax or abdomen, and instead these are entirely black and yellow. The abdomen has a series of yellow bands that are clear and generally uninterrupted, although the first does sometimes have

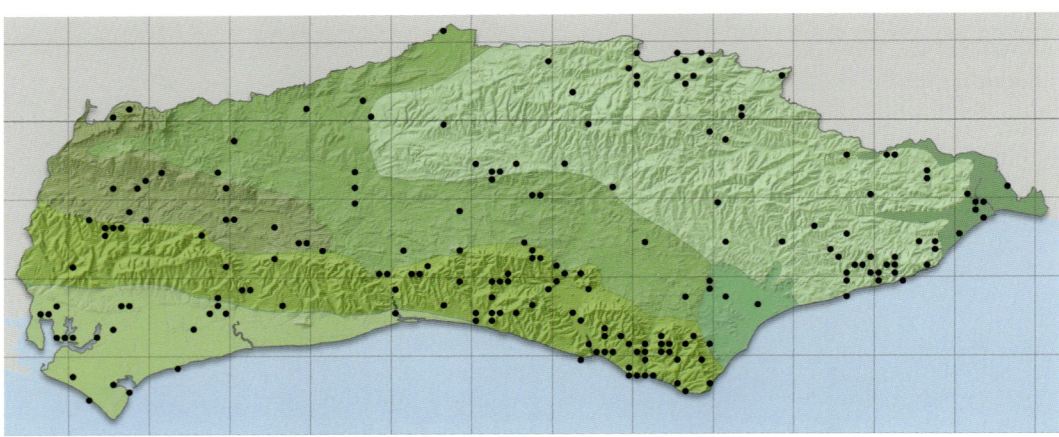

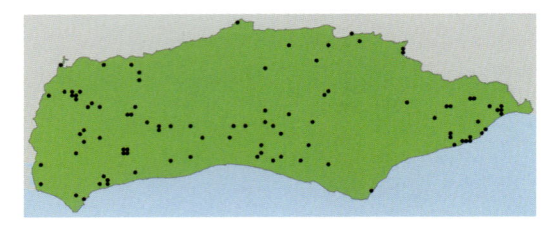

Large map: distribution 2000–2023
Small map: distribution 1844–1999

114 *The Bees of Sussex*

a narrow black band running lengthwise across the middle. Females can have yellow spots on the propodeum.

It most resembles *N. succincta*, a species that is currently restricted to the Channel Islands. In common with this last species, it has two to three curved spines on the hind tibiae that are very close together, a character that is shared with just three other species, *N. fucata* and *N. bifasciata* as well as *N. succincta*.

N. goodeniana is often found flying just above the ground, searching for its hosts' nests. These tend to be sited in areas of short vegetation or bare ground. Generally, just a handful of bees will be seen at one time, but males are sometimes seen flying rapidly in numbers over sunlit foliage in parks and gardens, often in the company of other widely distributed species such as *N. marshamella* and *N. flava*.

Behaviour and interactions
N. goodeniana parasitises at least five species of *Andrena*—*Andrena nigroaenea*, *A. nitida*, *A. pilipes*, *A. scotica* and *A. thoracica*. Other species that have been suggested as hosts are *A. cineraria* and *A. trimmerana*, but these relationships need confirmation.

This species has two flight seasons, with the second generation much less abundant than the first—since 2000 there has been just one Sussex record that falls within the later flight period.

On the wing
Sussex records run from the beginning of March until the end of June, and then from early July until mid September.

Andrena nigroaenea ♀, a host [SF]

Andrena nitida ♀, a host [PC]

Andrena thoracica ♀, a host [SF]

Feeding and foraging
Among a wide range of plant species that this bee has been seen visiting are Bluebell, bramble, members of the Cabbage family, Cat's-ear, Common Fleabane, Creeping Thistle, Creeping Willow, Cuckooflower, dandelion, Eared Willow, forget-me-not, gorse, Lesser Celandine, ragwort, Ramsons, Sea Campion, Smooth Sowthistle, Spear Thistle, White Dead-nettle and Wood Spurge.

Nomada guttulata Short-spined Nomad Bee
FIRST RECORD 1920, Hollington (William Butterfield) TOTAL RECORDS 10

Geography and history
Nomada guttulata is a very scarce bee (there are fewer than 80 records nationally) which until 2004 was restricted to southern England and south Wales. In 2004, however, it was recorded in Shropshire for the first time. It is much scarcer than its host *Andrena labiata*, and is usually known from flower-rich grasslands and open woodlands.

In Sussex, *N. guttulata* has only ever been recorded on ten occasions, with all but one of the records from the east of the county. It was most recently recorded in 2019 and 2020 at a complex of hay meadows and pasture near Warbleton in the High Weald. Earlier east Sussex records are from Hastings in 1920, Ditchling in 1943, near Ashurst in 2005 and 2007, Rye Harbour Nature Reserve in 2012 and Mayfield in 2017.

Hay meadow, Warbleton [NS]

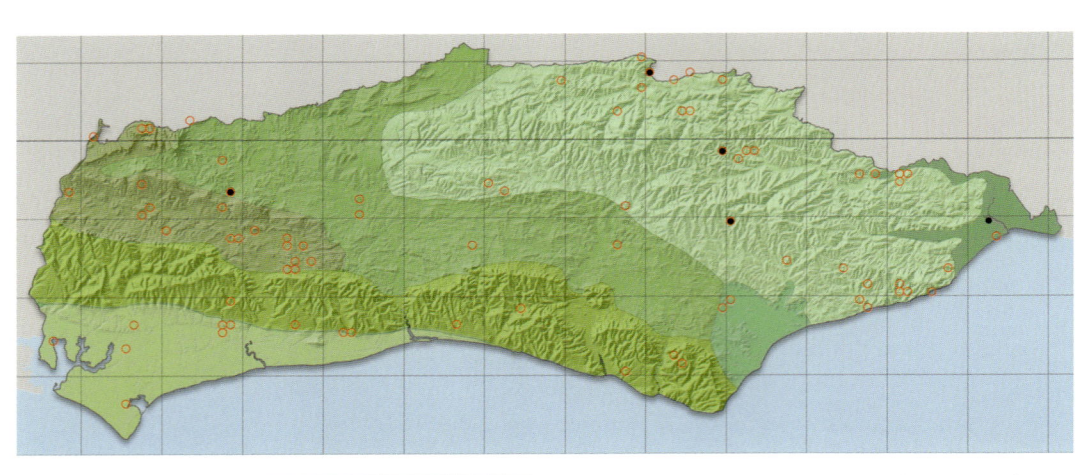

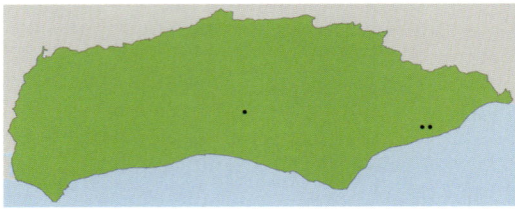

Large map: distribution 2000–2023, with orange circles showing distribution of *Andrena labiata*
Small map: distribution 1844–1999

116 *The Bees of Sussex*

Nomada guttulata ♀ [SF]

Nomada guttulata ♂ [SF]

There is just one record from west Sussex, collected from farmland north of Petworth in 2014. Unusually, this specimen was collected on a florally enhanced arable margin.

Out in the field and under the microscope

This is a small reddish-brown *Nomada* with small yellow spots on the sides of the abdomen. It is also one of two species with blunt-ended mandibles. The other is *N. striata*, which can occur in the same locations.

In females, there are a few characters that can be used to separate the species, including the spines on the hind tibiae. In *N. guttulata* there are three or four short straight black spines pressed very tightly together, while in *N. striata* these are separated from each other and are a little longer. These are not easy to see, even with the aid of a microscope. Unlike in *N. striata* males, the antennal scape in *N. guttulata* males is entirely black.

The first clue that *N. guttulata* might be present is often a sighting of *A. labiata* males and females either feeding or close to their nest sites. These can be scattered or in large aggregations, with reports indicating that they are sometimes sited on banks exposed to the sun. Just such a place is where Rosse Butterfield recorded *N. guttulata* in 1921, near Church-in-the-Wood, Hollington, a year after his brother William had first recorded it nearby.

N. guttulata often feeds from Germander Speedwell, and patches of this plant are particularly worth a close look, along with other food sources. On the Kent/Sussex county boundary, Ian Beavis reports that "several of both sexes were active over a flower-rich path-side bank in company with their host species *Andrena labiata* which was nesting in the bank. They were flying low and persistently over the bank, settling on sunlit foliage and visiting the flowers of Germander Speedwell."

Behaviour and interactions

This *Nomada* targets the nests of *A. labiata*, laying its eggs within brood cells that are being provisioned by the host. It has one flight season a year.

Andrena labiata ♀, the host [NO]

On the wing

The flight season for *N. guttulata* overlaps with that of its host, which has been recorded in Sussex between the beginning of April and the end of June.

Feeding and foraging

This bee has been recorded visiting buttercup, Germander Speedwell and Silverweed.

The Bees of Sussex

Nomada hirtipes Long-horned Nomad Bee
FIRST RECORD 1903, Slindon (Henry Guermonprez) TOTAL RECORDS 30

Geography and history
Until 2001 *Nomada hirtipes* was restricted to the downs between Slindon and Arundel but was then recorded from Hawkenbury Cemetery near Tunbridge Wells.

Subsequently, it has been recorded more widely and is now known from locations such as Lewes Cemetery, Friston Forest, West Dean Woods, Stoughton, Heyshott Down and Devil's Jumps. In 2021 and 2022 it was also recorded from Pulborough Brooks.

N. hirtipes is found in a range of different habitats, including chalk grassland, open woodland and on the edge of flower-rich grassland. It is also known from churchyards and cemeteries, where two of the Sussex records have been made.

Lewes Cemetery supports a number of scarce and threatened bee species, including *Nomada hirtipes* which was recorded here in 2020. [WP]

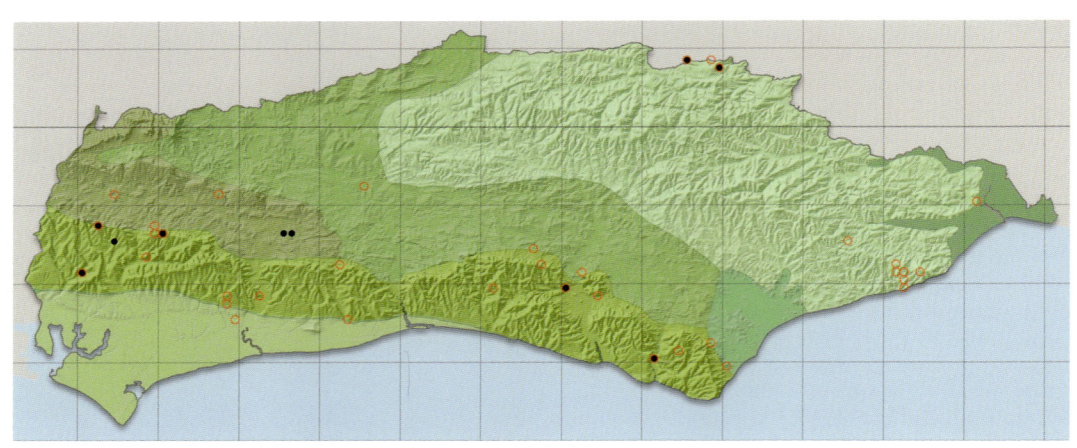

Large map: distribution 2000–2023, with orange circles showing distribution of *Andrena bucephala*
Small map: distribution 1844–1999

118 *The Bees of Sussex*

Out in the field and under the microscope

This is a reddish-brown and black *Nomada* with yellow markings on the abdomen. In females, the markings on tergite two are in the form of large, widely separated spots. Females also have a black band on tergite one, long pointed mandibles and comparatively long antennae. A key character is the presence of fine sparsely distributed hairs on the rear of the antennal segments. These are longer than those on the front face, but are only visible with the aid of a microscope.

Males have more extensive yellow markings on the abdomen than the females. These are widely separated by an extensive reddish brown marking on tergites two, three and four. Tergite one is black and reddish brown.

N. hirtipes is usually found flying in areas of scrubby species-rich grassland or in open woodland.

This *Nomada* is not easy to find, but searches within areas of suitable habitat could yield results—here females might be found as they search for the host's nest or as they fly close to the entrance. On a visit to Heyshott Down in 2022, Alice Parfitt reported that she "swept [a female] from the entrance to an *Andrena bucephala* nest". These nests are often sited on steep banks. Males can sometimes be seen at rest on sunlit foliage.

Nomada hirtipes ♀ [EP]

Nomada hirtipes ♂ [SF]

Behaviour and interactions

N. hirtipes parasitises nests of *A. bucephala*, a species where females share a nest entrance.

A set of observations from Devon in the early twentieth century suggests that up to ten per cent of the bees active around an *A. bucephala* nest can be this *Nomada*. On this occasion, most of the *Nomada* close to the nest were females, the males remaining within a range of some 50 yards, flying over scrub or vegetation and occasionally settling in hot sunshine.

Andrena bucephala ♀, the host [PB]

On the wing

Sussex records run from late April until early June. Elsewhere, it is reported to be active until early July.

Feeding and foraging

This bee has been seen visiting Bogbean, Cow Parsley, Cuckooflower, Cypress Spurge, dandelion, Field Maple and Wild Strawberry.

Nomada integra Cat's-ear Nomad Bee
FIRST RECORD 1937, Goring-by-Sea (Kenneth Guichard) TOTAL RECORDS 7

Geography and history
Until 2017 *Nomada integra* was treated as a single species. However, following the capture of a specimen in Lewisham that did not key out as *N. integra* but was otherwise very similar, investigations led to the realisation that it was *N. facilis*. This species had long been known from the continent, and examination of museum specimens revealed that it had in fact also been present in Britain with the earliest record now known to have been from 1802.

Of the 21 Sussex records that had previously been described as *N integra*, it has been possible to confirm that seven are correctly identified as this species - specimens for the remaining 14 records have not been traced. The seven confirmed records are from just three locations, Goring-by-Sea where it was recorded in 1937 and 1938, a sandpit near Ditchling where it was recorded on four occasions between 1943 and 1946, and, finally, Devil's Dyke where it was recorded in 1996.

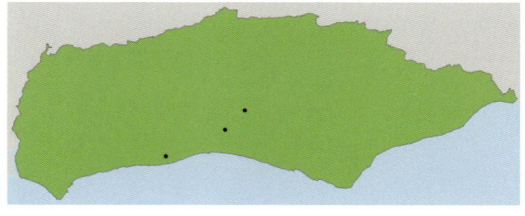

Large map: distribution 2000–2023, with orange circles showing distribution of *Andrena humilis*
Small map: distribution 1844–1999

120 *The Bees of Sussex*

Out in the field and under the microscope

N. integra is an elusive bee and is best looked for in the places where its host *Andrena humilis* is found. These are places with sandy soils and have included flower-rich meadows and pasture in the High Weald, species-rich grassland near the coast on the South Downs, and the heathlands of the Lower Greensand.

N. integra is entirely reddish brown and black on the abdomen and has no yellow or cream spots. It is very similar to *N. facilis*, and the differences between the two are very subtle. Female *N. integra* have a broad, almost rounded pygidium, while in *N. facilis* this is clearly pointed. Male *N. integra* have a small swelling halfway along the lower edge of each mandible, which is absent in *N. facilis*.

Behaviour and interactions

This bee lays its eggs within *A. humilis* nests and is generally only seen in low numbers. However, in 1937 Kenneth Guichard observed large numbers near Goring-by-Sea. Here it was flying with both *A. humilis* and *A. flavipes* in a mixed colony "some 20 yards long […] on a bank of stiff gravel facing south and only a short distance from high tide mark". As here, *N. integra* can be more abundant than its host.

Andrena humilis ♀, the host [SF]

On the wing

The flight season for *N. integra* overlaps with that of its host, which has been recorded in Sussex between early May and the beginning of August.

Feeding and foraging

There are records of this species visiting Bulbous Buttercup, Cat's-ear, dandelion, hawk's-beard, Sand Spurrey and stitchwort.

Sandy locations with yellow-flowered plants in the Daisy family such as Cat's-ear are the likeliest places to find *Nomada integra* and its host *Andrena humilis*. [PC]

Nomada lathburiana Lathbury's Nomad Bee
FIRST RECORD 2011, Iping Common (Graeme Lyons) TOTAL RECORDS 17

Geography and history
Nomada lathburiana is currently known from eleven widely separated locations in Sussex, stretching from Devil's Jumps in the west to Hastings in the east. It was first recorded in the county at Iping Common in 2011 and, more recently, has also been recorded from Lewes Cemetery, Truleigh Hill, Rewell Wood and Friston Forest.

The limited number of records is surprising given the frequency of records in adjoining counties such as Kent and Surrey, and the wide distribution of its main host *Andrena cineraria* in Sussex. *A. cineraria* occurs across the county and has been recorded from parks and gardens, woodland rides and clearings, chalk grasslands, heathlands, and areas of coastal grassland and scrub.

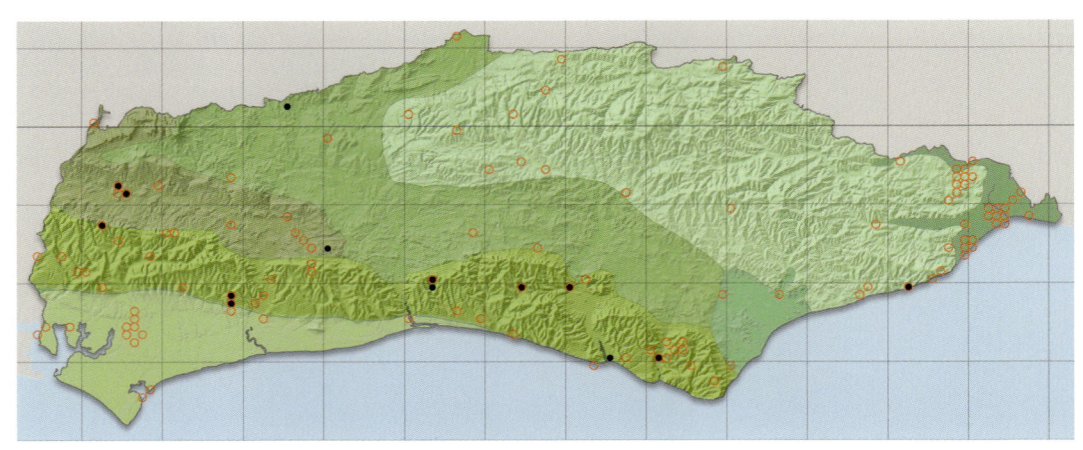

Large map: distribution 2000–2023, with orange circles showing distribution of *Andrena cineraria*
Small map: distribution 1844–1999

Its second host species is *A. vaga*, which has a much more limited distribution and is currently confined to the cliffs between Bexhill and Bulverhythe and to Rye Harbour Nature Reserve—there are no records of *N. lathburiana* in these places, although it has been recorded nearby in Hastings.

Out in the field and under the microscope

It is well worth checking any of the locations where either of its two hosts are known to be nesting, but especially *A. cineraria*. This last bee can form very large nesting aggregations and can therefore attract large numbers of the *Nomada*—at Ham Common in Surrey, for example, more than 5,000 *N. lathburiana* pairs have been reported.

N. lathburiana is one of the more readily distinguishable *Nomada*. Both males and females have black and yellow banding on the abdomen. The first tergite is black with a broad reddish-brown band, and fresh females have red hairs on the top of the thorax. This last character is unique to this species, but the colour does fade. Males are unique in the genus in that they have a projection on antennal segments 4–13.

Andrena cineraria ♀, a host [PC]

Behaviour and interactions

N. lathburiana generally has one generation a year, although there is a possibility that it may have a partial second generation.

In common with some other species of *Nomada*, during mating males wind their antennae around those of the female, using the small structures on the underside of the antennae to transfer pheromones. These may either be to reduce the attractiveness of the female to other males or to transfer a scent that helps her to gain access to a host's nest.

On the continent, *A. vaga* is described as the main host.

Andrena vaga ♀, a host [PG]

On the wing

Sussex records run from early April until late June.

Feeding and foraging

This bee visits plants such as cherry, Creeping Willow, dandelion, forget-me-not, Germander Speedwell, Gooseberry and Wood Spurge.

Nomada leucophthalma Early Nomad Bee
FIRST RECORD 1884, Fairlight (Edward Saunders) TOTAL RECORDS 188

Geography and history
Nomada leucophthalma is a widely distributed bee that is found throughout Britain. It has been recorded from a variety of locations across Sussex, including sites across the High Weald, the heathlands of the Lower Greensand, open-canopied woodlands on the western South Downs, and chalk grasslands between Falmer and Lullington Heath on the eastern South Downs.

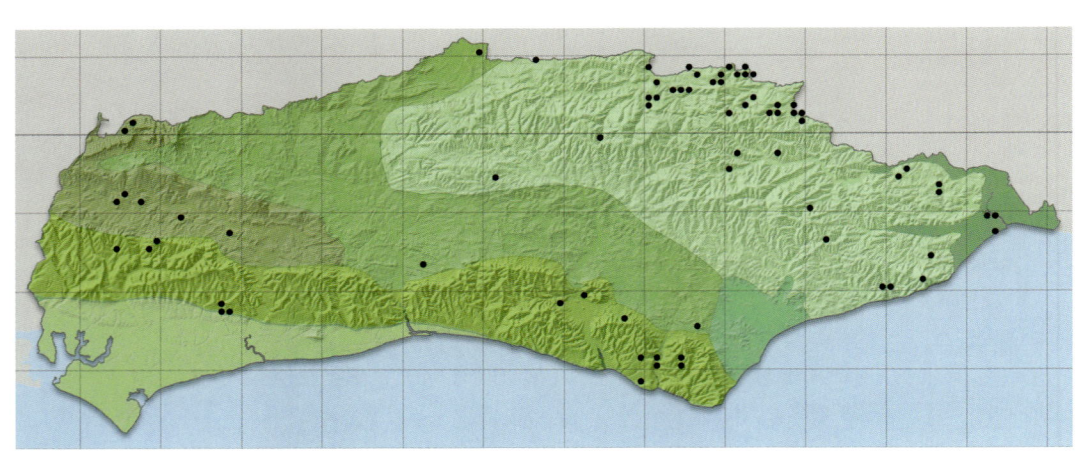

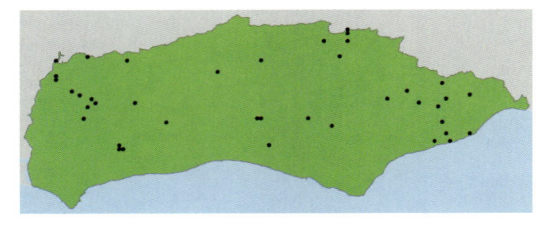

Large map: distribution 2000–2023
Small map: distribution 1844–1999

Out in the field and under the microscope

This bee is generally found during late March and April by searching places that are used by its hosts *Andrena apicata* and *A. clarkella*. Females are most likely to be found feeding or close to the hosts' nest sites. George Else, for example, has reported seeing a female *A. clarkella* fill her nest entrance, while two *N. leucophthalma* females were at rest nearby: "As soon as the nest's owner had flown away, one of these *Nomada* alighted at the nest entrance and immediately began to burrow into it, soon disappearing from view."

Like the females, males are often found feeding, but might also be seen patrolling in search of a female. Ian Beavis has found males feeding on Lesser Celandine, flying over and settling on sunlit leaflitter, or flying persistently around a sunlit gorse bush.

N. leucophthalma is one of the larger *Nomada* and is one of a small group that has black, yellow, and reddish-brown abdomens. Superficially, females can resemble *N. flava* or *N. panzeri*, but in Sussex they have a completely black scutum and propodeum, plus a pair of red markings on the scutellum. Males resemble the females but are smaller and generally more extensively marked with yellow.

Because *N. leucophthalma* is a parasite of two species of bee which specialise in collecting willow pollen, willow-rich habitats are the places to search for this bee, particularly locations supporting Goat Willow and Grey Willow.

Andrena apicata ♀, a host [PB]

Andrena clarkella ♀, a host [PC]

Behaviour and interactions

This is one of the first of the *Nomada* to appear each spring, the flight period coinciding with the emergence of its two host species, *A. apicata* and *A. clarkella*. It can be very abundant around nest sites.

It has one flight season a year.

On the wing

Sussex records run from early March until the end of May.

Feeding and foraging

N. leucophthalma has been recorded visiting Barren Strawberry, Bilberry, Colt's-foot, dandelion, forget-me-not and willow.

Nomada marshamella Marsham's Nomad Bee
FIRST RECORD 1894, Bognor Regis (Henry Guermonprez) TOTAL RECORDS 431

Geography and history
Nomada marshamella is found throughout the county—it is one of the most abundant and most widely distributed of the *Nomada* and can be found just about anywhere in Sussex. It is frequently found in gardens and parks but has also been found in woodlands, on heathlands, on chalk grassland, and in various coastal habitats, including soft-rock cliffs and grasslands.

Out in the field and under the microscope
N. marshamella is a very distinctive bee and one of the easier species in the genus to identify. It can often be recognised in the field, or from a photograph. It is also one of the most frequently recorded,

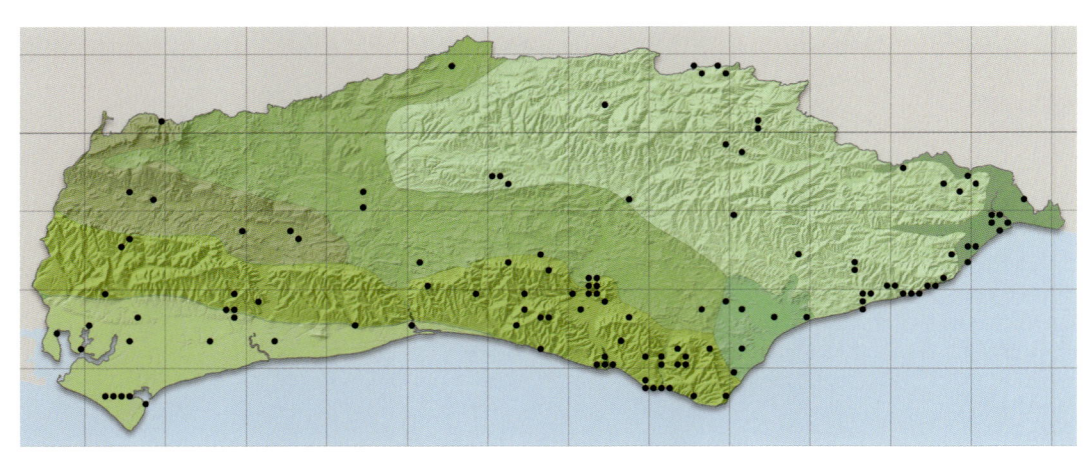

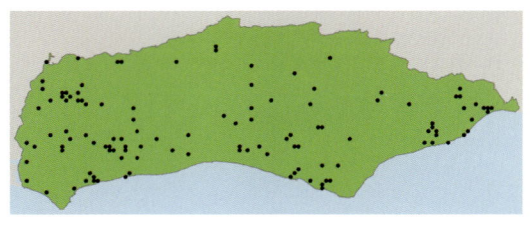

Large map: distribution 2000–2023
Small map: distribution 1844–1999

126 *The Bees of Sussex*

with 15 to 20 records per year not uncommon. The thorax and abdomen in both sexes are entirely marked in black and yellow (there is no reddish brown), while the yellow on the second tergite is divided in two by a black band. Generally, females have a pair of yellow spots on the scutellum, although these can be reduced in size or even absent. The antennae and legs are mostly orange.

Males are considerably smaller than the females and often lack a pair of yellow spots on the scutellum.

This *Nomada* is generally found while it is feeding or close to nesting aggregations of its main host, *Andrena scotica*. These aggregations can run to many hundreds of nests, so there is a strong likelihood of finding this *Nomada* once an aggregation is found. Several males can often be found flying rapidly together around sunlit foliage in parks and gardens, often with other common species such as *N. flava* and *N. goodeniana*.

Behaviour and interactions

It has been confirmed that *N. marshamella* attacks the nests of *A. scotica*, but it might also target other species. Those that have also been suggested are *A. ferox*, *A. haemorrhoa*, *A. rosae* and *A. trimmerana*.

In most parts of the country it has one generation a year, but the extended flight season in Sussex suggests that it has two here. The second generation is much less numerous than the first and may be linked to *A. trimmerana*.

Andrena scotica ♀, the confirmed host [PC]

Andrena haemorrhoa, a possible host ♀ [PC]

On the wing

This species has two flight seasons, the first running from mid March until late June and the second running from late June until early September.

Feeding and foraging

This species visits an enormous variety of spring and summer plants, including Alder Buckthorn, Bilberry, Blackthorn, bramble, members of the Cabbage family, members of the Carrot family, Common Ragwort, Creeping Thistle, Creeping Willow, Cuckooflower, currant, Daisy, dandelion, Eared Willow, forget-me-not, hawthorn, Hogweed, rhododendron, Sycamore, Thrift, White Willow, Wild Cherry and Wood Spurge.

Nomada obtusifrons Flat-ridged Nomad Bee
FIRST RECORD 1975, Chailey Common (Mike Edwards) TOTAL RECORDS 7

Geography and history
Nomada obtusifrons was last seen in Sussex in 1996, and this bee may have been lost from the county. Across Britain, this is a very local and scarce bee that has declined significantly. It was last recorded in Surrey in 1913, for example. Records are thinly scattered across the country, reaching as far north as the Inner Hebrides and as far west as Cornwall and Wales.

The handful of Sussex records are from the east of the county—the first being from Chailey Common in 1975 where a specimen was swept on ragwort by Mike Edwards. The remainder are from the Beckley and Robertsbridge area. Its hosts are more widely distributed (though never common) and are generally found in open woodlands and heathy areas.

The last Sussex record is from August 1996 when one individual was swept by Peter Hodge from vegetation in an area of wet and dry grassland close to the River Brede near Westfield. This was the day after a specimen of *Andrena coitana*, one of its hosts, had been collected from the same location.

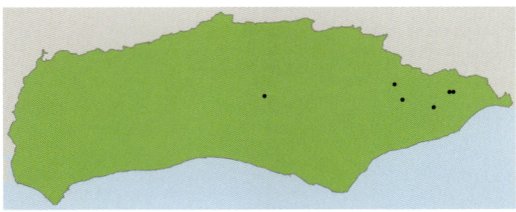

Nomada obtusifrons ♀ [PS]

Nomada obtusifrons ♂ [TF]

Large map: distribution 2000–2023, with orange circles showing distribution of *Andrena coitana* and *A. tarsata*
Small map: distribution 1844–1999

128 *The Bees of Sussex*

Out in the field and under the microscope
This is a small *Nomada*. Both males and females have a reddish-brown and black abdomen with small cream-coloured markings. This species can be readily separated from all the other *Nomada* by the presence of a flat-topped or slightly domed area between the antennae—every other species has a sharp ridge here.

Behaviour and interactions
This *Nomada* attacks the nests of *A. coitana* and *A. tarsata*.

Observations of a Scottish colony by Andrew Jarman and Brian Little in 2010 describe females crawling or flying close to *A. tarsata* nests sited in Bracken. They would visit several nest entrances before "a female found one of interest. At that time she stopped at the nest entrance with antennae pointed forwards towards the burrow and constantly fanned her wings. After a short time she would either leave to assess another burrow or enter the nest for 5–10 seconds."

N. obtusifrons has one generation a year.

Andrena coitana ♀, a host [PS]

Andrena tarsata ♀, a host [PS]

On the wing
The flight season for *N. obtusifrons* overlaps with that of its hosts, which have been recorded in Sussex between early June and early September.

Feeding and foraging
This bee visits bramble, Creeping Thistle, goldenrod, Nipplewort, Oxeye Daisy, ragwort, Sheep's-bit, Smooth Hawk's-beard, Sweet Chestnut, Tormentil and Wild Angelica.

Nomada panzeri Panzer's Nomad Bee
FIRST RECORD 1894, Eartham (Henry Guermonprez) **TOTAL RECORDS** 142

Geography and history
Nomada panzeri is a widely distributed bee but is more frequent in northern and western areas of Britain. It has been found across Sussex and has been recorded in woodlands, hay meadows and pasture, chalk grasslands and heathlands. Locations that have supplied recent records include Rye Harbour Nature Reserve, Ebernoe Common and Great Dixter.

In 2022, DNA analysis and a review of physical characters confirmed that *N. panzeri* is in fact two closely related species, *N. panzeri* and *N. glabella*. It is also possible that the forms of *N. panzeri* associated with the hosts *Andrena fulva* and *A. fucata* will also prove to be distinct species.

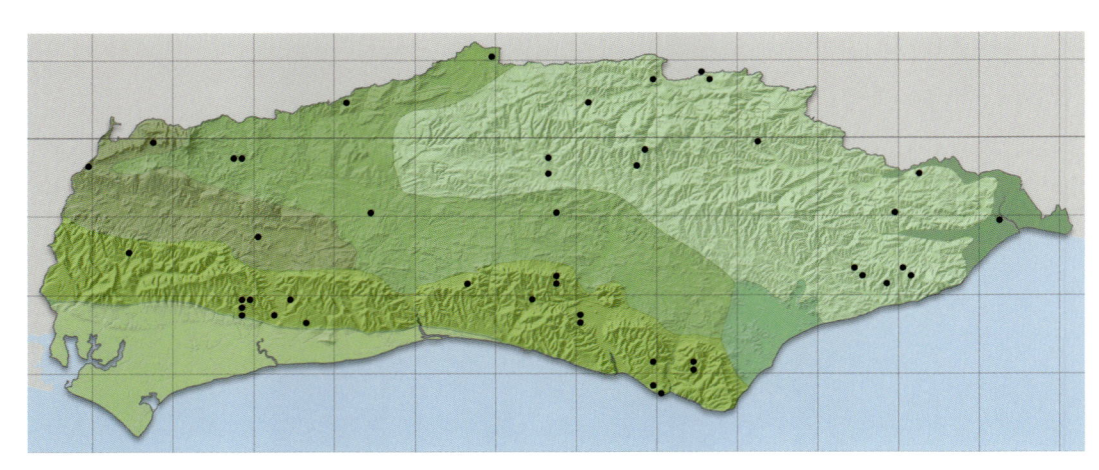

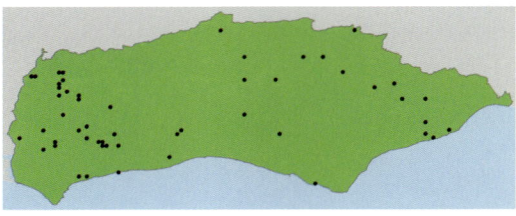

Nomada panzeri ♀ [PB]

Nomada panzeri ♂ [SF]

Large map: distribution 2000–2023
Small map: distribution 1844–1999

130 *The Bees of Sussex*

Out in the field and under the microscope

N. panzeri is particularly close to *N. glabella* but also to *N. flava*—the differences between these three species are very subtle. Female *N. panzeri* have red markings on the top and sides of the thorax, as well as black, reddish-brown and yellow colouring on the abdomen. *N. flava* is more extensively marked with yellow.

A key difference between *N. panzeri* and *N. glabella* females is that the bristles on sternites two, three and four are much shorter in *N. panzeri* (they are about half the length of those on *N. glabella*). In female *N. panzeri*, the punctures on the clypeus vary more in size, becoming less well formed closer to the lower edge of the clypeus.

To separate *N. panzeri* females from *N. flava*, one key character to look for is the colour of the bristles on the face. These are black in *N. panzeri* and yellow in *N. flava*.

Many males of the three species are indistinguishable from each other, but, like the females, male *N. panzeri* appear to have short bristles on sternites two, three and four.

Females might be seen exploring bare ground or leaf litter, searching for the hosts' nest sites, or settled on sunlit foliage.

Andrena fucata ♀, a host [SF]

Behaviour and interactions

This *Nomada* attacks several species of closely related bees—*A. fucata*, *A. fulva*, *A. helvola*, *A. synadelpha* and *A. varians*. It had been thought to also attack *A. lapponica*, but this species' parasite is now known to be *N. glabella*.

N. panzeri overwinters as an adult, emerging in the spring. It has one generation a year and a lengthy flight period. In one Welsh study, as many as 18 per cent of the *A. fulva* brood cells had been parasitised by this bee.

Andrena fulva ♀, a host [PC]

Andrena synadelpha ♀, a host [SF]

On the wing

Sussex records run from the beginning of April until the beginning of July.

Feeding and foraging

A wide range of plant species are visited. These include Bilberry, Buckthorn, Alder Buckthorn, buttercup, cherry, dandelion, Eared Willow, hawthorn, ragwort, Raspberry, Red Campion, Rough Chervil, strawberry, cinquefoil and Wood Spurge.

Nomada ruficornis Fork-jawed Nomad Bee
FIRST RECORD 1905, Hastings (Esam) TOTAL RECORDS 196

Geography and history
This bee is widely distributed throughout the county. It is most often found in open woodlands, but also on heathlands and chalk grasslands, as well as in gardens and cemeteries. Sites that have produced recent records include Ebernoe Common, Flatropers Wood, Lullington Heath, Sheffield Park and Woods Mill.

Out in the field and under the microscope
In the field, *Nomada ruficornis* can easily be confused with three other three-coloured *Nomada*—*N. flava*, *N. panzeri* and *N. glabella*. Under magnification, however, two teeth on the mandibles are

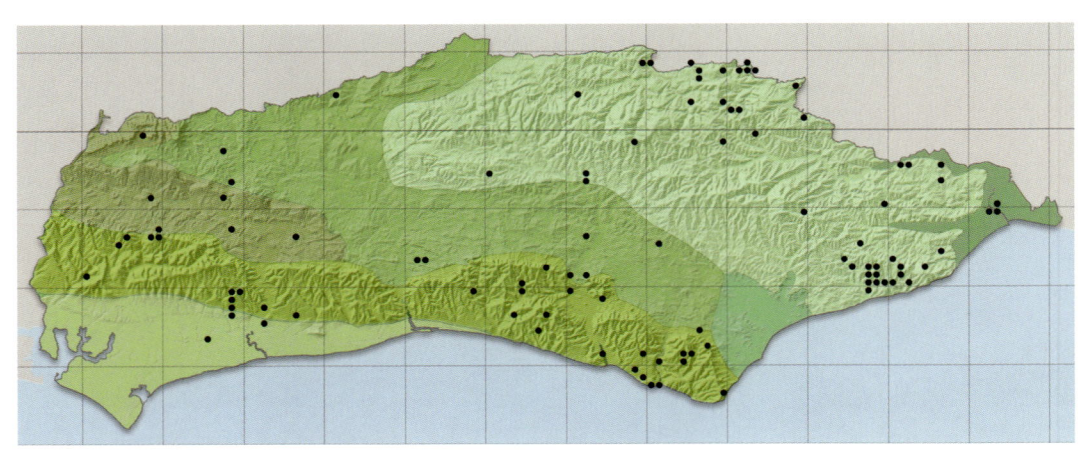

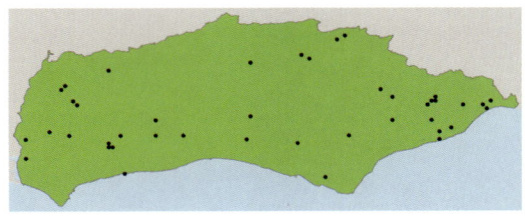

Large map: distribution 2000–2023
Small map: distribution 1844–1999

visible in both sexes of this *Nomada*, while these other species have just one. *N. fabriciana* is the only other species with two teeth here. Unlike *N. ruficornis*, this last bee has a predominantly reddish-brown abdomen and small yellow spots, while both male and female *N. ruficornis* are extensively marked with yellow.

Its host *Andrena haemorrhoa* is a very abundant spring bee often seen on flowering shrubs such as Hawthorn, so any shrub being visited by this *Andrena* should be checked for this *Nomada*. Males and females might also be found feeding on plants such as Wood Spurge growing along woodland rides, or flying rapidly among tall vegetation in areas with light shade.

A. haemorrhoa nests are thinly scattered in areas with a short sward, on the edge of tracks or on banks, and females of the *Nomada* are often seen searching in these areas, flying to-and-fro just above the ground.

Andrena haemorrhoa ♀, the host [PC]

Behaviour and interactions
This *Nomada* attacks the brood cells of *A. haemorrhoa*, a widely distributed spring-flying species.

On the wing
Sussex records run from early April until the beginning of July.

Feeding and foraging
Plants visited include Berberis, Bilberry, Bogbean, bramble, buttercup, cherry, Creeping Willow, dandelion, forget-me-not, Grey Willow, Guelder-rose, Hawthorn, Hogweed, Sycamore, water-dropwort, White Willow and Wood Spurge.

Wood Spurge [NS]

Nomada rufipes Black-horned Nomad Bee
FIRST RECORD 1870, Plashett, Lewes (William Unwin) TOTAL RECORDS 314

Geography and history
Nomada rufipes is most closely associated with heathlands, especially those on the Lower Greensand and in the High Weald. Locations where it is regularly found include Iping and Stedham Commons, Ashdown Forest and Broadwater Warren. It was also re-found on Chailey Common in the Low Weald in 2018 after an absence of 23 years.

Away from Ashdown Forest and Broadwater Warren, *N. rufipes* is thinly distributed throughout the High Weald, including on a complex of hay meadows and pasture near Warbleton. Here, the grassland contains pockets of heather. Very occasionally *N. rufipes* is recorded in other locations—in 2007 it was found at Seaford Head, and it has previously been found in Rewell Wood and at Lewes Railway Land.

On the continent, it has been suggested that *N. rufipes* may prove to be more than one species.

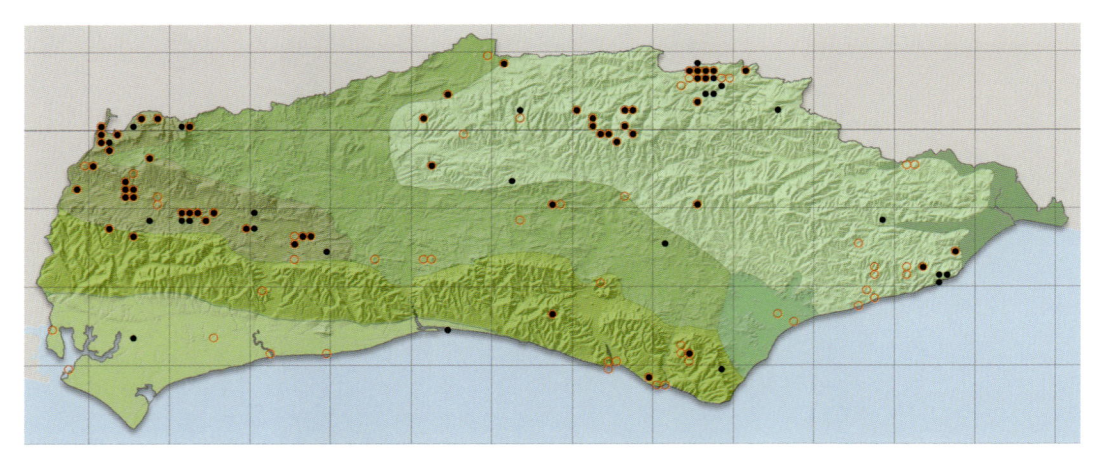

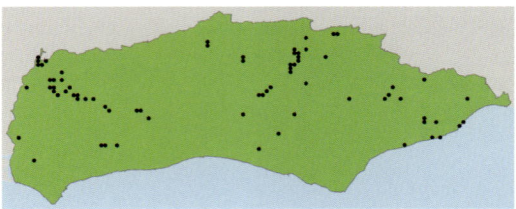

Large map: distribution 2000–2023, with orange circles showing distribution of *Andrena denticulata, A. fuscipes, A. nigriceps* and *A. nitidiuscula*
Small map: distribution 1844–1999

134 *The Bees of Sussex*

Out in the field and under the microscope

N. rufipes is a distinctive *Nomada*. It is the only species in which females have antennae that are almost completely black and a single yellow patch on the scutellum. This patch of yellow is bright and roughly in the shape of a rectangle. There are other species with a single yellow marking here (*N. bifasciata* and *N. fucata*, for example), but their antennae are not as dark. The abdomen can either be black, yellow and reddish brown, or just black and reddish brown, while the propodeum is very smooth and shiny with large punctures.

Nomada rufipes ♀ with orangey-red markings on the lower face. Yellow-marked females are also found. [PB]

Nomada rufipes ♂, showing yellow markings on the face [PB]

Males, like the females, have black antennae and a single yellow spot on the scutellum, and are usually just black and yellow on the abdomen.

This is one of a handful of *Nomada* that flies late in the season. It can be very abundant on heathland sites, either feeding on Heather flowers alongside its main host *Andrena fuscipes*, or flying low above the ground searching for suitable nest sites.

Behaviour and interactions

This *Nomada* attacks at least five closely related species of *Andrena*—*A. denticulata*, *A. fuscipes*, *A. nigriceps*, *A. nitidiuscula* and *A. simillima*. This last species is confirmed as a host on Salisbury Plain, and this may be the case in Sussex too. In places where it parasitises *A. fuscipes*, it can be more abundant than this host.

Andrena fuscipes ♀, a host [PC]

Andrena denticulata ♀, a host [SF]

On the wing

Sussex records run from mid June until early October.

Feeding and foraging

N. rufipes has been recorded visiting a range of different plant species, including Heather, Bell Heather, Wild Parsnip, thyme, Sheep's-bit, Creeping Thistle, Yarrow and Common Ragwort.

Nomada sheppardana Sheppard's Nomad Bee
FIRST RECORD 1905, Brighton (William Unwin) **TOTAL RECORDS** 99

Geography and history
A high proportion of the records for *Nomada sheppardana* are from across the High Weald, and particularly from Broadwater Forest. Away from here, records are very thinly scattered across the county.

Given the range of habitats used by its various hosts, *N. sheppardana* could turn up just about anywhere in Sussex. These habitats include chalk grassland, flower-rich meadows and pasture, heathland, sandpits and woodland. It has also been recorded from gardens.

Guestling Wood. South-facing woodland banks are good locations to search for *Nomada sheppardana*. [PG]

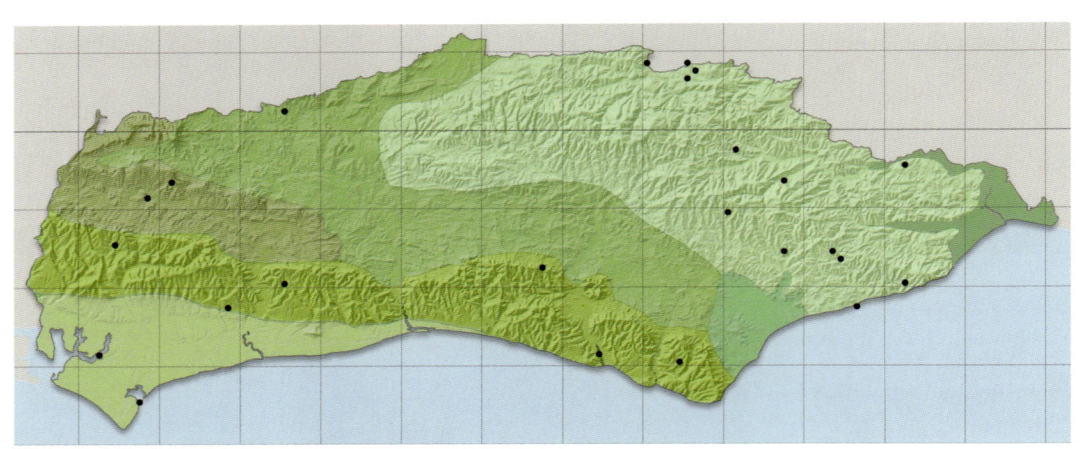

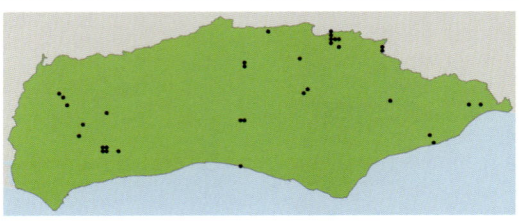

Large map: distribution 2000–2023
Small map: distribution 1844–1999

136 *The Bees of Sussex*

Out in the field and under the microscope
At between 4 mm and 6 mm long, *N. sheppardana* is the smallest of the *Nomada* and extremely easy to overlook. The females are very dark with reddish-brown bands across tergites one and two, and no yellow markings on the abdomen. They most resemble small *N. flavoguttata*, but this species usually has yellow spots on the abdomen.

The males are almost black but do have small cream spots and dashes on the sides of the abdomen, plus indistinct reddish-brown bands on the first and second tergites.

Finding this bee is not easy—searching for the nest sites of one of its hosts is one approach. At Batemans in 2022, for example, a female was observed flying slowly along a wall within the gardens, checking crevices in crumbling mortar, while it has also been recorded in some numbers on a low, south-facing earth bank in Fore Wood. Ian Beavis has also reported that males and females are "often seen flying around bare ground, whether large expanses, […], or smaller areas beside heathland rides or woodland paths". Most records are from May, June and July.

Open areas of wooded habitat are particularly worth checking.

Nomada sheppardana ♀ [PB]

Nomada sheppardana ♂ [SF]

Behaviour and interactions
N. sheppardana is unique among the British *Nomada* in that it attacks various species of Halictid bee. These include *Lasioglossum nitidiusculum*, its close relative *L. parvulum*, and potentially *Halictus tumulorum*, *L. smeathmanellum* and *L. villosulum*. The record from Batemans, for example, was from within an area frequented by large numbers of *L. smeathmanellum*. On the continent *L. sexstrigatum*, a species that was first recorded in Britain in 2006, is also suggested as a host.

N. sheppardana has a single flight season in Britain but can be bivoltine on the continent.

On the wing
Sussex records run from the beginning of April until late July.

Feeding and foraging
There is very little information on the flowers visited by this bee, with just Cat's-ear, Daisy, dandelion, cinquefoil and spurge listed in various sources for Britain. Stitchwort and Germander Speedwell are suggested from the continent.

Lasioglossum parvulum ♀, a host [NO]

Lasioglossum nitidiusculum ♀, a host [SF]

Nomada signata Broad-banded Nomad Bee
FIRST RECORD 1894, Hastings (Edward Saunders) **TOTAL RECORDS** 32

Geography and history
Despite the abundance of its host *Andrena fulva*, *Nomada signata* is a scarce bee and is found much less frequently than its host.

Prior to it being recorded on the Kent/Sussex border at Tunbridge Wells Common in 2000, *N. signata* had not been recorded in Sussex for 77 years. Here, it was found close to the location of the previous record from 1923. Since its re-discovery, it has been recorded regularly in the county, with recent records from places such as Mount Caburn, a garden in Lewes, Abbot's Wood, woodland in Brighton and St Leonards. It was most recently recorded in the county in 2021 at Buchan Country Park, and in 2022 and 2023 in Lewes Cemetery.

Its host is often found in gardens, parks and flower-rich meadows and pasture.

Out in the field and under the microscope
N. signata is a very distinctive bee, and females can be identified in the field by the clear, very broad yellow bands on the abdomen which are bordered by straightish narrower black bands. The females are also the only *Nomada* that consistently have a pair of yellow patches over a

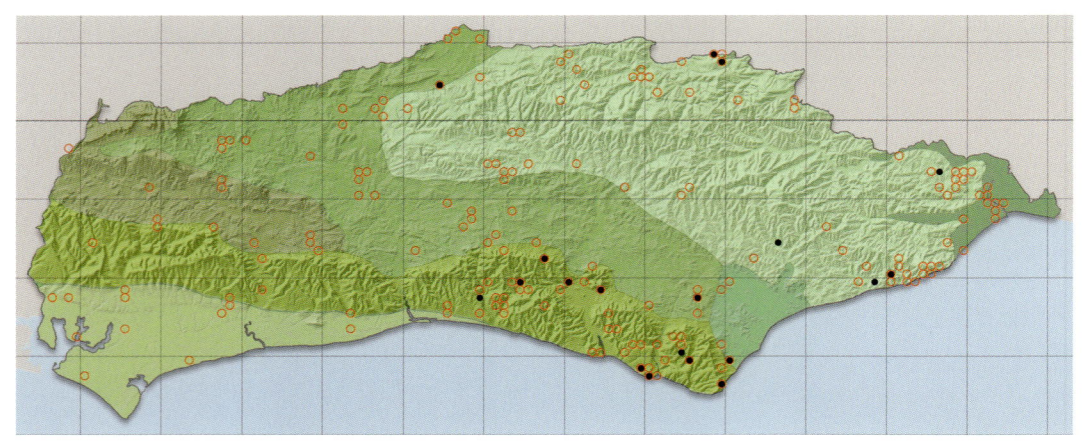

Nomada signata ♀ [MM]

Large map: distribution 2000–2023, with orange circles showing distribution of *Andrena fulva*
Small map: distribution 1844–1999

138 *The Bees of Sussex*

large portion of the propodeum (some *N. goodeniana* also have some yellow here). In common with a handful of other species, *N. signata* also has reddish-brown markings on tergite one.

Some males resemble *N. flava*, *N. panzeri* and *N. glabella* but, unlike in these other species, *N. signata* males have clear yellow bands on tergites two and three. They also have yellow markings that extend a short way up the side of each eye, whereas the yellow markings in these other species are longer, reaching a point that is almost in line with the antennal sockets.

This *Nomada* is most frequently found close to the nest sites of its host, *A. fulva*. Ian Beavis is responsible for a cluster of records from the Tunbridge Wells area and reports seeing females flying above leaf litter alongside a wooded path and males at rest on foliage, feeding or flying around a bush. There is also a report from Surrey of a female "basking in the sun on Cherry Laurel".

Behaviour and interactions

N. signata targets *A. fulva* brood cells. This last species nests in warm, sunny locations with light soils, and nests are often sited on the edges of paths. Nest sites can be quite extensive, but are otherwise in small, loose aggregations.

One set of observations from Jeremy Early in Surrey is of a female *N. signata* that flew slowly and methodically across an *A. fulva* nest site, briefly entering two of the nests. The female returned at 5 o'clock the following afternoon, and he reports that "she immediately went into one of the nests inspected the day before, staying inside for a couple of minutes". On each of the days that the female *N. signata* was seen at the nest site, she entered the nests ten to fifteen minutes after the resident *A. fulva* had left.

On the wing

Sussex records run from the beginning of April until mid May.

Feeding and foraging

Plants visited include dandelion, willow, Gorse and Wood Spurge.

Nomada striata Blunt-jawed Nomad Bee
FIRST RECORD 1888, Hastings (Frisby and W.H. Bennett) TOTAL RECORDS 62

Geography and history
Nomada striata is found in the same places as its host *Andrena wilkella*. This last bee sources much of its pollen from plants in the Pea family such as bird's-foot-trefoil, clover and vetch, so it is often found in hay meadows and pasture, on chalk grassland and in open woodlands. It has also been recorded from the heathlands of the Lower Greensand. *N. striata* is much scarcer than its host.

Nomada striata can be found in flower-rich meadows and pasture in the High Weald. [PG]

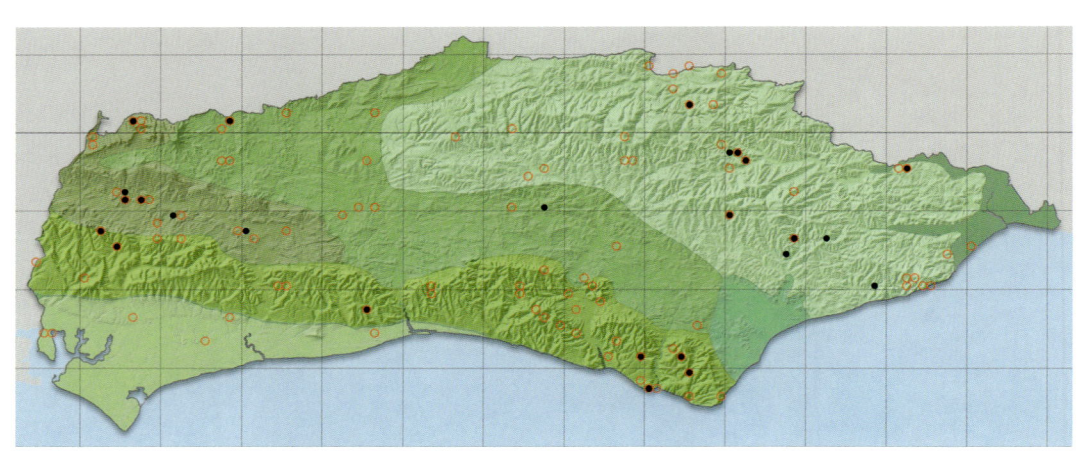

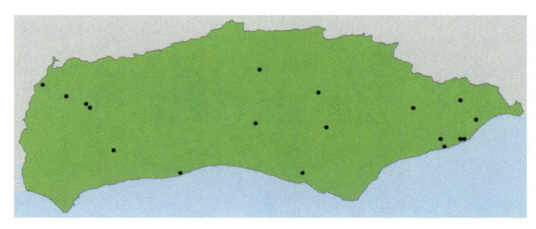

Large map: distribution 2000–2023, with orange circles showing distribution of *Andrena wilkella*
Small map: distribution 1844–1999

The Bees of Sussex

Nomada striata ♀ [SF]

Nomada striata ♂ [SF]

Out in the field and under the microscope
Both males and females are reddish brown and black, with small and widely separated yellow spots on the abdomen, and they are tricky to distinguish from other similarly marked *Nomada*. Females, however, have truncated tips to the mandibles, and this separates them from all other species of *Nomada* except for *N. guttulata*. This last species is usually slightly smaller and has three to four tightly packed short black spines at the base of the hind tibiae, while *N. striata* has longer and more widely spaced spines here.

Males also have truncated tips to the mandibles and are more extensively marked with yellow.

N. striata is often found feeding on a flower head or searching for its host's nests. In common with other species of *Nomada*, the females can be more visible as they fly close to the nest entrances. The nests can be either scattered thinly through an area of grassland or in very large and dense aggregations.

Behaviour and interactions
This *Nomada* lays its eggs within *A. wilkella* brood cells. Several other species of *Andrena* have been suggested as hosts on the continent, including two species found in Britain, *A. fucata* and *A. russula*.

Andrena wilkella ♀, the host [PB]

On the wing
Sussex records run from early May until the end of June.

Feeding and foraging
This bee visits Bell Heather, bird's-foot-trefoil, Bogbean, Bulbous Buttercup, Germander Speedwell, Raspberry, White Bryony, Wood Avens, Wood Spurge, and plants with yellow flowers in the Daisy family such as dandelion and ragwort.

Nomada zonata Variable Nomad Bee
FIRST RECORD 2018, Seaford Head (Steven Falk) TOTAL RECORDS 45

Geography and history
In recent years *Nomada zonata* has increased its range across western Europe. It reached Britain in 2010, which is when it was first recorded in Kent.

The first Sussex record is from April 2018 when a number were seen flying together by Steven Falk at Seaford Head. Since then, *N. zonata* has continued to spread, and it reached west Sussex in 2020. Given that its main host *Andrena dorsata* is now a very widely distributed bee, *N. zonata* could be found just about anywhere in the county.

Seaford Head, where *Nomada zonata* was first recorded in Sussex in 2018 [NS]

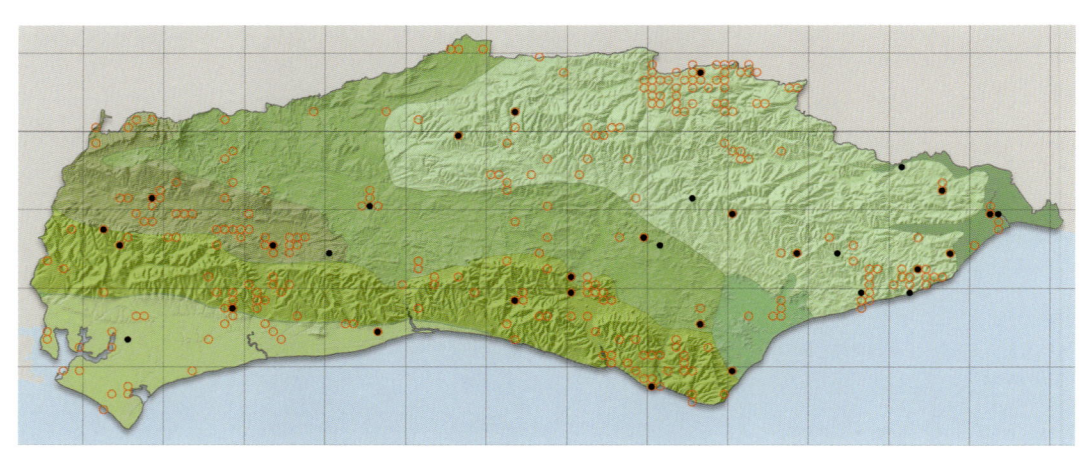

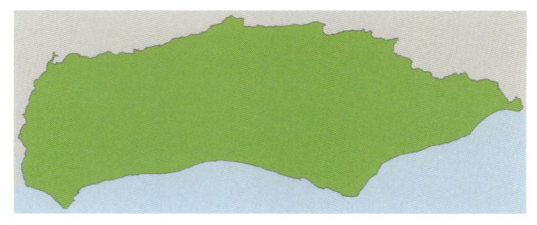

Large map: distribution 2000–2023, with orange circles showing distribution of *Andrena dorsata*
Small map: distribution 1844–1999

142 *The Bees of Sussex*

Nomada zonata ♀ [PC]

Nomada zonata ♂ [SF]

Out in the field and under the microscope
A. dorsata is found in a wide range of habitats, including heathland, chalk grassland and scrubby coastal grasslands, as well as in gardens, and these are all places where *N. zonata* might be found.

N. zonata females have a black, yellow and reddish-brown abdomen and are very variable. Seen from above, the antennae in females have pale tips which contrast strongly with the adjacent dark segments. One other character to look for (although it is not constant) is the presence of reddish-brown markings along the midline of the second and third tergites. Here, these markings often form the inner edge to the large yellow spots.

Typical males have black and yellow abdomens, without the reddish brown found on females.

Behaviour and interactions
This *Nomada* parasitises *A. dorsata* nests. On the continent, it is also reported that *N. zonata* attacks *A. confinis*, another species found in Britain.

Andrena dorsata ♀, the host [SF]

On the wing
This species has two flight seasons a year. These overlap with that of its host, which in Sussex is on the wing between early March until the end of May, and again from the beginning of July until late August.

Feeding and foraging
Plants in the Buckthorn, Daisy, Dead-nettle, Pea, Rose, Stonecrop and Willow families have been reported as sources of nectar on the continent.

Xylocopa – Large Carpenter Bees

This is a genus of large robust solitary bees that are sometimes confused with bumblebees. The females' mandibles are modified to enable the excavation of nest tunnels in wood, including solid timber. Two species have been recorded in Britain, neither of which is permanently resident. One of these, *Xylocopa virginica*, is very occasionally accidentally imported from eastern North America, while the second, *X. violacea*, is an occasional vagrant from Europe as well as being an occasional accidental import. There is evidence to suggest that in some years this last species has overwintered in Britain. Of the two species, only *X. violacea* has been recorded in Sussex.

Xylocopa violacea Violet Carpenter Bee
FIRST RECORD 1996 (Worthing, Mike Edwards) TOTAL RECORDS 3

Geography and history
Xylocopa violacea has been recorded with increasing frequency in Britain in recent years, and this coincides with a rapid expansion in its range in western Europe. It is possible that it will establish a breeding population in Britain—including in Sussex—in the not-too-distant future.

There are three confirmed records for *X. violacea* in Sussex—from Worthing in 1996, East Grinstead in 2003, and Great Dixter in 2023. The first record is of a female excavating a nest in the leg of a bird table in Worthing, evidence of an early attempt to establish a nest here (the bee was killed by the houseowner before the nest was completed, to save the bird table).

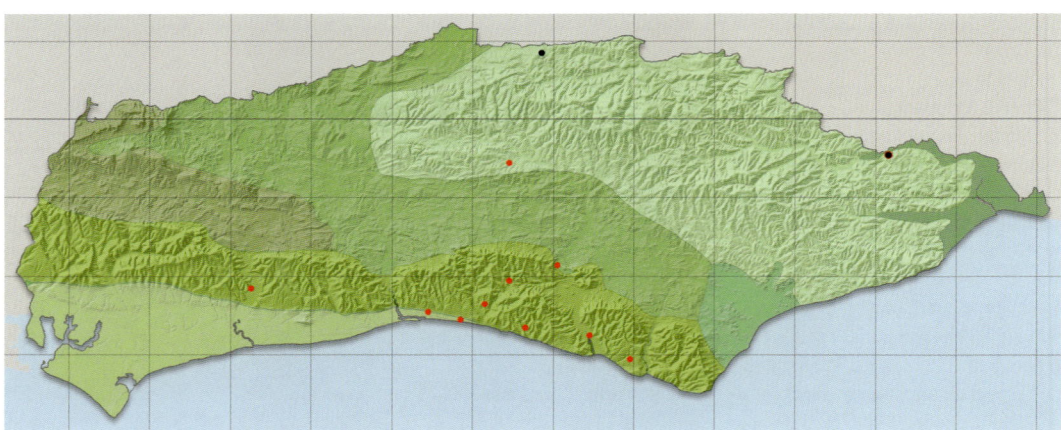

Xylocopa violacea ♀ [PB]

Large map: distribution 2000–2023, with black dots showing distribution of *Xylocopa violacea* and red dots showing distribution of *Xylocopa* sp.
Small map: distribution 1844–1999

The most recent record, from March 2023, is of a male that was recorded at Great Dixter a year after an unconfirmed sighting of a female from the same location—an indication that this species may have successfully bred in Sussex.

While there are just three confirmed records for *X. violacea* in Sussex, there have been a further eleven sightings of bees in this genus from across the county. While it is most likely that these sightings are also of *X. violacea*, in the absence of specimens it has not been possible to confirm that this is the case. These might have been observations of the closely related—and very similar—*X. valga* which may also reach Britain in the near future. Like *X. violacea*, *X. valga* has increased its range across Europe, although not to the same extent.

On the continent, *X. violacea* is frequently recorded in orchards, gardens and parks.

Out in the field and under the microscope
The combination of its large size and black body, as well as the long darkened wings with violet reflections, means that this bee is difficult to confuse with any species resident in Britain.

Males of *X. violacea* and *X. valga* can be separated in the field. The key feature to focus on is the colour of the antennae. In *X. valga* these are entirely dark, while in *X. violacea* they are reddish yellow towards the tip. Females are much more difficult. A key feature is the number of rows of teeth present on the hind tibiae—two rows in *X. violacea*, several in *X. valga*—but this character is very hard to see even with the aid of a microscope.

Xylocopa violacea ♂, showing the reddish-yellow tips to the antennae [PW]

Behaviour and interactions
Males and females overwinter as adults, the adult bees living for more than a year. The females have low productivity, one report showing that they each produce between seven and eight offspring.

On the wing
Males are active between April and August, while females could be on the wing in any month of the year.

Feeding and foraging
X. violacea visits plants in the Borage, Daisy and Pea families, plus acanthus, buddleia, daffodil, lavender, mahonia, phlomis, rosemary, tree-mallow and verbena. The 2023 record from Great Dixter is of a male feeding on Blackthorn.

Anthidium – Wool Carder Bees

These are broad, parallel-sided solitary bees. The body is usually black with yellow markings. Females gather pollen in a brush beneath the abdomen and establish nests made of plant hairs in pre-existing cavities. This is an unusual genus in that the males are often larger than the females and are very territorial. There are some 30 species in Europe, some of which are expanding their range. Just one species is resident in Britain, *Anthidium manicatum*, which is also known from Sussex.

Anthidium manicatum Wool Carder Bee

FIRST RECORD 1877, Brighton (A. Brazenor) TOTAL RECORDS 337

Geography and history

Anthidium manicatum is a very widely distributed bee and is found throughout Sussex. It is most often found in gardens, especially if suitable plants are present, but it has also been found in a very wide

Males are often seen aggressively holding a territory around garden plants such as Caucasian Germander. [WP]

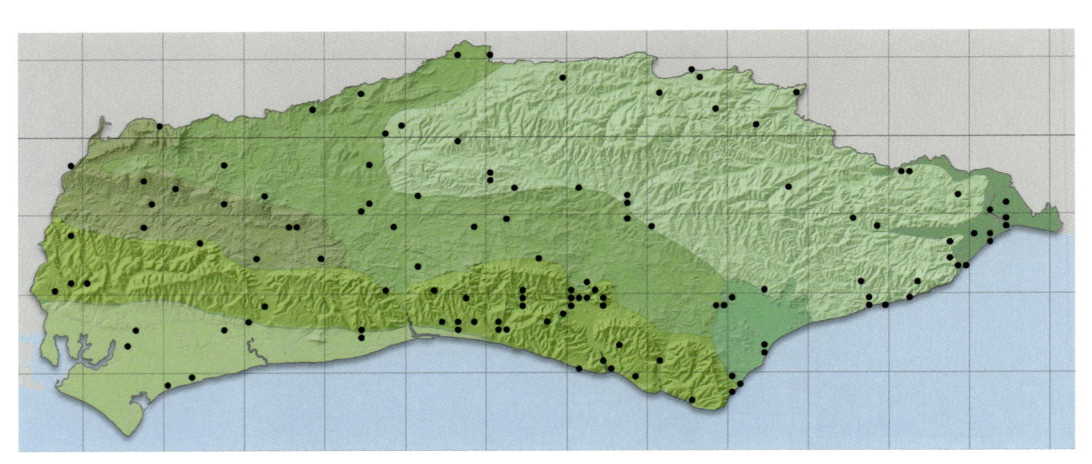

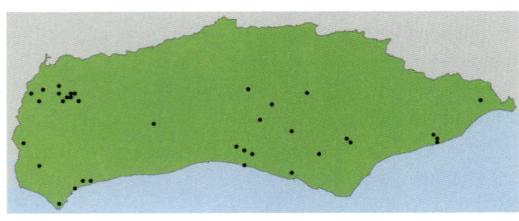

Large map: distribution 2000–2023
Small map: distribution 1844–1999

146 *The Bees of Sussex*

Anthidium manicatum ♀ visiting Lamb's-ear [PB]

Below *Anthidium manicatum* ♀ gathering plant hairs from Lamb's-ear [PB]

Anthidium manicatum ♂ [PC]

range of habitats. These include chalk grassland, flower-rich meadows and pasture, heathland, open woodland, soft-rock cliffs and vegetated shingle.

Out in the field and under the microscope
This bee is unmistakable and can usually be confidently identified from a photograph or on the wing. The males are particularly striking. As with the females, they have yellow spots on the sides of the abdomen, but they also have a set of short spines at the tip.

There is some variation in this species, and it is possible to occasionally encounter males and females with short bands of yellow on the abdomen, instead of spots, or with very dark forms where the yellow spots are much reduced.

Behaviour and interactions
To maximise their chances of mating, the males guard patches of a particular plant and reserve access to the pollen and nectar for visiting females. If necessary, the males will ram into intruders to drive them off, sometimes using the spines at the tip of the abdomen to crush them. Smaller males do not hold territories, adopting a more opportunistic mating strategy.

As well as gathering plant hairs, females gather secretions from certain plants, adding these to the plant hairs that form the partitions between their brood cells.

The bee *Stelis punctulatissima* targets nests of this species.

On the wing
Sussex records run from late May until September.

Feeding and foraging
A wide range of plants is visited for pollen and nectar, with those in the Dead-nettle, Speedwell and Pea families being particularly important sources for pollen.

Chelostoma – Scissor Bees

Chelostoma are elongated bees with a black body. Males have two teeth that project downwards from the tip of the abdomen. Females transport pollen with the aid of a brush of hairs on the underside of the abdomen and can be very efficient at gathering pollen. One species from the continent is reported to gather over 95 per cent of the available pollen grains while foraging. All species nest in pre-existing cavities, creating a row of brood cells that are separated from each other by partitions made of soil or sand grains. Potentially all species in the genus specialise in collecting pollen from a narrow range of plants. Two species have been recorded in Britain, both of which are present in Sussex, *Chelostoma campanularum* and *C. florisomne*.

Chelostoma campanularum Small Scissor Bee
FIRST RECORD 1896, Worthing (Edward Saunders) TOTAL RECORDS 160

Geography and history
Across Sussex, this is a widely distributed bee but one that is often overlooked because of its very small size. As *Chelostoma campanularum* is closely associated with bellflowers, there are many records from habitats that support these plants, including acid grasslands in the Wealden Greensand and species-rich chalk grasslands on the South Downs. It is very frequently recorded in gardens.

Out in the field and under the microscope
This is a very small dark and slender bee up to 4.5 mm long. Despite its small size, it is not difficult to find in gardens and can often be seen flying around a bellflower or flying to and from small tunnel entrances in bee hotels. Males are frequently seen flying persistently around bellflowers and nest sites in search of a mate.

It is a little more difficult to find in the wider countryside, but watching flowering bellflowers or fence posts with small beetle holes can yield results. Even on cool days, it is possible to find this bee within a bellflower.

Any garden that contains bellflowers is likely to support this bee, particularly if there is deadwood nearby. At Great Dixter, for example, Andy Phillips reported that a large number of nests had been

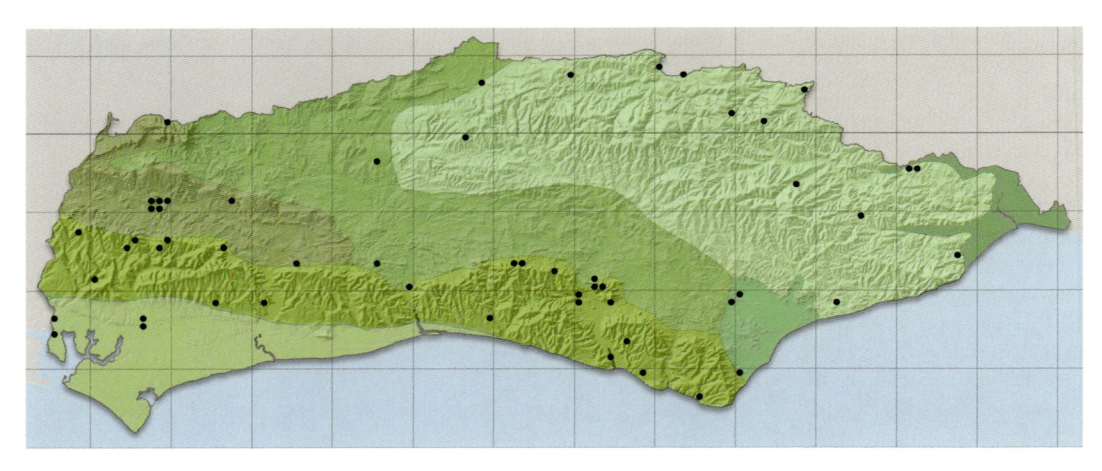

Large map: distribution 2000–2023
Small map: distribution 1844–1999

Chelostoma campanularum ♀ [NO]

Chelostoma campanularum ♂ [SF]

established in woodworm exit tunnels in some of the old barns. This bee is also often seen around trees with dead or decaying branches, and in 1996 David Porter reported finding this bee "flying around old beetle holes in an old apple bough". This was in his garden in Hailsham.

Females have a brush of white hairs on the underside of the abdomen, and males have a pair of downward-pointing teeth at the tip of the abdomen.

Behaviour and interactions
The brood cells are arranged in a line within a small existing tunnel in a piece of timber or reed. Females nest gregariously within these tunnels.

On the wing
Sussex records run from the beginning of June until the end of August.

Feeding and foraging
Pollen is mostly gathered from plants in the Bellflower family, although plants species from other families are also visited. These include geranium, stitchwort, buttercup and members of the Cabbage family.

Chelostoma campanularum visiting a garden bellflower [LL]

The Bees of Sussex 149

Chelostoma florisomne Large Scissor Bee
FIRST RECORD 1905, Brighton (William Unwin) TOTAL RECORDS 67

Geography and history
Most Sussex records are from the High Weald just to the south of Tunbridge Wells and are otherwise thinly scattered across the county. *Chelostoma florisomne* is generally seen along flowery woodland rides or in meadows and pasture close to woodland.

It has also been recorded from a garden in Lewes. Here, the buttercup-rich lawn and a nearby area of woodland mimic its typical habitats.

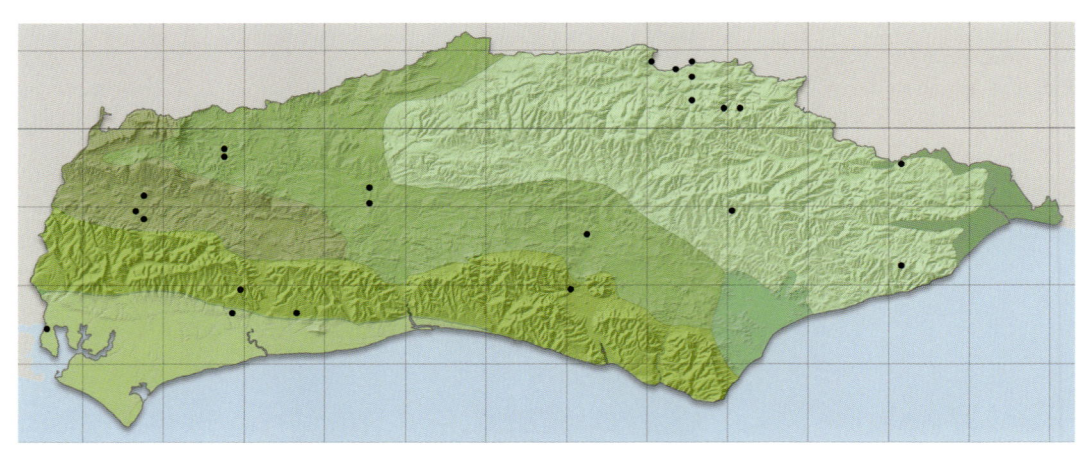

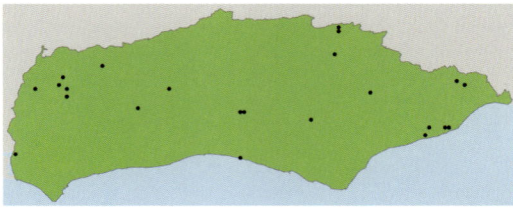

Large map: distribution 2000–2023
Small map: distribution 1844–1999

150 *The Bees of Sussex*

Out in the field and under the microscope
This is a long and slender dark bee. Females are unmistakable and have a brush of white hairs along the underside of the abdomen, an obviously blocky head, and, like the males, very long, scissor-like mandibles.

The males can be distinguished by the two blunt downward-pointing teeth at the tip of the abdomen, visible to the naked eye. They also have noticeably knobbly antennae.

The most successful way of finding this bee is to search buttercup flowers within sheltered areas close to or within woodland. Males can also be seen flying persistently around sunlit timber and coursing along adjacent foliage in search of females.

Behaviour and interactions
C. florisomne is one of the few bee species to specialise in collecting buttercup pollen.

Brood cells are established within pre-existing tunnels in decaying timber, fence posts or even reed stems that are being used as roof thatch. Typically there are two to three brood cells per nest, although there can be as many as eight. The cells are arranged linearly, with female cells towards the back and males towards the front. Cell walls are lined with moistened clay or sand, and a similar mix is used to seal the entrance with the addition of small stones. This hardens as it dries.

The club-horned wasp *Monosapyga clavicornis* is a significant predator of *C. florisomne*, laying its eggs in brood cells that have been stocked with pollen. As the wasp is able to push its abdomen through a cell wall to lay an egg within a sealed cell, the bee may also construct empty cells to deceive the wasp. If the wasp eggs are laid before the brood cells have been sealed, the female bee may remove these.

Monosapyga clavicornis ♀ [PC]

On the wing
Sussex records run from the beginning of May until the end of July. Elsewhere, *C. florisomne* has been reported as occasionally flying in August.

Feeding and foraging
C. florisomne largely forages for pollen from buttercup, particularly Bulbous Buttercup, Creeping Buttercup and Meadow Buttercup. Plants that it visits for nectar include Common Valerian, Germander Speedwell, and yellow-flowered plants in the Daisy family, including dandelion, sow-thistle and hawk's-beard. Other plants visited include Lesser Stitchwort, rose, Small-leaved Lime and Wood Crane's-bill.

Coelioxys – Sharp-tail Bees

The *Coelioxys* are unlike any other bee genera in that the females have strongly tapered cone-like abdomens. They also have abundant hairs on their eyes, a character shared only with the Honey Bee. Although the shape of the abdomen is distinctive in females, separating out to species can be a challenge. Males have more rounded abdomens which can end in a variable number of spines. Both males and females are largely black with patches or bands of white hairs on the abdomen. Females of many species of *Coelioxys* cut a slit in the wall of the host's brood cell with the tip of the abdomen, before inserting an egg that then protrudes into the cell. Others lay their eggs either within or on top of the food provisions. Various species of *Megachile* and *Anthophora* are targeted, although the precise relationships are not always understood. Seven species have been recorded in Britain, with six known from Sussex.

Coelioxys conoidea Large Sharp-tail Bee
FIRST RECORD 1845, Littlehampton (unknown)　TOTAL RECORDS 94

Geography and history
Coelioxys conoidea has a very restricted distribution and is found on areas of dry heathland on the Lower Greensand, as well as in coastal locations. Examples include Camber Sands, East Head, Newhaven, and the sand dunes at Littlehampton. It has also been recorded from a steep chalk grassland bank near Lewes.

There are other recent records from The Crumbles, an area of vegetated shingle in Eastbourne, although this population may since have been lost to development.

Out in the field and under the microscope
C. conoidea is the largest of the *Coelioxys* and, like the other species in the genus, is a fast-flying insect. The first sign that it might be present on a site can be a sighting of its host, the very large *Megachile maritima*.

Although this is probably the easiest of the species in this genus to identify (with practice, it can be identified in the field), it is still tricky. Males and females have broad and widely separated patches

Large map: distribution 2000–2023, with orange circles showing distribution of *Megachile maritima*
Small map: distribution 1844–1999

Coelioxys conoidea ♀ on Sea-holly [PB]

Coelioxys conoidea ♂ on Sea-holly [CG]

Megachile maritima ♀, the host [PB]

of white hairs on the tergites and on the sternites. All other species have narrower, uninterrupted or scarcely interrupted bands here. In females, the underside of the final segment of the abdomen is broad, narrowing evenly towards the tip.

Behaviour and interactions
In Britain, *C. conoidea* targets the nests of *M. maritima* and, unusually for bees in this genus, it can be abundant close to nest sites of the host.

On the wing
Sussex records run from late June until late August.

Feeding and foraging
C. conoidea visits a wide range of plants, including knapweed, bramble, Common Sea-lavender, Creeping Thistle, mallow, ragwort, Sea-holly, Sea Bindweed, Sea Rocket and thyme.

Common Sea-lavender [PC]

Sea Bindweed [PC]

The Bees of Sussex **153**

Coelioxys elongata Dull-vented Sharp-tail Bee
FIRST RECORD 1896, Bognor Regis (Henry Guermonprez) TOTAL RECORDS 52

Geography and history
Despite the abundance of its main host, this is an elusive bee with few Sussex records. Between 1919 and 1974, for example, it was unrecorded in the county, while over the last decade there have been just sixteen records.

These more recent records are from widely separated locations across the county, including Lewes, Eastbourne, Great Dixter, Graffham Common and Rye Harbour Nature Reserve. The most recent record is from near Etchingham in 2022.

Coelioxys elongata ♀ resting on a bee hotel [PC]

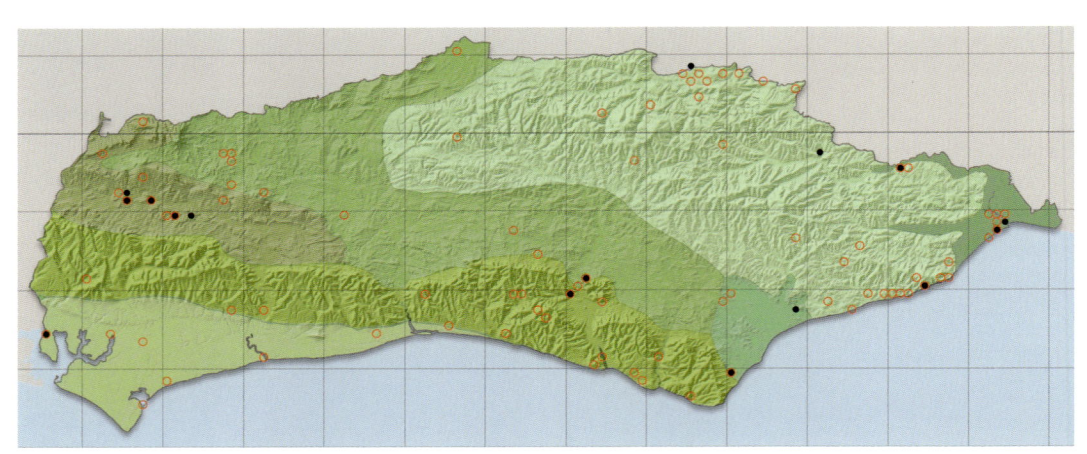

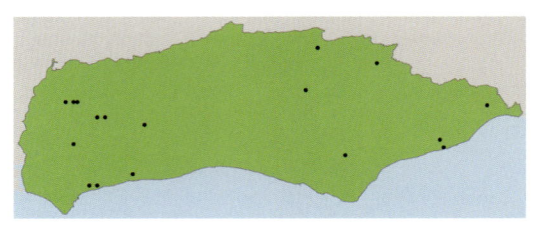

Large map: distribution 2000–2023, with orange circles showing distribution of *Megachile circumcincta* and *M. willughbiella*
Small map: distribution 1844–1999

154 *The Bees of Sussex*

Coelioxys elongata ♀ [PC]

Coelioxys elongata ♂ [SF]

Out in the field and under the microscope
This is not an easy bee to find, but searching areas where its main host *Megachile willughbiella* is known to be active could yield results. Possible locations include parks and gardens where *M. willughbiella* is frequently found, perhaps nesting in fence posts or in bee hotels. In these situations, *Coelioxys elongata* might be seen searching for a suitable nest to parasitise.

Female *C. elongata* are similar in appearance to *C. inermis* and *C. mandibularis*, with all three species having tiny acute teeth towards the tip of the abdomen. *C. mandibularis*, however, can be separated from the others by the right-angled shape of its mandibles.

Female *C. elongata* and *C. inermis* are more of a challenge to separate but can be distinguished from each other by subtle differences, including how dull or shiny the midline of sternite four is and how dense the punctation is here. In *C. elongata* this area is very dull and has dense punctation, whereas in *C. inermis* it is clearly shiny and has much sparser punctation.

Separating the males requires examination of the genital capsule.

Behaviour and interactions
C. elongata lays its eggs within sealed brood cells of *M. circumcincta* and *M. willughbiella*. Other possible hosts include *Anthophora bimaculata* and *M. maritima*, with additional species also cited on the continent.

Ian Beavis has reported males "flying fast and low among grass and herbs in flower-rich grassland […] and visiting buttercups and bird's-foot-trefoil in company with *Megachile willughbiella*, *Andrena wilkella* and *Osmia caerulescens*". He has also recorded single females in allotments and visiting bird's-foot-trefoil in grassland by the River Grom.

Megachile willughbiella ♀, the principal host [ME]

On the wing
Sussex records run from mid May until late August.

Feeding and foraging
This bee visits bellflower, bramble, buttercup, Common Fleabane, Greater Bird's-foot-trefoil and knapweed.

Coelioxys inermis Shiny-vented Sharp-tail Bee
FIRST RECORD 1879, Hastings (Edward Saunders) TOTAL RECORDS 50

Geography and history
There have been just fifteen Sussex records for *Coelioxys inermis* over the last 20 years—it is a very scarce bee even though its main hosts *Megachile centuncularis* and *M. versicolor* are relatively abundant. The more recent records are from the Wealden Greensand, from the downs between Falmer and Eastbourne, from Chichester Harbour, Littlehampton and from the coast at Fairlight and Rye Harbour Nature Reserve.

Out in the field and under the microscope
Gardens are good places to search for *C. inermis*, although it is not an easy bee to find, especially as it rarely occurs in numbers. Checking possible nest sites for its hosts is one approach— in a garden in Lewes, for example, it has been seen flying close to a tree stump being used for nesting by a species of *Megachile* and, over the course of several days, close to a bee hotel. It is also occasionally found feeding.

It is a challenging species to identify as it is very similar in appearance to *C. elongata* and *C. mandibularis*.

Females of all three species have tiny sharp teeth towards the tip of the final sternite.

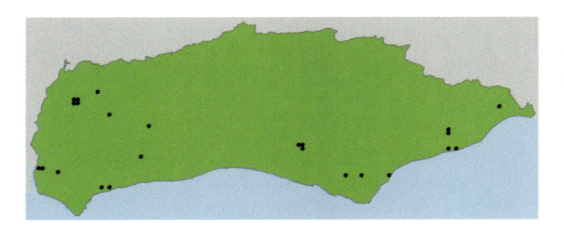

Coelioxys inermis ♀ at rest close to nest of *Megachile ligniseca* [JP]

Large map: distribution 2000–2023, with orange circles showing distribution of *Megachile centuncularis* and *M. versicolor*
Small map: distribution 1844–1999

156 *The Bees of Sussex*

Coelioxys inermis ♀ [NO]

Coelioxys inermis ♂ [NJ]

However, *C. mandibularis* has very distinctively shaped, right-angled mandibles, while with practice *C. inermis* and *C. elongata* can be separated by checking the underside of the abdomen. In *C. inermis*, the punctures on the fourth sternite either side of the midline are large and relatively widely separated, while in *C. elongata* they are small and close together. The surface of the sternite around these punctures is smooth in *C. inermis* and roughened in *C. elongata*.

Males are even more challenging, and examination of the genital capsule is necessary to distinguish between them.

Behaviour and interactions

C. inermis lays its eggs in brood cells of at least four species of *Megachile*. It has long been known to parasitise the nests of *M. centuncularis* and *M. versicolor*. In July 1933 and again in 1934, Alfred Brazenor reported that he reared females from "cells of a *M. versicolor* in a dead branch of willow". As recently as 2019, a new host was confirmed for this bee, *M. leachella*, while a relationship was also confirmed with *M. ligniseca* in 2020.

Megachile centuncularis ♀, a host [SF]

It appears that host-parasite relationships vary across the country. In Norfolk, for example, *C. inermis* parasitises *M. leachella*, while in South Wales and Kent it is *C. mandibularis* which parasitises *M. leachella*.

On the wing

Sussex records run from the end of May until late August. Elsewhere, it is reported to fly until the beginning of September.

Feeding and foraging

There are few records of this bee visiting plants. Those listed in the literature are Cross-leaved Heath, bird's-foot-trefoil, bramble, spurge and ragwort.

Coelioxys mandibularis Square-jawed Sharp-tail Bee
FIRST RECORD 1981, The Crumbles (Gerald Dicker) TOTAL RECORDS 1

Geography and history
Across Britain, *Coelioxys mandibularis* is a scarce bee known from the east Kent coast, south and north-west Wales, and Merseyside. In these places, it is found in coastal sand dunes and sandy areas.

In Sussex, it has only been recorded once, at The Crumbles in 1981. This is despite the abundance of its probable hosts along sections of the coast.

Out in the field and under the microscope
Females can be separated from the other species of *Coelioxys* by the very distinctive mandibles—these are bent almost to a right-angle. The males are trickier, with the abundance of hairs within a

Large map: distribution 2000–2023, with orange circles showing distribution of *Megachile leachella* and *M. maritima*
Small map: distribution 1844–1999

groove on the second tergite and the colour of spines on the hind tibiae among the characters to look for. In this species, the groove should be filled with short hairs, while the spines are generally black or a dark reddish brown.

C. mandibularis is possibly extinct in Sussex but might yet be found around the nests of its hosts. In the case of one of these, *Megachile leachella*, these tend to be in aggregations sited on areas of sandy soil.

Behaviour and interactions

C. mandibularis is a parasite of various species of *Megachile*. In Britain, a host-parasite relationship has been confirmed with *M. leachella* and *M. maritima*. Individual bees from populations in Kent—associated with *M. leachella*, the smallest of the *Megachile*—tend to be much smaller than those found in places such as Merseyside, where *M. maritima*—one of the largest *Megachile*—is the host.

On the continent, other species of *Megachile* suggested as hosts are *M. centuncularis*, *M. circumcincta* and *M. versicolor*.

Megachile leachella ♀, a host [PC]

Megachile maritima ♀, a host [SF]

On the wing

The flight season for *C. mandibularis* overlaps with that of its hosts, which have been recorded in Sussex between late May and the end of August.

Feeding and foraging

In Britain, Sea-holly and thyme are the only plant species reported as being visited by this bee. On the continent, the following are some of the species given: Common Bird's-foot-trefoil, Field Scabious, Goldenrod, Hop Trefoil, Rosebay Willowherb, Sheep's-bit, Small Scabious and Wood Crane's-bill.

Sea-holly ♀ [PC]

Coelioxys quadridentata Grooved Sharp-tail Bee
FIRST RECORD 1896, Hastings (Edward Saunders) TOTAL RECORDS 5

Geography and history
The five Sussex records for *Coelioxys quadridentata* are from very scattered locations across the county. These are Hastings in 1896, Slindon in 1905, Harting Down in 1975, Deep Dean in 1988, and then from the undercliffs at Hastings Country Park in 2007—111 years after the first county record from the same area.

Nationally, this bee is now very scarce with very few modern records, all restricted to the southern half of Britain.

Out in the field and under the microscope
Searching its hosts' nest sites is one way of locating this species. One Sussex record is from a specimen reared from an *Anthophora furcata* nest, which was within a piece of rotten wood collected on the

Coelioxys quadridentata ♀ [PW]

Coelioxys quadridentata ♂ [SF]

Large map: distribution 2000–2023
Small map: distribution 1844–1999

downs, while elsewhere it has been found close to an *A. quadrimaculata* nesting aggregation in an old stone wall. On the continent, it has been found close to areas of bare sandy soil being used as a nest site by *A. bimaculata*.

Females can be readily separated from other species in the genus by the shape of the underside of the abdomen towards the tip, where there is an obvious constriction.

Males represent more of a challenge. Among other characters, the details of the furrow that runs across the second tergite will help to identify this species. This furrow should run uninterrupted all the way across the tergite, unlike in *C. rufescens*.

Behaviour and interactions

C. quadridentata attacks the brood cells of different species of *Anthophora* and *Megachile*. Of these, *Megachile circumcincta* is possibly the main host, and the decline of this bee may be one reason for the scarcity of *C. quadridentata*.

A. furcata is also a confirmed host, based in part on observations made by Mike Edwards on Harting Down in 1975. Here he reported how he "bred *Coelioxys quadridentata* from a stem containing *Anthophora furcata* cells".

Other hosts suggested on the continent are *A. bimaculata*, *M. leachella* and *M. willughbiella*.

Megachile circumcincta ♀, possibly the main host [SF]

Anthophora furcata ♀, a host [IT]

On the wing

The flight season for *C. quadridentata* overlaps with that of its hosts, which have been recorded in Sussex between late May and early September.

Feeding and foraging

Common Bird's-foot-trefoil and White Bryony are the only associated plant species reported from Britain.

Coelioxys rufescens Rufescent Sharp-tail Bee
FIRST RECORD 1895, Slindon (Henry Guermonprez) TOTAL RECORDS 49

Geography and history
This is one of the more frequently recorded of the *Coelioxys*, but it is still not seen every year. Most modern Sussex records are from the heathlands of the Lower Greensand and from the coast near Hastings. There are also isolated records from places such as Seaford Head, Lullington Heath, Eastbourne, Great Dixter and Poynings.

Out in the field and under the microscope
The underside of the final segment of the female's abdomen is unlike any other *Coelioxys* species as it has 'shoulders' just before a triangular tip. This makes it a more straightforward species to identify than most others in the genus.

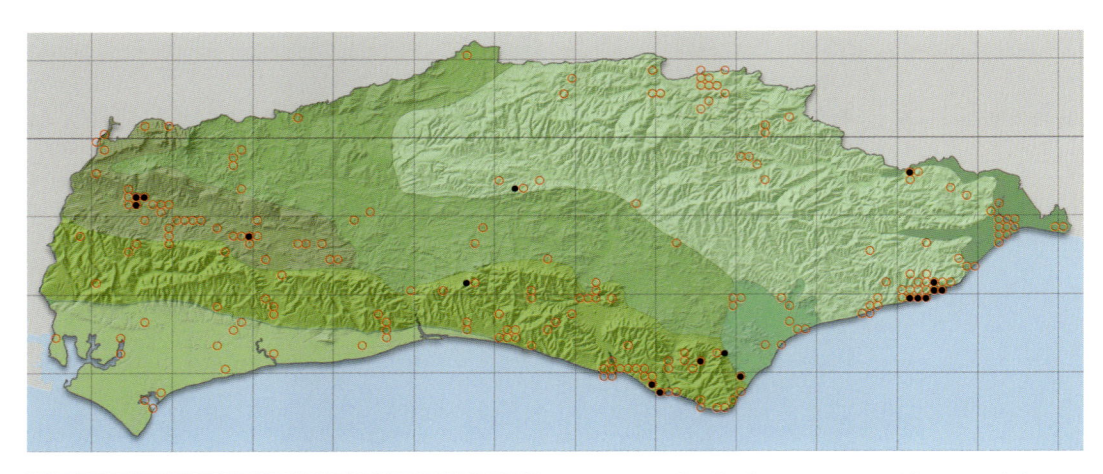

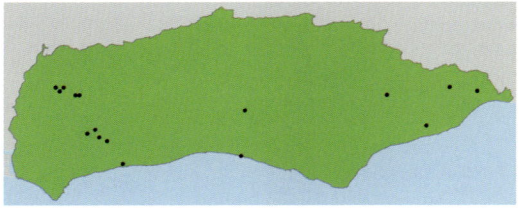

Large map: distribution 2000–2023, with orange circles showing distribution of *Anthophora bimaculata, A. furcata* and *Megachile centuncularis*
Small map: distribution 1844–1999

Males have continuous bands of hairs on the tergites and sternites, and most resemble *Coelioxys quadridentata*. *C. rufescens* can be separated from *C. quadridentata* by checking the furrow near the front of the second tergite. This becomes weak or indistinct in the middle in *C. rufescens* but is continuous in *C. quadridentata*.

This bee is most likely to be seen visiting flowers. Near Tunbridge Wells, Ian Beavis has recorded females visiting bramble, Wild Marjoram and Hebe, and males "flying persistently around low foliage and settling on sunlit fences". Searching close to the nest sites of its hosts could also yield results. A 2017 record from Great Dixter, for example, was from near a log pile.

Behaviour and interactions

C. rufescens is known to lay its eggs in the brood cells of various species of *Anthophora*, especially *Anthophora bimaculata* and *A. furcata*, and it has been reared from nests of both. *A. quadrimaculata* has been suggested as a host on the continent.

It has also been reared from *Megachile centuncularis* brood cells, and *M. circumcincta*, *M. ligniseca* and *M. willughbiella* are other possible hosts.

Anthophora bimaculata ♀, one of the hosts [SF]

Anthophora furcata ♀, one of the hosts [PB]

On the wing

Sussex records run from late May until late August.

Feeding and foraging

Visits to bramble, buttercup, Common Mallow, Creeping Thistle, Hebe, Wild Marjoram, ragwort, scabious, Tufted Vetch and willowherb are all reported in the literature.

Heriades – Resin Bees

The slender bees in this genus are found in many parts of the world, including Asia, Europe, and North and Central America. The females carry pollen in a brush of hairs on the underside of the abdomen and establish their linear nests in pre-existing tunnels in wood or within the pithy stems of plants such as bramble. One of the features that distinguishes this genus from others is the presence of a small but very distinct ridge at the front of the abdomen. Two species have been recorded in Britain, one of which, *Heriades truncorum*, is known from Sussex. The second species, *H. rubicola*, is possibly an accidental introduction. This was recorded in Dorset in 2006 and again in Greenwich in 2016. On the continent, it is largely restricted to the Mediterranean region.

Heriades truncorum Small-headed Resin Bee
FIRST RECORD 1974, Cocking (Mike Edwards) TOTAL RECORDS 119

Geography and history
Heriades truncorum is much more widely distributed than the records show as its population has expanded rapidly in recent years, and yet it is often overlooked.

After it was first found in Sussex in 1974 at Cocking, there was just a handful of records over the next 20 years, mostly from the Wealden Greensand. However, since the 1990s the number of records has risen, and there are now records from locations across the county.

Out in the field and under the microscope
A bee hotel or a piece of timber with small pre-drilled holes sited in a garden is a great place to

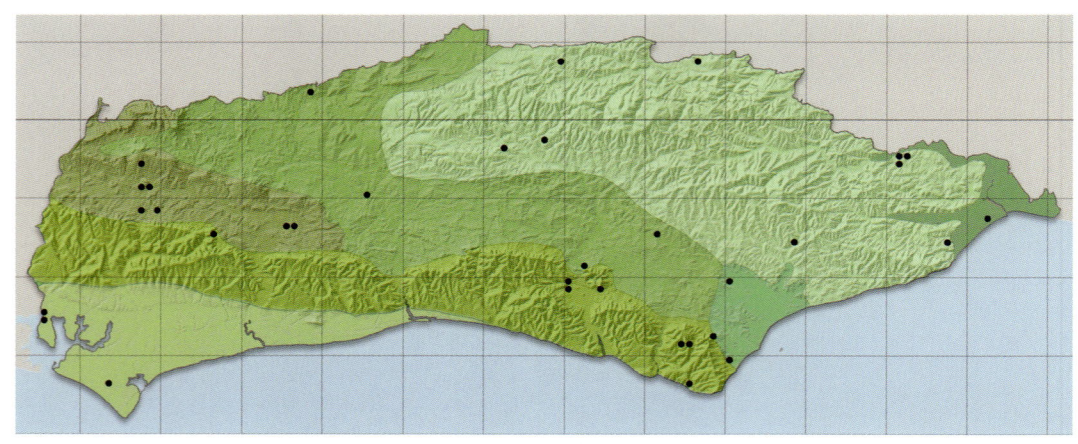

Heriades truncorum ♀ [BY]

Large map: distribution 2000–2023
Small map: distribution 1844–1999

Heriades truncorum ♀ **carrying resin** [BY]

Heriades truncorum ♂ [SF]

watch for this bee. Here, females are often seen flying to their nest burrows with yellow pollen, or carrying pieces of resin in their mandibles. They are also frequently found feeding, particularly on plants with yellow flowers in the Daisy family, while males might be seen resting on flower heads.

This is a small, broad-headed blackish bee. Key features to look for are the narrow ridge at the front of the abdomen in both sexes and an orange pollen-collecting brush beneath the abdomen of the female.

A gate post with pre-drilled holes is a good place to watch *Heriades truncorum*. [WP]

Exposed nest within hollow stem showing cells divided by resin and *Heriades truncorum* **larvae feeding on pollen** [PW]

Behaviour and interactions
As well as in bee hotels and pieces of pre-drilled timber, *H. truncorum* nests in dead trees, fence posts and bramble stems. Resin from broadleaves and conifers is used to create partitions between the unlined brood cells and to seal the nest entrance.

H. truncorum is targeted by the cuckoo bee *Stelis breviuscula*, which lays its eggs within open brood cells while they are still being provisioned.

On the wing
Sussex records run from the end of May until mid September. Elsewhere, it is reported to be active into early October.

Feeding and foraging
H. truncorum largely specialises in collecting pollen from plants with yellow flowers in the Daisy family such as Cat's-ear, fleabane, hawkbit, and especially ragwort. It also regularly visits knapweed.

Hoplitis – Lesser Mason Bees

Bees in this genus are very similar in appearance and behaviour to those in the genus *Osmia*. As with this other genus, nests are established in wood, in soil and in cracks in rock, often using chewed sections of leaf to line the brood cells and to seal the entrance to the nest. Three species have been recorded in Britain. These are *Hoplitis adunca* (first recorded at Greenwich in 2016), *H. leucomelana* (recorded once in Britain, in 1920), and *H. claviventris*, which is the only one to have been recorded in Sussex. *Osmia spinulosa* was also included within this genus until recently.

Hoplitis claviventris Welted Mason Bee
FIRST RECORD 1896, Hastings (Edward Saunders) and Bexhill (unknown) TOTAL RECORDS 157

Geography and history
Although a high proportion of Sussex records for *Hoplitis claviventris* are from chalk grasslands on the South Downs, there is also a good number of records from the Lower Greensand. Elsewhere in the county, records are thinly scattered and are from a variety of habitats, including orchards, sand dunes, undercliffs, vegetated shingle and woodland. It is also known from gardens in the county.

There are recent records from Stedham Common, Ashburnham Place and Cow Gap.

Out in the field and under the microscope
This is not an easy bee to find as populations are generally very localised. The best approach is to search sheltered spots in any of the habitats that it uses, checking foodplants such as Common Bird's-foot-trefoil. Another technique is to collect dead, hollowed-out bramble stems in late winter to see if adult bees emerge once they have been brought indoors.

In females, the white hair bands on the tergites form a strong contrast to the shiny black abdomen and are visible in the field. They resemble an *Osmia*, especially *Osmia spinulosa*. However, female *H. claviventris* have an all-white brush for gathering pollen on the underside of the abdomen (it is bright orange in *O. spinulosa*).

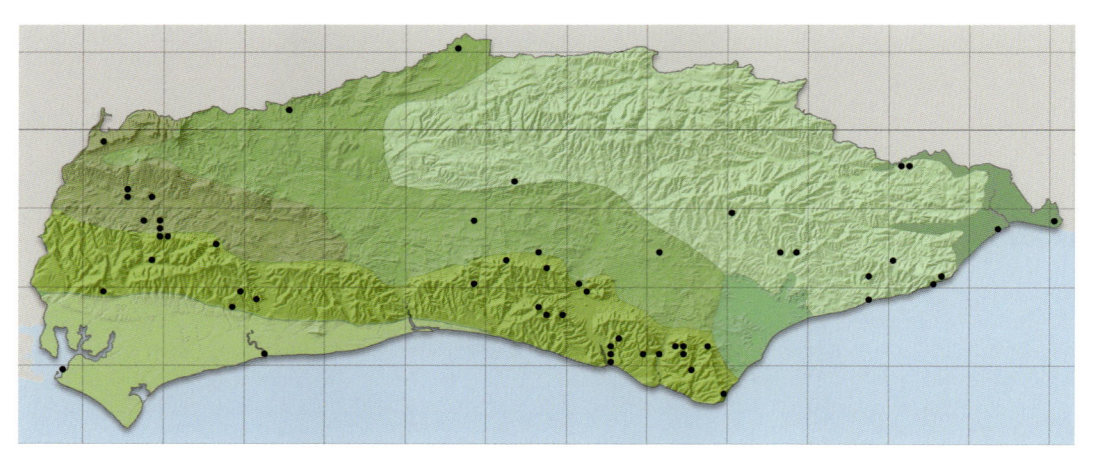

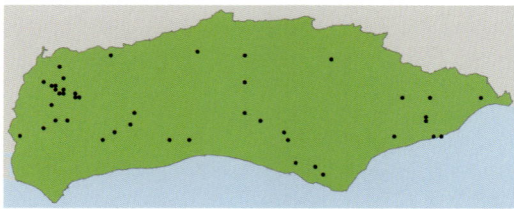

Large map: distribution 2000–2023
Small map: distribution 1844–1999

The south-facing chalk grassland at Cow Gap is a good location to search for *Hoplitis claviventris*. [EJ]

The males of these two species are a little trickier to separate, but in *H. claviventris* a broad, chunky plate protrudes downwards on sternite two and is visible under magnification, while *O. spinulosa* has a slender spine here. *O. spinulosa* also has backward pointing spines on plates adjacent to the sides of the scutellum.

Hoplitis claviventris ♀ [PC]

Hoplitis claviventris ♂ [IT]

Behaviour and interactions
H. claviventris generally makes its nests in dead stems of plants with a pithy centre. These include bramble, ragwort and rose. Typically, four brood cells are arranged in a line within the stem and separated from each other by partitions made of chewed pieces of leaf. The cell walls themselves are unlined. Males tend to be near the front and emerge first.

Stelis ornatula lays its eggs within brood cells of this species.

On the wing
Sussex records run from mid May until late August.

Feeding and foraging
This species visits a range of plants, including Wild Angelica, bramble, Common Bird's-foot-trefoil, dandelion, buttercup, scabious, hawk's-beard, *Erica*, vetch and clover.

Megachile – Leafcutter Bees

Megachile bees are known as 'Leafcutter Bees' because most species transport leaf material to the nest with which to line and seal their cells. In Britain, all but one species use pieces of soft fresh leaves, while the exception—*Megachile ericetorum*—uses mud. This last species is assumed to be extinct, although a male was recorded in Kent in 2022, possibly as a result of an accidental introduction. The females transport pollen with a brush of hairs on the underside of the abdomen—the pollen is gathered by pressing the abdomen against the anthers. Nine species have been confirmed from Britain, with a further two of doubtful status. Seven species have been recorded in Sussex.

Megachile centuncularis Patchwork Leafcutter Bee
FIRST RECORD 1891, Bognor Regis (Henry Guermonprez) TOTAL RECORDS 257

Geography and history
Megachile centuncularis is a widely distributed bee in Sussex and could occur just about anywhere. It is very commonly seen in gardens, but also occurs on heathlands, in open woodlands, on scrubby grasslands, along hedgerows, and along the coast.

Out in the field and under the microscope
This is quite a distinctive bee, and it should be possible to identify females in the field. Seen from above, the orange hairs of the pollen brush on the underside of the abdomen are visible, forming a ring of reddish orange around the otherwise darkened abdomen. These hairs extend all the way to the tip of the abdomen,

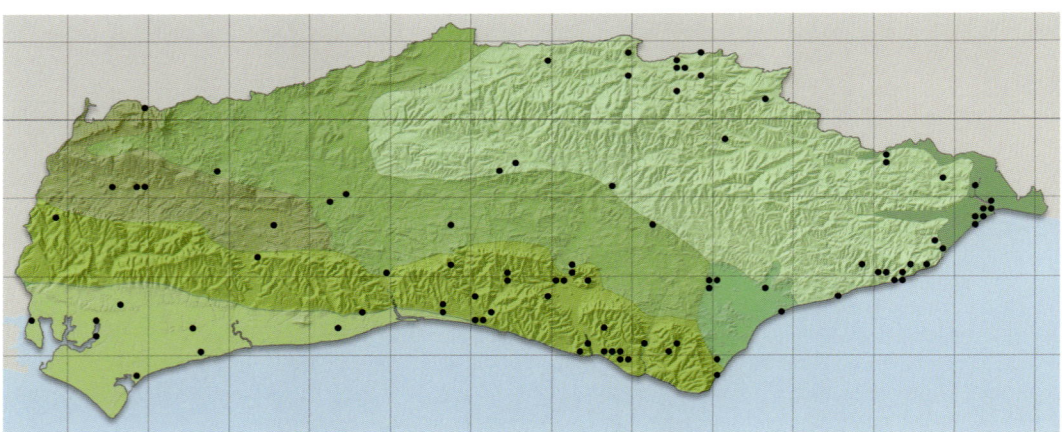

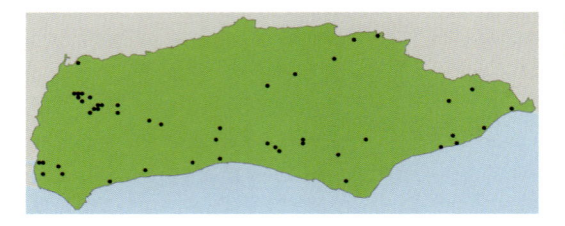

Megachile centuncularis ♀ [PC]

Large map: distribution 2000–2023
Small map: distribution 1844–1999

The Bees of Sussex

whereas the otherwise similar *M. versicolor* females have black hairs nearest the tip on sternite six.

Male *M. centuncularis* resemble male *M. versicolor*, with differences between the two very subtle and only visible under magnification. One of the key characters to check is the hind margin of the fourth sternite—in *M. centuncularis* this is dull and covered in minute hairs which are particularly visible along the margin of the sternite, while in *M. versicolor* this area is shiny and almost completely hairless.

Gardens are the most reliable places to search for *M. centuncularis*.

Megachile centuncularis ♂ [NO]

Behaviour and interactions
Nests are established in pre-existing cavities in dead trees, woody plant stems and walls. Each brood cell is lined with sections of leaf cut from a range of different plants. Leaf material is also used to seal the entrance. Each female might construct as many as 30 brood cells.

M. centuncularis is attacked by different species of *Coelioxys*, and *C. inermis* and *C. rufescens* have both been reared from cells of this bee. On the continent, *C. elongata* and *C. mandibularis* are also cited as parasites.

Megachile centuncularis ♀ cutting a rose leaf [PC]

On the wing
Sussex records run from mid May until late September.

Feeding and foraging
This bee visits a very wide range of plants for pollen and nectar. The Daisy and Pea families, for example, are both important sources of pollen.

Megachile circumcincta Black-headed Leafcutter Bee
FIRST RECORD 1911, Graffham (Henry Guermonprez) TOTAL RECORDS 9

Geography and history
When Ian Beavis found a female *Megachile circumcincta* in Fore Wood in 2019, it was the first time this species had been recorded in Sussex since 1994.

M. circumcincta has seemingly disappeared from many other parts of southern England and was last recorded in Kent in 1918 and in Surrey in 1973. There are still strong colonies in other parts of Britain, including in coastal areas of Scotland.

This bee is restricted to sandy habitats, particularly heathland and sand dunes, and in Sussex was recorded from Graffham in 1911, Hastings in 1919, Ambersham Common in 1976, Camber Sands in 1986 and 1994, and Hastings again, also in 1994.

Fore Wood, where *Megachile circumcincta* was recorded in 2019, is an unlikely location to find this bee. [IB]

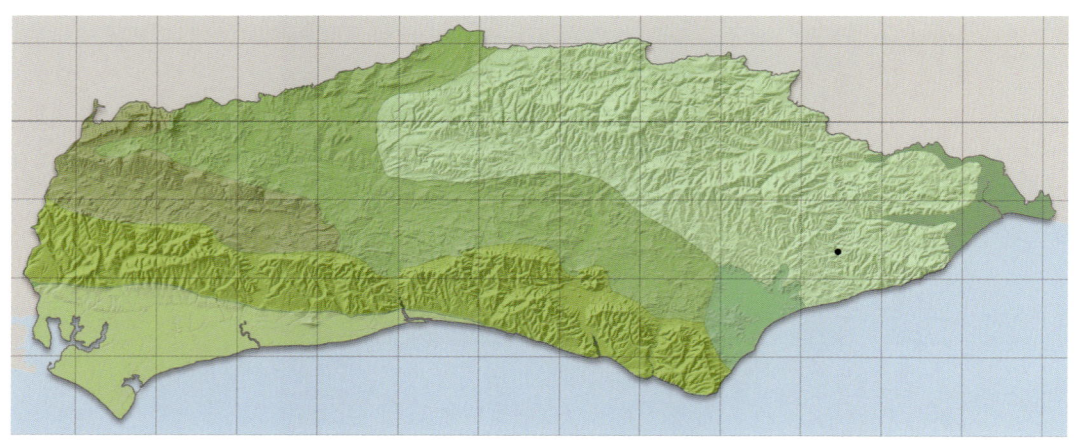

Large map: distribution 2000–2023
Small map: distribution 1844–1999

170 *The Bees of Sussex*

Megachile circumcincta ♀ [NO]

Megachile circumcincta ♂ [SF]

Out in the field and under the microscope
This is a distinctive bee. The females have large black-haired heads, and the colour of the hairs here contrasts strongly with the reddish-brown hairs on the top of the thorax. The hairs on the sides of the thorax are pale, while the brush of hairs on the underside of the abdomen is mostly a reddish colour but black at the tip.

In males, the tarsi on the front legs are expanded, and this is one of just three species where this is the case. In *M. circumcincta*, however, this is not to the same extent as in the other two species, *M. maritima* and *M. willughbiella*.

The 2019 record from Fore Wood is of an individual that had settled on a sunlit wooden bench in a coppiced clearing.

Behaviour and interactions
M. circumcincta generally excavates very shallow nests that are just 2 to 3 cm below the ground. These are sited in areas of sandy ground, especially where the ground is firm and facing south or under stones. It will also use pre-existing cavities. Brood cells are lined with sections of leaf that are principally collected from roses and birch trees. On the continent, other sources of leaf matter cited are lime and Hornbeam.

This bee is possibly parasitised by *Coelioxys elongata*, *C. mandibularis*, *C. quadridentata* and *C. rufescens*.

On the wing
Typically, *M. circumcincta* is on the wing between late May and early August.

Feeding and foraging
M. circumcincta visits bramble, broom, Common Bird's-foot-trefoil, knapweed, restharrow and thistle.

Megachile leachella Silvery Leafcutter Bee
FIRST RECORD 1855, Littlehampton (Frederick Smith) **TOTAL RECORDS** 158

Geography and history
Megachile leachella is largely a coastal species but is also occasionally recorded from inland locations where there are sandy soils and little vegetation. In Sussex, it has been recorded just twice away from the coast—on both occasions this was at a sandpit near Midhurst, most recently in 2019.

There have been recent records from Camber Sands, Chichester Harbour, sand dunes near Littlehampton, clifftop grassland at Newhaven, vegetated shingle at Norman's Bay and Galley Hill.

Clifftop vegetation, Newhaven [WP]

Large map: distribution 2000–2023
Small map: distribution 1844–1999

172 *The Bees of Sussex*

Out in the field and under the microscope
This is the smallest of the *Megachile* and one of the easiest to identify in the field. Females have a white brush of hairs covering the underside of the abdomen, very distinctive bands of white hairs on the tergites, and two small white hair patches on the tip of the abdomen.

The males are smaller than the females and have greenish eyes and, like the females, have white hair bands on the tergites. With care, it is possible to see an area of flattened white hairs that covers the tip of the abdomen, a character that is not shared with any other *Megachile*. Both sexes have a silvery appearance.

Megachile leachella ♀, transporting a leaf fragment [PB]

Megachile leachella ♂ showing the pale green eyes and white hairs on the clypeus [PC]

Megachile leachella ♂ [PC]

This species is usually seen flying just above areas of sparsely vegetated ground, and males are sometimes seen patrolling just above ground-level, moving rapidly between nest entrances and investigating other insects in flight. It is also possible to see males at rest on vegetation before they dart off to investigate passing insects.

In 2016 George Else recorded this species "mainly nesting in sand heaped into a long pile" in a sandpit near Midhurst.

Behaviour and interactions
Female *M. leachella* excavate their nest cavities in sandy soils, forming either loose or dense aggregations. These can extend over an area of several hundred metres. On the continent, it is reported that this bee also nests above ground, often in hollow plant stems.

Brood cells are lined with sections of leaf cut from a wide range of plants including alder, Traveller's-joy, rose and willow. Females emerge from the innermost cells, with males emerging from cells closer to the nest entrance.

Small forms of *Coelioxys mandibularis* and *C. inermis* lay their eggs in the brood cells of this species. It is thought, however, that the precise relationships vary and that these two species parasitise *M. leachella* brood cells in different parts of the country.

On the wing
Sussex records run from late May until the end of August.

Feeding and foraging
M. leachella forages for pollen from the Dead-nettle, Pea and Stonecrop families. Other plants visited include Bog Pimpernel, bramble, Creeping Cinquefoil, Viper's-bugloss and White Bryony.

Megachile ligniseca Wood-carving Leafcutter Bee
FIRST RECORD 1899, Felpham (Henry Guermonprez) **TOTAL RECORDS** 155

Geography and history
Megachile ligniseca is widely distributed across Sussex. It has frequently been recorded from gardens, but also from brownfield sites, chalk grasslands, meadows and pasture, woodlands, plus a variety of locations along the coast.

Out in the field and under the microscope
Females of this *Megachile* are most often found as they gather pollen from plants such as Spear Thistle or Teasel in areas of scrubby grassland. As this species is also often seen in gardens, it might be found provisioning a nest in a bee hotel.

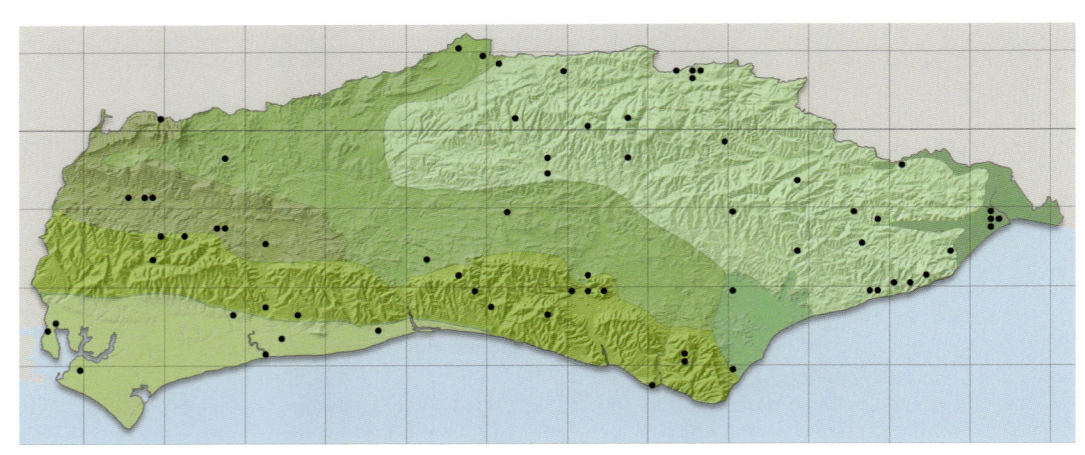

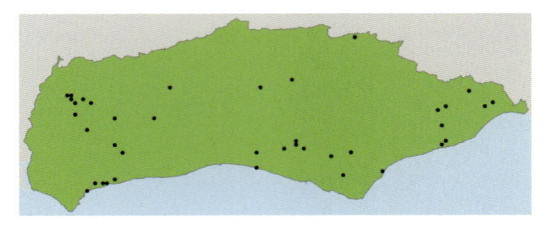

Megachile ligniseca ♀ [PC]

Megachile ligniseca ♂ [NO]

Large map: distribution 2000–2023
Small map: distribution 1844–1999

This is one of three species of *Megachile* in which the females are large and robust. With care, it can be separated from the other two, *M. maritima* and *M. willughbiella*, by its comparatively large head and the pale, light-brown brush of hairs on the underside of the abdomen—both these other species have a brighter, reddish-orange pollen brush here.

Males are also large, though much smaller than the females and, unlike *M. maritima* and *M. willughbiella*, do not have expanded segments on the front legs. They can be identified by looking for a very distinctive V-shaped notch in the middle of the hind edge of tergite six. As in other males in the genus, this tergite continues onto the underside of the abdomen.

Behaviour and interactions

M. ligniseca establishes its nests in dead trees, tree stumps and fence posts, as well as in bee hotels and robust plant stems. Brood cells are lined with pieces of leaf. Pollen is gathered by pressing the brush of hairs on the underside of the abdomen against the anthers.

Observations in Norfolk have shown that *Coelioxys inermis* is a parasite of this species, with males having emerged from two of the ten brood cells within a trap-nest. *C. elongata* is also thought to lay its eggs in the brood cells of this species, although this is yet to be confirmed.

On the wing

Sussex records run from late May until mid September.

Feeding and foraging

Pollen is collected from a wide range of plants from a limited range of families, and especially from members of the Daisy family such as Spear Thistle, knapweed and burdock. Members of the Deadnettle, Pea, Plantain and Teasel families are also visited for pollen, and Ian Beavis reports seeing this species collecting pollen from bramble.

Teasel [WP]

Megachile maritima Coast Leafcutter Bee
FIRST RECORD 1876, Littlehampton (Frederick Smith) TOTAL RECORDS 121

Geography and history
Megachile maritima is the least widely distributed of the three large and robust *Megachile* found in Sussex, and it has an unusual distribution in the county. It is found in many locations along the Sussex coast but otherwise is largely restricted to heathlands on the Lower Greensand. Here, it has been recorded sporadically over the years.

There are four records from other inland locations, namely Rewell Wood in 1979 and 1982, Lewes in 2016, and a site in the Low Weald near Lewes in 2021. One of the earliest records is of males and

Megachile maritima nest site, Lewes [WP]

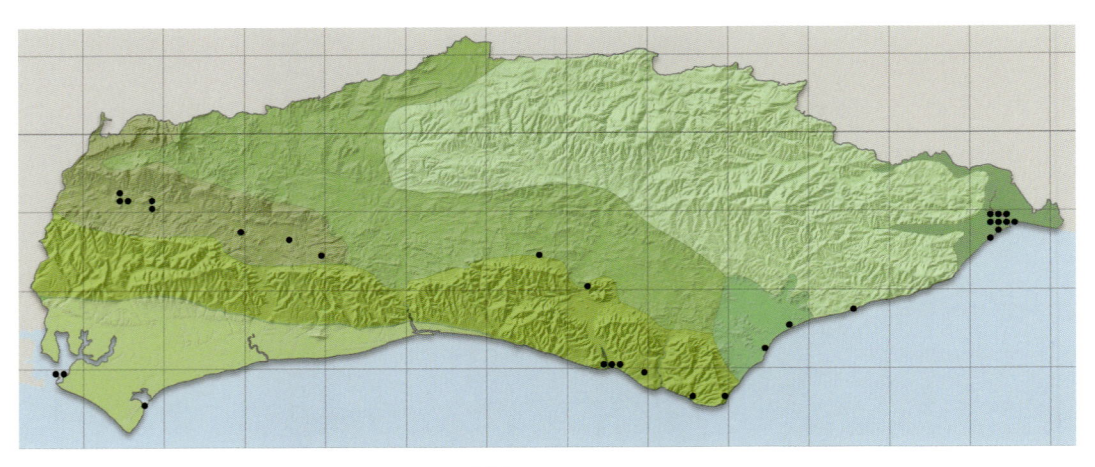

Large map: distribution 2000–2023
Small map: distribution 1844–1999

females which were collected from a colony close to the grandstand at Brighton Racecourse in June 1879 by C. Brazenor.

Out in the field and under the microscope

M. maritima is best searched for by checking flowering plants along the coast or inland on the heathlands of the Lower Greensand. Sandpits are also worth investigating.

M. maritima females are difficult to separate from *M. willughbiella* females, but *M. maritima* has very narrow, obvious whitish hair bands along the edges of tergites two to five and a much paler brush of reddish hairs on the underside of the abdomen, particularly towards the front. Fresh specimens also have a longer fringe of hairs on the back of the front basitarsi, although these can rub off with age.

Like *M. willughbiella* males, *M. maritima* males have enlarged segments on the front legs. *M. maritima* can, however, be readily separated from *M. willughbiella* by the number of teeth on the mandibles—*M. maritima* has three, while *M. willughbiella* has four.

Megachile maritima ♀ [PB]

Megachile maritima ♂ [NO]

Behaviour and interactions

M. maritima females usually nest singly but do sometimes establish small aggregations. The Lewes record from 2016, for example, was of a small number of females nesting close together on a steep, rocky and sparsely vegetated south-facing chalk bank.

Nests are excavated in the ground, with one study on the continent indicating that typically between one and two brood cells are established per nest, although as many as five is possible.

Each brood cell is lined with pieces of leaf collected from various trees, shrubs and flowering plants. These include Hound's-tongue and willow, with Field Maple, Sycamore, birch and Dog-rose also reported from the continent. Tunnel entrances are filled with sand.

Coelioxys conoidea is known to lay its eggs within brood cells of this species, while a large form of *C. mandibularis* is suspected of this behaviour in south Wales.

On the wing

Sussex records run from early June until late August.

Feeding and foraging

Pollen is collected from plants in the Daisy, Borage, Pea, Willowherb and Plantain families. This species has also been reported visiting bramble, Nettle-leaved Bellflower, Sea-holly and Hogweed.

Megachile versicolor Brown-footed Leafcutter Bee
FIRST RECORD 1920, Eridge (Charles Nurse) TOTAL RECORDS 221

Geography and history
Megachile versicolor is widely distributed across Sussex. This bee uses a wide range of habitats and has recently been recorded in places as diverse as Broadwater Warren, Mount Caburn, Selwyns Wood, Stedham Common and Thorney Island. It is also often recorded from gardens.

Mount Caburn [ME]

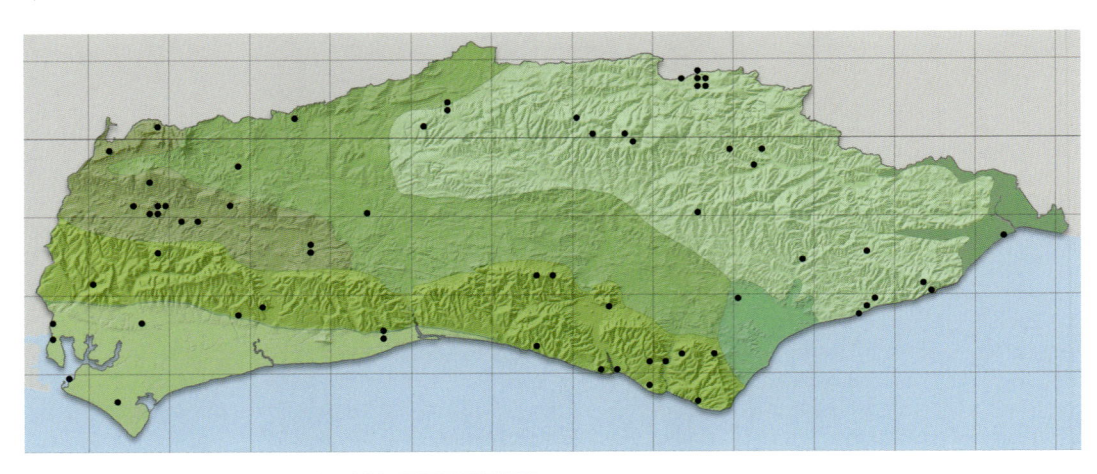

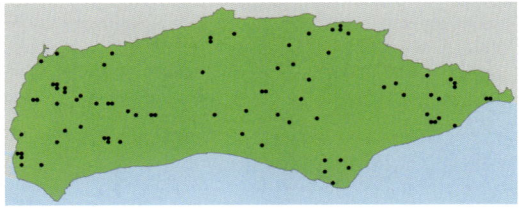

Large map: distribution 2000–2023
Small map: distribution 1844–1999

Megachile versicolor ♀ [PB]

Megachile versicolor ♂ [SF]

Out in the field and under the microscope
This species is tricky to separate from *M. centuncularis*. The females are dark when viewed from above, with no hair bands on the abdomen. A key character to check are the hairs on the surface of the final tergite—these all lie flat in this species, but some are always upright in *M. centuncularis*. On the underside of the abdomen, *M. versicolor* females have a bright, golden-orange brush of hairs with black hairs on the final two sternites. In *M. centuncularis*, the orange hairs extend all the way to the tip of the abdomen.

In males, the key character to look for is on the fourth sternite. Here, there should be a shiny, virtually hairless hind margin to the sternite. In the otherwise very similar *M. centuncularis*, this area is dull and covered in tiny microscopic hairs which are particularly visible along the hind margin.

Males and females are often seen on plants such as dandelion, thistle, Common Bird's-foot-trefoil, and females are frequently seen establishing nests in bee hotels.

Behaviour and interactions
Although this *Megachile* usually nests in crevices in deadwood and in large hollow stems of robust plants such as bramble, rose and mullein, it does sometimes nest in the ground in sandy soil. Cells are often lined with leaf sections from roses, while on the continent it is known that it uses Blackthorn.

Coelioxys inermis has been confirmed as a parasite of this species in Britain, with *C. mandibularis* reported from the continent.

M. versicolor is thought to generally have a single flight period. However, it is possible that it has a second generation during late summer, as is the case in Germany.

On the wing
Sussex records run from the beginning of May until mid September.

Feeding and foraging
Pollen is collected from the Daisy, Borage, Teasel, Pea and Plantain families. The Daisy family may be particularly important, with plants such as fleabane, thistle, dandelion and ragwort all reported from the High Weald.

Megachile willughbiella Willughby's Leafcutter Bee
FIRST RECORD 1896, Bognor Regis (Henry Guermonprez) TOTAL RECORDS 255

Geography and history
Megachile willughbiella is widely distributed and is found across all parts of the county. As it uses a wide range of habitats, it can be found just about anywhere in Sussex that has plants in flower and suitable nest sites. It is very often seen in towns and villages, and is a frequent visitor to gardens.

Out in the field and under the microscope
This large and robust bee is most often seen either foraging, provisioning or lining a nest in a bee hotel. Sometimes females can be seen pushing their way inside the flowers of a plant like Sweet-pea to reach the pollen inside, using their hind legs to gain extra traction. Males are often seen persistently patrolling patches of flowers frequented by the females.

Megachile willughbiella ♀ lining a nest in a bee hotel [PC]

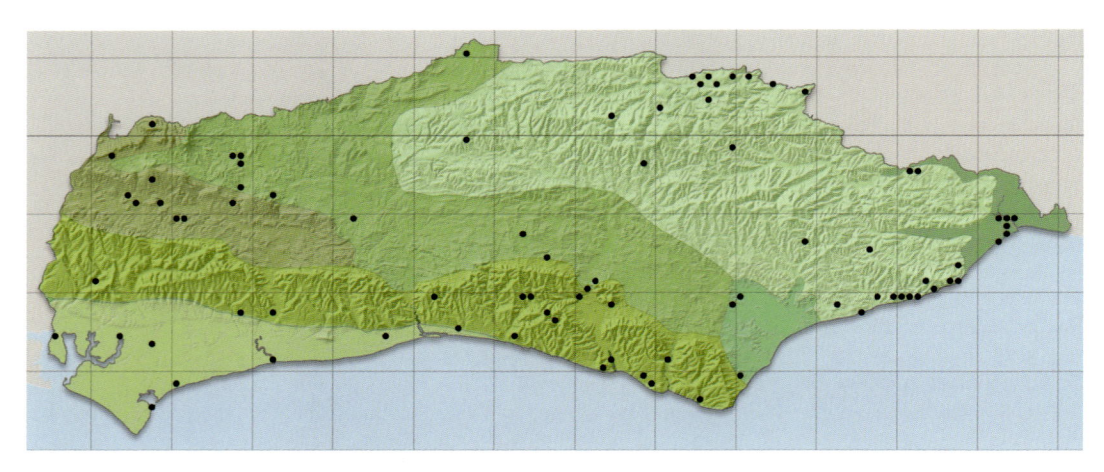

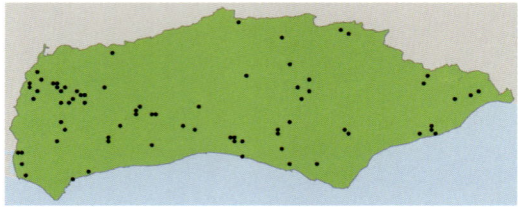

Large map: distribution 2000–2023
Small map: distribution 1844–1999

180 *The Bees of Sussex*

Megachile willughbiella ♀ [SF]

Megachile willughbiella ♂ [PC]

M. willughbiella males have very distinctive front tarsi. These are flattened and expanded, are covered in short white hairs, and have a fringe of long white hairs. Expanded front tarsi are only otherwise found in *M. maritima* and the smaller *M. circumcincta*. Of these two, however, only the front tarsi on *M. maritima* males closely resemble those on *M. willughbiella*. These last two species can be readily distinguished from each other by the number of teeth they have on the mandibles—four for *M. willughbiella*, and three for *M. maritima*.

Females have an orange brush of hairs beneath the abdomen that becomes black at the tip, and they closely resemble *M. maritima* females. The length of the hairs on the front basitarsi is one key distinguishing feature—these are much shorter in *M. willughbiella*—while the colour of the brush of hairs on the underside of the abdomen is also a good guide. In *M. willughbiella*, this is a more consistent reddish colour throughout and is not as pale towards the front of the abdomen, as in *M. maritima*.

Behaviour and interactions
This is a very adaptable species which takes advantage of a wide range of different nest sites. It usually nests in deadwood but will also use holes in walls, and on occasion it is known to nest in compost in flower pots. One example from Surrey, reported by David Baldock, is of three females establishing their nests in grow bags within a greenhouse. In this case, five nests were completed, with between three and six brood cells per nest. It took the females some 20 minutes to collect a pollen load, each one completing eleven loads in one day.

Males have several unusual adaptations that are used during mating. The most striking is the structure of the front tarsi. As well as being flattened and having a fringe with long hairs, these also contain a linear groove. This has special pores within it that release pheromones onto the female's antennae as they are held in the groove during courtship and mating. At the same time, the flattened tarsi and the long hair fringes are placed over the female's eyes.

Coelioxys elongata and *C. quadridentata* target the brood cells of this species.

On the wing
Sussex records run from mid May until the end of September.

Feeding and foraging
This bee collects pollen from the Bellflower, Daisy, Pea, Stonecrop and Willow families.

Osmia – Mason Bees

This is a large genus of robustly built bees. Many of the species found in Britain construct their nests within existing cavities, including within abandoned insect burrows in wood, in crevices or in abandoned snail shells. One species, *Osmia inermis*, constructs its cells on the underside of rocks, while a second, *O. pilicornis*, excavates its own nest chamber. The brood cells are constructed from mud or chewed leaf fragments which are sometimes mixed with mud or resin. Pollen is transported within a brush of hairs on the underside of the abdomen. Eight species are known from Sussex, with one of these, *O. xanthomelana*, extinct in the county. A further five species have been confirmed from Britain, and one of these, *O. cornuta*, was found for the first time in 2012, in Petersham, Surrey.

Osmia aurulenta Gold-fringed Mason Bee
FIRST RECORD 1877, Brighton (A. Brazenor) TOTAL RECORDS 147

Geography and history
Osmia aurulenta has two main centres of population in Sussex. One is on the species-rich chalk grasslands of the South Downs running east from the River Adur as far as Beachy Head, while the second is concentrated between Lydd Ranges and Rye Harbour Nature Reserve. There have also been two unusual records from species-rich meadows and pasture in the High Weald, most recently in 2021 when it was swept on a farm near Warbleton by Chris Glanfield.

Out in the field and under the microscope
The best way of finding this bee is to search for foraging females as they gather pollen. These are frequently seen working their way from flower to flower, using special hairs on the face to remove the pollen. Alternatively, they might be seen harvesting pieces of leaf from patches of Common Rock-rose or Wild Strawberry. As well as feeding, males might be seen holding a territory on a snail shell.

Females are very distinctive; no other species of *Osmia* is quite like it. Fresh specimens have a dense covering of long orange-red hairs on the thorax and a thinner covering of shorter orange-red

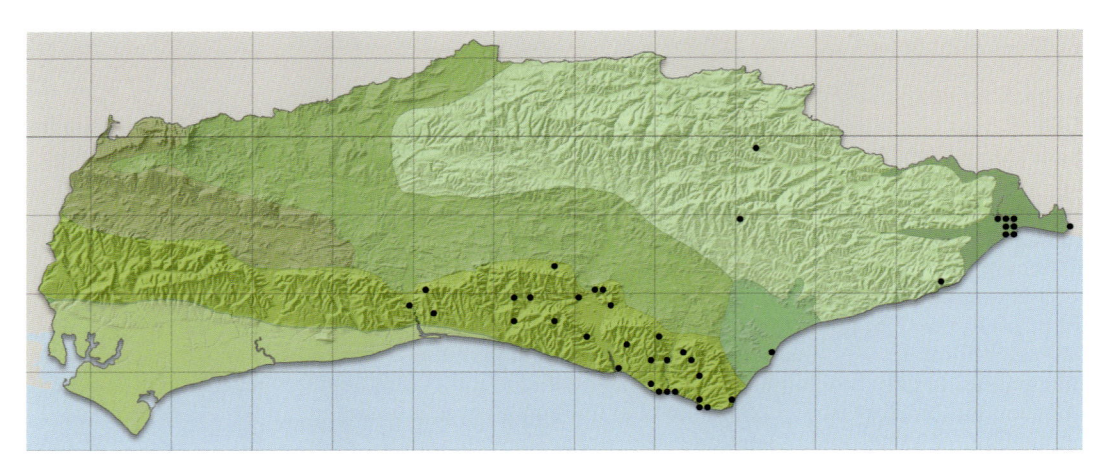

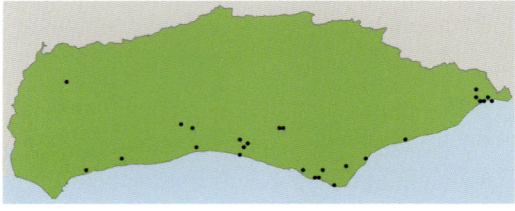

Large map: distribution 2000–2023
Small map: distribution 1844–1999

Osmia aurulenta ♀ [PC]

Osmia aurulenta ♂ [NO]

hairs on the abdomen. There is also a dense band of very short pale hairs along the hind margins of each of the tergites.

Males are less straightforward, especially once the orangey-red colour has faded from the hairs on the abdomen. At this point, they resemble *O. bicolor* males. However, the hind margin of tergite six has a central notch in this species whereas the margin is evenly rounded in *O. bicolor*. *O. aurulenta* also has a pair of side notches, one each side of the central notch, although these can be obscured by hairs and difficult to see.

Behaviour and interactions

After mating, females establish their nests within abandoned snail shells, with up to ten brood cells within each shell. In 1898, Alfred Brazenor found three cells within a Garden Snail shell collected at Bevendean. The brood cells are separated from each other with partitions that are made of chewed leaf matter. This is also used to seal the entrance to the shell and is daubed on the outside. Here, it soon turns brown, helping to camouflage the nest.

Osmia aurulenta ♀ **investigating an abandoned snail shell** [PC]

Osmia aurulenta **brood cells and larvae inside a snail shell** [PW]

On the wing
Sussex records run from early April until late August.

Feeding and foraging
Although a wide range of plants is visited at different times, pollen is collected from just Common Bird's-foot-trefoil, Horseshoe Vetch, Kidney Vetch and restharrow.

The Bees of Sussex

Osmia bicolor Red-tailed Mason Bee
FIRST RECORD 1877, Brighton (A. Brazenor)   TOTAL RECORDS 231

Geography and history
Although *Osmia bicolor* has been recorded sporadically from the Lower Greensand, it is rare to find this bee far from the South Downs where the abundant snail shells provide ample opportunities for the females to nest. It is widely distributed throughout this landscape and could be found in just about any area of species-rich chalk grassland. It has also been recorded from open woodlands within the South Downs.

Away from the Lower Greensand and the South Downs, there are a handful of additional records, including two from the High Weald. These include one record from the Haywards Heath area in 1974 and one from a farm near Warbleton. Here, a female was swept by Chris Glanfield in an area of damp poorly drained, species-rich hay meadows and pasture in 2021. This unlikely location encompasses a small area of alkaline soils that seemingly supports a sufficient population of suitable snails to sustain *O. bicolor*.

Out in the field and under the microscope
A female *O. bicolor* is a very striking bee: the head and thorax are black-haired, contrasting strongly with the red-haired abdomen and red pollen brush. Males are smaller than the females and are black with reddish-ginger hairs on the tip of the abdomen, although the colour here soon fades. They most closely resemble male *O. aurulenta* but have a smooth, un-notched hind margin on tergite six.

O. bicolor is best looked for in the spring and early summer on areas of chalk grassland. Here, the red and black females might be seen flying rapidly close to the ground or foraging. A female might also be seen carrying a grass stem. Males might be seen feeding, at rest, or holding a territory on a snail shell.

Behaviour and interactions
This is the earliest of the *Osmia* to emerge each year, with the males appearing on sunny days from March onwards. They generally appear some two weeks ahead of the females. Mated females lay between one and five eggs within a shell, separating each brood cell with a partition made from

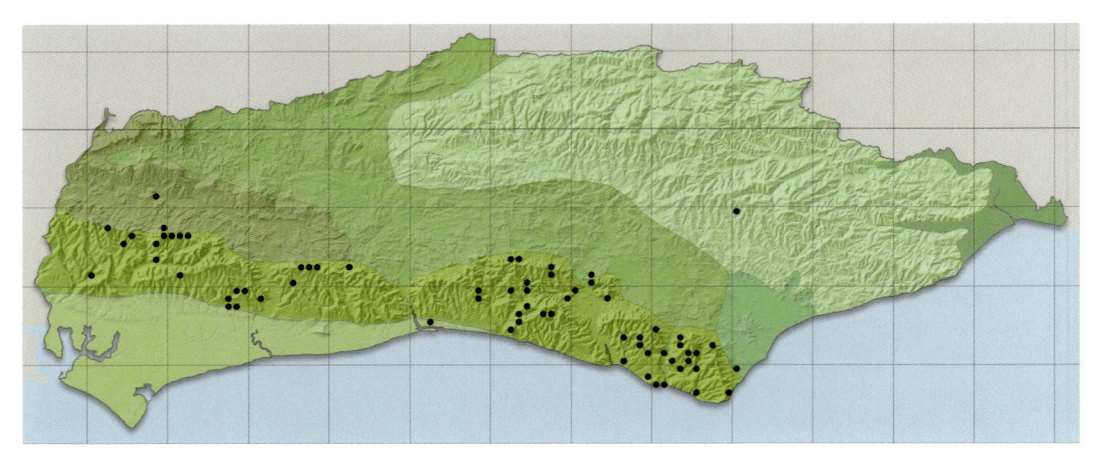

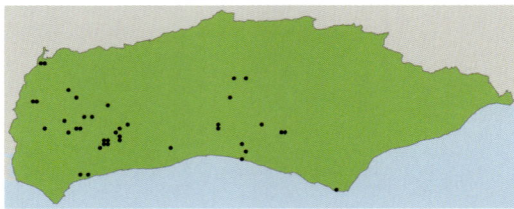

Large map: distribution 2000–2023
Small map: distribution 1844–1999

Osmia bicolor ♀ [PC]

Osmia bicolor ♂ [PC]

chewed leaf fragments, which are transported to the nest as a ball of pulp in the mandibles. The entrance to the shell is blocked with small stones and shells, before being closed with more chewed leaf fragments.

Females are also known to excavate a shallow depression beneath the shell, which is then rotated so that the entrance faces downwards and sits within the hollow. The outside of the shell is daubed with chewed leaf fragments and covered with hundreds of grass stems of between 2 and 15 cm in length, as well as with other pieces of vegetation. This process can take two to three days to complete. Nests are generally thinly spread out, but small aggregations are known. In the combe on Malling Down, for example, several females were seen in 2012, nesting near each other in a sheltered area of short-cropped grassland.

Osmia bicolor ♀ emerging from a snail shell [all PC]

Osmia bicolor ♀ carrying a grass stem [PW]

On the wing
Sussex records run from early March until mid July.

Feeding and foraging
Common Bird's-foot-trefoil, Horseshoe Vetch and Kidney Vetch are important pollen sources, with Cowslip, Creeping Willow and Wood Anemone also visited. Violets are an important source of nectar.

The Bees of Sussex

Osmia bicornis Red Mason Bee
FIRST RECORD 1878, Brighton (A. Brazenor) TOTAL RECORDS 715

Geography and history
Osmia bicornis is one of the most frequently recorded solitary bees in the county—it occurs in all parts of Sussex and is very abundant in gardens, especially around bee hotels.

Out in the field and under the microscope
Females have a relatively large head with a pair of prominent horns on the face (hence the scientific name *bicornis*), a character shared only with *O. cornuta*. This species has been present in Britain since at least 2012, and while it is not currently known from Sussex, it is expanding its range. Like *O. bicornis*, it is commonly found in gardens.

O. bicornis females have black hairs on the head, a mix of black and golden hairs on the top of the thorax, and reddish-orange hairs on the top of the abdomen. The pollen brush on the underside of

Both *Osmia bicornis* ♀ left [PC] and *O. cornuta* ♀ below [PW] have a pair of horns on the face

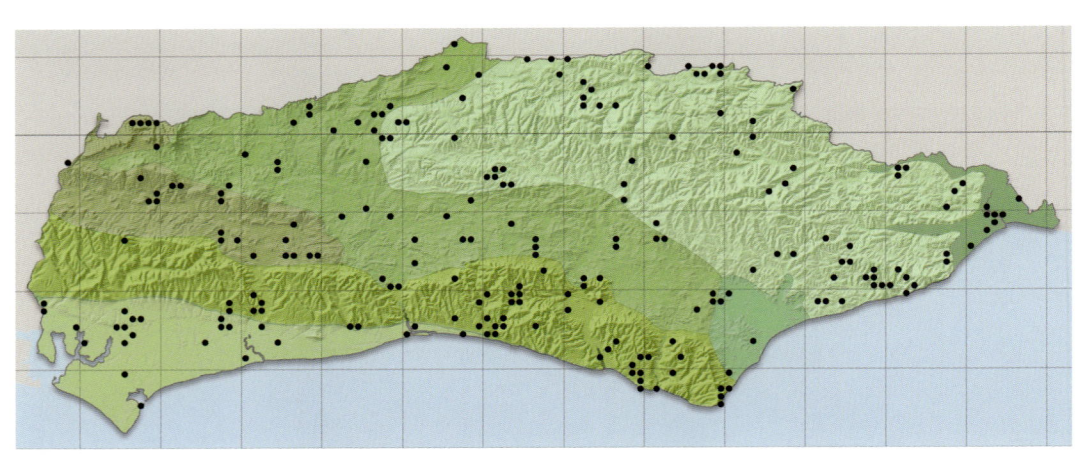

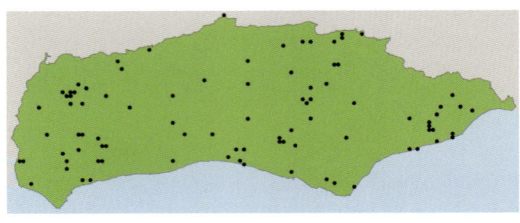

Large map: distribution 2000–2023
Small map: distribution 1844–1999

the abdomen is also reddish orange. *O. cornuta* is generally a much bigger insect and in colouring is closer to *O. bicolor*, with more obviously red and black hairs.

Male *O. bicornis* are smaller than the females and have particularly long antennae. They also have white hairs on the face and pale spurs on the hind tibiae. They have a thin covering of reddish-orange hairs, although the colour soon fades.

Females are generally seen close to their nest sites but can also be seen foraging or gathering mud from the edges of streams, ponds or puddles. Males will also be busy close to nest sites and are frequently seen coursing around likely nest locations.

Osmia bicornis ♂ [PC]

Behaviour and interactions

O. bicornis often forms very large nesting aggregations in pre-existing cavities but will sometimes excavate its own tunnel if the soil is loose and friable. It establishes its nests within holes in masonry, holes in timber, and within reed and other hollow plant stems. However, it is an opportunist and will take advantage of any number of locations, including hollow bamboo canes or pre-drilled holes in bee hotels.

Females transport the wet mud used to form their brood cells in their mandibles and then shape and position it with their horns. Each brood cell is separated from the others by a partition made from damp soil. The cells are not normally lined unless the female is using a wide chamber, in which case the cells can lie side by side and are separated from each other by mud walls. Males outnumber females about two to one. After mating, they insert a plug to prevent a female from mating again.

The fly *Cacoxenus indagator* is a common sight around bee hotels that are being used by *O. bicornis* for egg laying. The female fly waits close to the nest entrance while a female bee is provisioning a brood cell and once this process is complete—but before the cell wall is built—the fly enters and lays several eggs on the pollen and nectar provisions. When the fly larvae hatch, they feed on the provisions, reducing the likelihood of the bee larvae reaching adulthood.

Cacoxenus indagator ♀ [PC]

To escape from the brood cell, an adult fly relies on holes created in the cell walls by bees emerging from their cells, or on small openings created by the fly itself as it applies pressure on the cell wall with an inflatable sac on its head.

On the wing

Sussex records run from mid March until mid August, with April to June the main flight period.

Feeding and foraging

O. bicornis females collect pollen from a very wide range of plant families (19 are cited in one continental source). However, if a particular species is producing abundant pollen, females tend to be very faithful to this one species.

Plant species visited for pollen in Britain include oak and buttercup, plus Hawthorn, Crab Apple, Sycamore, Horse-chestnut, White Clover and strawberry. The Mint and Borage families are well used in gardens, as are fruit trees and shrubs such as apple, pear, plum, cherry and Raspberry.

Osmia caerulescens Blue Mason Bee
FIRST RECORD 1877, Worthing (Edward Saunders)　TOTAL RECORDS 266

Geography and history
Osmia caerulescens is widely distributed across Sussex and is found in a range of different habitats. It is frequently found on the heaths of the Lower Greensand, in areas of flower-rich grassland in the High Weald, on chalk grasslands on the South Downs, and in woodlands. It is also often found in gardens.

Out in the field and under the microscope
This *Osmia* is most often seen foraging, although females may also be seen searching for a suitable nest site or provisioning a brood cell. Males might be seen flying persistently in small groups around likely nest sites. Nests are often established in bee hotels, and these can be good places to search for this bee.

Osmia caerulescens ♀ nectaring on Ivy-leaved Toadflax [PC]

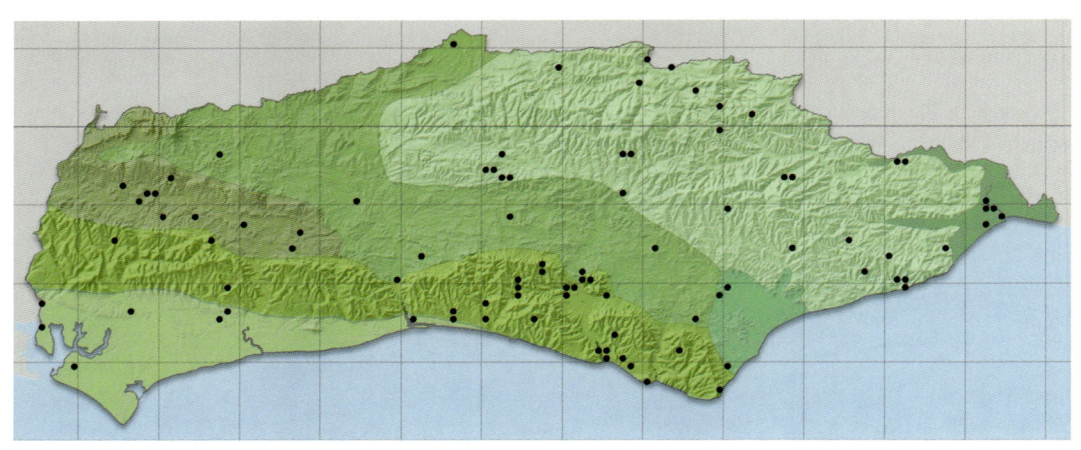

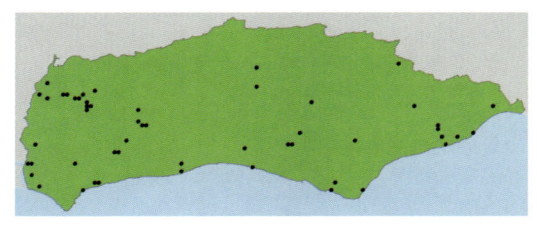

Large map: distribution 2000–2023
Small map: distribution 1844–1999

188　*The Bees of Sussex*

The females are very distinctive and have a large blocky head, a tinge of blue to the otherwise dark abdomen, and a brush of black hairs on the underside of the abdomen.

The males are a metallic bronze colour and most resemble *O. leaiana* and *O. niveata*, the last of which is not known from Sussex. One of the best characters to separate *O. caerulescens* from *O. leaiana* is the smooth and shiny front face to the first tergite. In *O. leaiana*, as in *O. niveata*, this area is slightly roughened. *O. caerulescens* also has a roughly serrated hind margin to tergite six either side of a central notch, whereas *O. leaiana* has a completely even hind margin either side of a notch. In the field, eye colour can also help to separate males of *O. caerulescens* and *O. leaiana*—the eyes are a bluey green in *O. caerulescens* and dark in *O. leaiana*.

Osmia caerulescens ♂ [PC]

Behaviour and interactions

Nests are established in pre-existing cavities in dead wood (such as those created by beetles), hollow plant stems and masonry. *O. caerulescens* generally creates linear nests, such that the brood cells are in a line, separated from each other by pieces of chewed leaf. The tunnel entrance is also sealed with pieces of chewed leaf.

Osmia caerulescens larvae within their individual cells feeding on their pollen and nectar provisions [PW]

One study has shown that it can take a female about 20 flights to provision a brood cell, and that she completes between one and two brood cells per day. Pollen is collected with the aid of specially adapted hairs and bristles on the face.

O. caerulescens is thought to generally be single brooded but may produce a second generation in southern counties such as Sussex. Most records have been made during April, May, June and July, but over the years there has also been a handful from August. There is also one exceptionally late record from Worthing, made on 20th October 1877. This record was made by Edward Saunders, who also lists a remarkable number of other species from this late date and comments, "It certainly was an exceptional day."

Although no cuckoo bees have been identified in Britain, on the continent *Stelis ornatula* is cited.

On the wing

Sussex records run from late March until mid August.

Feeding and foraging

Pollen is sometimes collected from plants in the Cabbage and Daisy families but especially from plants in the Dead-nettle and Pea families. Species visited for nectar include Ivy-leaved Toadflax, knapweed and speedwell.

Osmia leaiana Orange-vented Mason Bee
FIRST RECORD 1905, Ore (Edward Saunders) and Bognor Regis (Henry Guermonprez) TOTAL RECORDS 182

Geography and history
Osmia leaiana is widely distributed but rarely abundant. It might be found in parks and gardens (where it will use bee hotels), in orchards, on grasslands with scrub, and along woodland rides. A key requirement is the presence of plants in the Daisy family.

Out in the field and under the microscope
A good way of searching for this bee is to check the flower heads of plants such as knapweed, dandelion and thistle. It is also frequently found close to nest sites, or at rest either on foliage or fence posts.

The females have a blocky head, a shiny, sparsely haired black abdomen and a reddish-orange brush of hairs that runs to the tip of the abdomen. They are very close in appearance to female

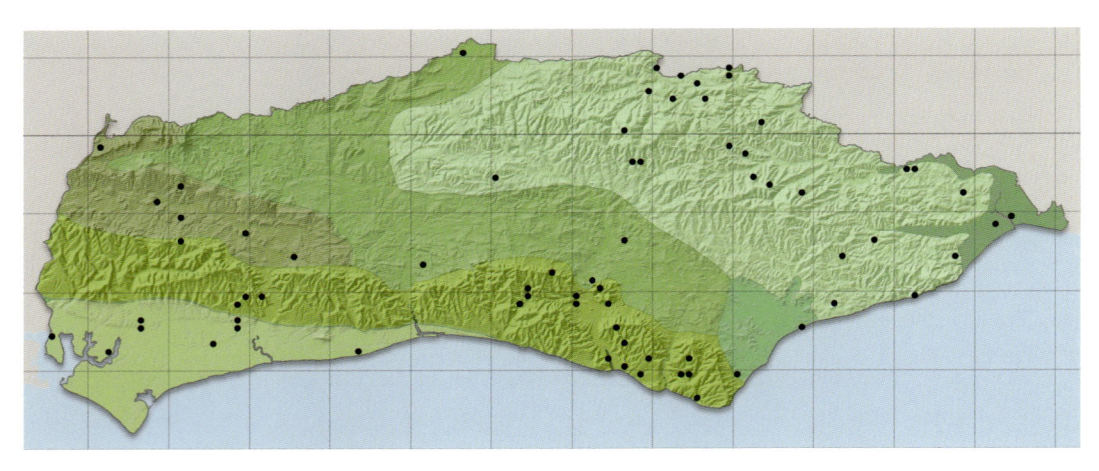

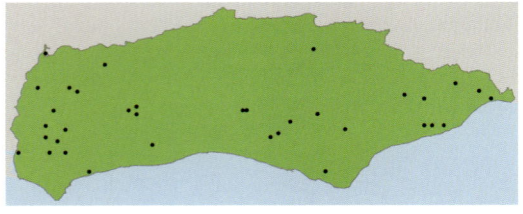

Osmia leaiana ♀ [CG]

Osmia leaiana ♂ [PC]

Large map: distribution 2000–2023
Small map: distribution 1844–1999

190 *The Bees of Sussex*

O. niveata, a species that is not known from Sussex. A key distinguishing feature is the presence of two small rounded teeth on the bottom edge of the clypeus (these are absent in *O. niveata*).

The males are about two thirds the size of the females. They most closely resemble *O. caerulescens* and *O. niveata*. Like *O. caerulescens* males, *O. leaiana* males have bluish reflections on the thorax. To separate this species from *O. caerulescens*, two characters to check are the front face of the abdomen and the hind margin of tergite six. The front face of the abdomen is slightly roughened in *O. leaiana* and smooth and shiny in *O. caerulescens*, while *O. leaiana* males have a smooth and even hind margin to tergite six either side of a central notch. This contrasts with the roughly serrated hind margin either side of a central notch in *O. caerulescens*. Live males of the two species also differ in their eye colour. This is dark in *O. leaiana* and bluey green in *O. caerulescens*.

Behaviour and interactions

O. leaiana establishes its nests within pre-existing cavities in dead wood, within holes in masonry, and in hollow canes and plant stems. Each brood cell is separated from the other cells by chewed pieces of leaf, and the tunnel entrance is also sealed with pieces of chewed leaf.

A study has shown that a single brood cell can be built and provisioned in a day. The females collect nectar and pollen on each trip, combining them so that the pollen 'loaf' is thoroughly mixed. In this study, more than 99 per cent of the pollen collected had come from plants in the Daisy family.

It is possible that some individuals in a population have a two-year lifecycle, remaining in the nest as adults for a second winter.

Stelis phaeoptera lays its eggs within the brood cells of this species, and a second species, *S. punctulatissima*, is also suspected of this behaviour. There are also two species of wasp that target *O. leaiana* (as well as other species of bee) and that are regularly seen around *O. leaiana* nests—*Sapyga quinquepunctata* and *Chrysura radians*. *S. quinquepunctata* lays an egg within a brood cell and the large-jawed larva feeds initially on the bee's egg before then feeding on the pollen deposited in the cell. In contrast, a *C. radians* larva waits for the *O. leaiana* larva to hatch before attaching itself to feed on the bee larva itself.

Sapyga quinquepunctata ♀ [PC]

Chrysura radians ♀ [SF]

On the wing
Sussex records run from mid April until late August.

Feeding and foraging
Pollen is almost entirely collected from plants in the Daisy family such as hawk's-beard, knapweed and thistle, although small quantities of pollen from other plant families may also be gathered.

Osmia pilicornis Fringe-horned Mason Bee

FIRST RECORD 1892, Guestling (Francis Morice) TOTAL RECORDS 60

Geography and history

The very first Sussex record for *Osmia pilicornis* is from Guestling in 1892 and implies that it was a well-known species at the time. This is when Francis Morice expressed frustration that "at Guestling I took what I fondly hoped was a ♀ of *Osmia xanthomelana*, but it proved to be only a *pilicornis*". Since then, it has become increasingly scarce, its decline mirrored by a reduction in the area of ancient woodland that is coppiced annually, and by a decline in woodland management more generally. Even on sites where it was once frequently found, it is now very elusive.

This species is found in open-structured woodlands. Historically, it has had two principal centres of population in the county, one focused on the woodlands and copses of the western South Downs, and the other based on the eastern High Weald. The only places to provide Sussex records this

Osmia pilicornis ♀ **feeding on Bugle** [NF left] **and collecting Tormentil fragments** [RBI below]

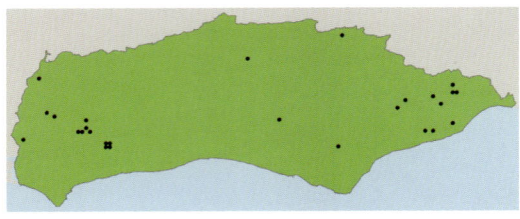

Large map: distribution 2000–2023
Small map: distribution 1844–1999

192 *The Bees of Sussex*

century are woodlands on the downs above Cocking in 2000, Rewell Wood in 2013, and Fore Wood in 2013, 2015, 2017 and 2018. There have also been records in this period from Tudeley Woods near Tunbridge Wells, just over the county boundary in Kent.

As this bee lives at a low population density, it could well still be present within these landscapes.

Out in the field and under the microscope

To find this bee it is necessary to search warm, sunny clearings and rides within areas of broadleaved woodland which are undergoing regular, annual management. In these situations, *O. pilicornis* has often been seen at rest on leaf litter or tree stumps that are exposed to sunlight. It is well suited to coppiced woodland and uses the same habitat as the Pearl-bordered Fritillary butterfly, so any recently cleared areas with flowering Bugle, Common Dog-violet and Ground-ivy are worth a close look. Here, females might be seen foraging for pollen and nectar, and males might be seen dashing from flower to flower as they feed or patrolling likely flowers in search of females.

Osmia pilicornis ♂ [RBI]

The hairs on the thorax and at the front of the abdomen are reddish brown in females, while the underside of the abdomen has a brush of black hairs. Males resemble worn *O. bicolor* males but have very long antennae with a sparse covering of long hairs on the rear of the segments.

Behaviour and interactions

O. pilicornis males follow a regular beat when patrolling flowers in search of a mate. After mating, females excavate their nests in rotting wood within a metre of the ground, using old coppice stools, tree stumps or fallen branches. They will even nest within conifer wood such as larch.

In Fore Wood, Ian Beavis reports that he and Rosie Bleet were able to observe "a female gathering leaf pulp (biting off and chewing up small strips of leaf) for nesting from Tormentil foliage, making repeated visits from the plants to her nest in a small, decayed log a few yards away. The nest and the leaf pulp source were all in the same coppiced clearing."

Osmia pilicornis nest entrance at Fore Wood [RBI]

When foraging, *O. pilicornis* females use the hooked hairs on the tongue to gather pollen from very narrow flower tubes, before mixing this with nectar to provision their brood cells. Twenty-five provisioning flights per cell have been reported.

On the wing

The few modern Sussex records run from late March until early May, with older records running until the end of June or, occasionally, mid July.

Feeding and foraging

While pollen is collected from plants in the Borage, Dead-nettle, Pea and Rose families, this bee is particularly closely associated with Bugle and Ground-ivy.

Visits to Common Dog-violet, comfrey, Hairy Violet, Selfheal and willow have been reported.

Osmia spinulosa Spined Mason Bee
FIRST RECORD 1884, Hastings (Edward Saunders) TOTAL RECORDS 206

Geography and history
Osmia spinulosa is closely associated with places that support abundant snail populations and is therefore most often found on dry calcareous grasslands and other calcium-rich habitats. In Sussex, the main populations are centred on the chalk grasslands of the South Downs and the sandy soils found on the coast near Rye Harbour.

It has also been found in other locations with sandy soils, including the cliffs at Glyne Gap, sand dunes at Littlehampton, and vegetated shingle at Norman's Bay.

Sandy habitats along the coast such as the cliffs at Glyne Gap are worth searching for *Osmia spinulosa*. [AP]

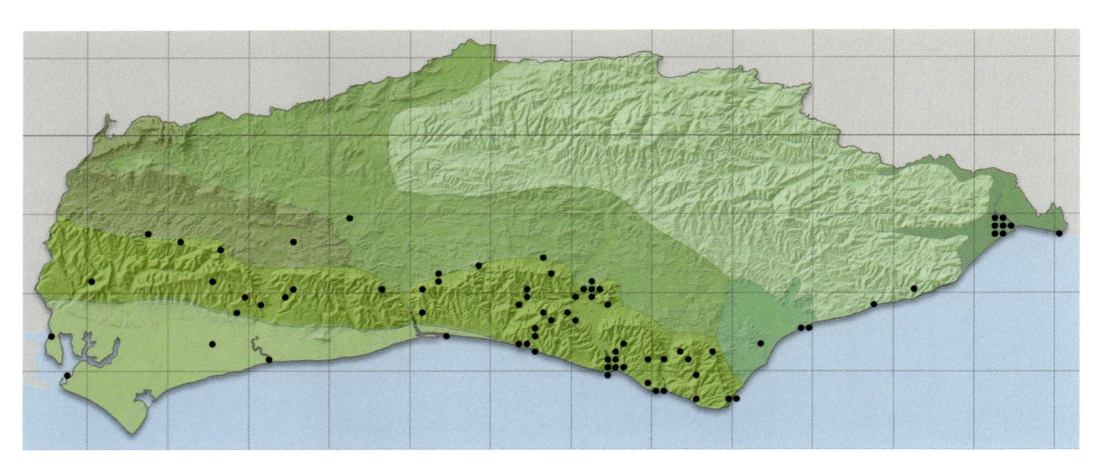

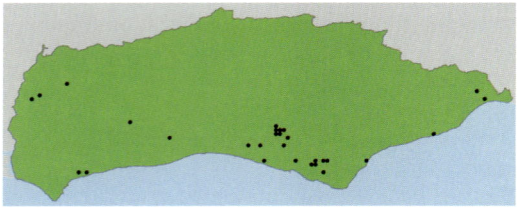

Large map: distribution 2000–2023
Small map: distribution 1844–1999

Osmia spinulosa ♀ [PC]

Osmia spinulosa ♂ [PC]

Out in the field and under the microscope
This is the smallest *Osmia*, and its status has been the source of confusion over the years—until recently it was thought that it belonged to a different genus and was described as *Hoplitis spinulosa*. One character that distinguishes it from all other *Osmia* and the superficially similar *H. claviventris* is the presence in both sexes of a pair of pointed plates to the sides of the scutellum.

The females have an orange brush of hairs on the underside of the abdomen and most resemble small examples of *O. leaiana*.

The males can also be readily distinguished by the slender spine that protrudes from the underside of the abdomen on sternite two. The only other species with a similar structure here is *H. claviventris*, but in that species the spine is much chunkier and blunt.

O. spinulosa is best searched for by checking the flowers of plants in the Daisy family for foraging females—in good habitat they can be abundant. Sites with sandy soils are all worth investigating.

Behaviour and interactions
O. spinulosa nests within empty shells of a wide range of different snail species, including Garden Snail, Heath Snail and White-lipped Snail. Brood cells are created within the spiral, with each cell separated from the others by quite robust partitions. The entrance to the shell is also sealed, a process which can take as long as an hour.

The cell partitions and the entrance seal are made from chewed leaves gathered from plants such as cinquefoil and Salad Burnet. Unlike other species of *Osmia* which nest in snail shells, this species does not daub the surface of the shell with patches of chewed leaf.

Males have been seen with Twayblade pollinia attached to the underside of the abdomen.

Stelis odontopyga lays its eggs within brood cells of this bee. This species is a relatively new arrival in Britain—the earliest record is from 2017—and it was first found in Sussex in 2020.

On the wing
Sussex records run from late May until early September. Late May, June and July are the main periods of activity.

Feeding and foraging
Pollen is collected from a wide range of plants within the Daisy family, including Common Fleabane, Ox-eye Daisy and Rough Hawkbit.

There are also reports of *O. spinulosa* visiting a variety of other plant species including buttercup, Field Scabious, Small Scabious and Wild Marjoram.

Stelis – Dark Bees

This is a genus of very elusive bees that are mostly black (one species, *Stelis ornatula*, has white markings, while some continental species have yellow markings). In Britain, they target *Anthidium*, *Heriades*, *Hoplitis* and *Osmia* bees, laying their eggs in open cells that are being provisioned by the host. On finding a suitable nest, a female *Stelis* may return repeatedly to the same nest, laying an egg in a series of cells as each is provisioned by the host. After emergence from the egg, a *Stelis* larva will locate and kill the host larva, using its dagger-like upper jaws. Five species are known from Britain, all of which have been recorded in Sussex.

Stelis breviuscula Little Dark Bee
FIRST RECORD 1984, Iping Common (George Else) TOTAL RECORDS 50

Geography and history
Stelis breviuscula was first discovered in Britain at Iping Common in 1984. This is when George Else came across "a small bee at rest on a flower of [*Jacobaea vulgaris*] on the perimeter of the Iping Common car park". A comparison with material in the Natural History Museum confirmed that this specimen was *S. breviuscula*. It remained a scarce species for some years and, in Sussex, was initially confined to a handful of sites close to Midhurst. Then in the mid 1990s it was found at several locations at Chichester Harbour as well as in Warnham.

Its host *Heriades truncorum* also used to be a very scarce bee but has rapidly expanded its range in recent years and has become relatively widespread. *S. breviuscula* has followed suit and, despite the paucity of records, is believed to be widely distributed across south-east England, including across Sussex. In 2016 it was recorded at Rye Harbour Nature Reserve, in 2017 at Great Dixter, in 2020 in Lewes, and in 2021 at Broadwater Warren.

Out in the field and under the microscope
A high proportion of records for *S. breviuscula* are of males and females close to *H. truncorum* nests in deadwood, where they are often seen at rest or investigating holes rather than provisioning a nest.

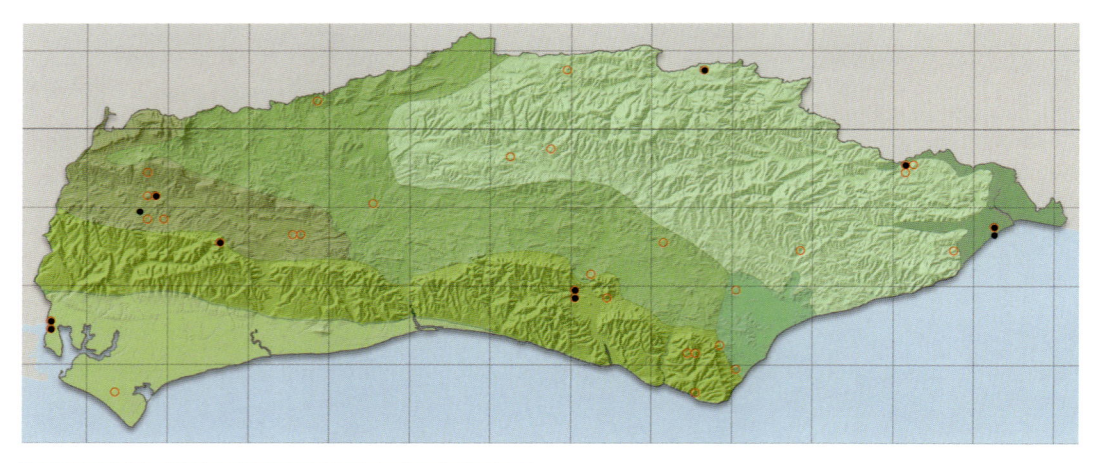

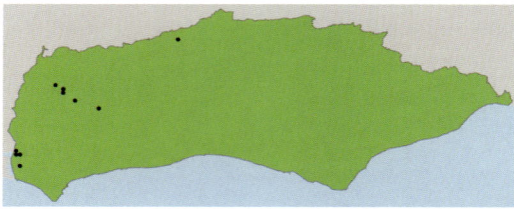

Large map: distribution 2000–2023, with orange circles showing distribution of *Heriades truncorum*
Small map: distribution 1844–1999

Any plant with yellow flowers in the Daisy family is also worth checking.

Males and females are very alike. This is a small black bee that has narrow bands of white hairs on the hind margins of the tergites. The hairs are clearly separated from each other, and these, as well as the bee's small size, help to distinguish this species from the two other all-black species in the genus—the much larger S. odontopyga and S. phaeoptera.

S. breviuscula closely resembles its host, H. truncorum, but, unlike that species, it does not have a ridge across the front of the abdomen.

Behaviour and interactions
S. breviuscula lays its eggs within the brood cells of the bee H. truncorum, which nests in small pre-existing cavities in wood and occasionally in bramble stems.

On the wing
Sussex records run from mid June until early August.

Feeding and foraging
S. breviuscula is most often seen feeding on plants in the Daisy family such as Common Fleabane, Common Ragwort, hawkweed and thistle.

A pre-drilled gate post [WP]

Stelis breviuscula ♀ [SF]

Heriades truncorum ♀, the host [SF]

Stelis odontopyga Smooth-saddled Dark Bee
FIRST RECORD 2020, Brighton (Graeme Lyons) **TOTAL RECORDS** 1

Geography and history
The only time that *Stelis odontopyga* has been recorded in Sussex was in 2020, which is when it was found by Graeme Lyons during a survey in Brighton. This was two years after it had been recognised as a resident British species from specimens caught in Kent and Oxfordshire. It subsequently transpired that a specimen taken in Kent in 2017 was also of this species, and it is possible that it is more widely distributed in Britain than is currently known. On the continent, this is a very scarce bee.

Brighton 'Bee Bank' where *Stelis odontopyga* was first recorded in Sussex in 2020 [GL]

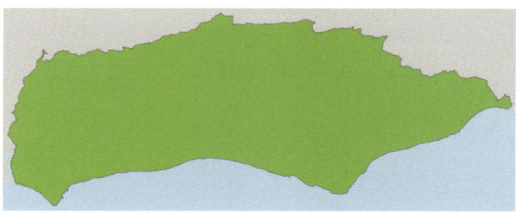

**Large map: distribution 2000–2023, with orange circles showing distribution of *Osmia spinulosa*
Small map: distribution 1844–1999**

198 *The Bees of Sussex*

S. odontopyga's host *Osmia spinulosa* is a species of calcareous soils with abundant snail populations. In Sussex populations of this *Osmia* are concentrated on the South Downs and on the coast at Rye Harbour Nature Reserve. It also has outlying populations at scattered locations with sandy soils.

The specimen swept in Brighton was on an area of bare chalk that is being managed as a 'Bee Bank'. The Kent and Oxfordshire specimens are from very different brownfield sites, but in both cases there was also a calcareous substrate.

Out in the field and under the microscope

Searching known foodplants in locations that support its host, *O. spinulosa*, is likely to be the most effective way of finding this bee.

S. odontopyga is all black and closely resembles its host in both colouring and size. It is also very similar to another species in the same genus, *S. phaeoptera*. To separate females of these two species, it is necessary to check the surface of the propodeal triangle. In *S. odontopyga* this area is polished and lacks punctures in the middle, while in *S. phaeoptera* it is dull and covered in punctures. The tip of the abdomen in male *S. odontopyga* ends in a tiny point, a feature missing in *S. phaeoptera*. The only other all-black species in the genus, *S. breviuscula*, is much smaller and has bands of white hairs across the hind margins of the tergites.

Stelis odontopyga ♀ [SF]

Osmia spinulosa ♀, the host [PC]

Behaviour and interactions

S. odontopyga lays its eggs within brood cells that are being provisioned by its host, *O. spinulosa*. This last bee establishes its nests within empty snail shells of a variety of species, including Garden Snail, Heath Snail and White-lipped Snail.

On the wing

The flight season for *S. odontopyga* overlaps with that of its host, which has been recorded in Sussex between late May and early September. Late May, June and July are the main periods of activity for the host.

Feeding and foraging

Species cited in the literature are Common Ragwort, hawkweed, Elecampane, Rough Hawkbit, stonecrop, Tansy, Viper's-bugloss and Wild Carrot.

Stelis ornatula **Spotted Dark Bee**
FIRST RECORD 1977, Ambersham Common (Mike Edwards) TOTAL RECORDS 23

Geography and history
Records for *Stelis ornatula* are very thinly scattered across the county and are closely tied to the very localised populations of its host *Hoplitis claviventris*. Most records for this last bee are from chalk grasslands on the South Downs, with additional records from heathlands, orchards, sand dunes, undercliffs, vegetated shingle, woodlands and gardens. This pattern is mirrored in the distribution of *S. ornatula*, which was last recorded in Sussex at Rye Harbour Nature Reserve in 2021.

Across Britain, most records are concentrated in the south-east and are otherwise thinly scattered as far north as the Sefton coast in Lancashire.

Ambersham Common where *Stelis ornatula* was first recorded in Sussex in 1977 [ME]

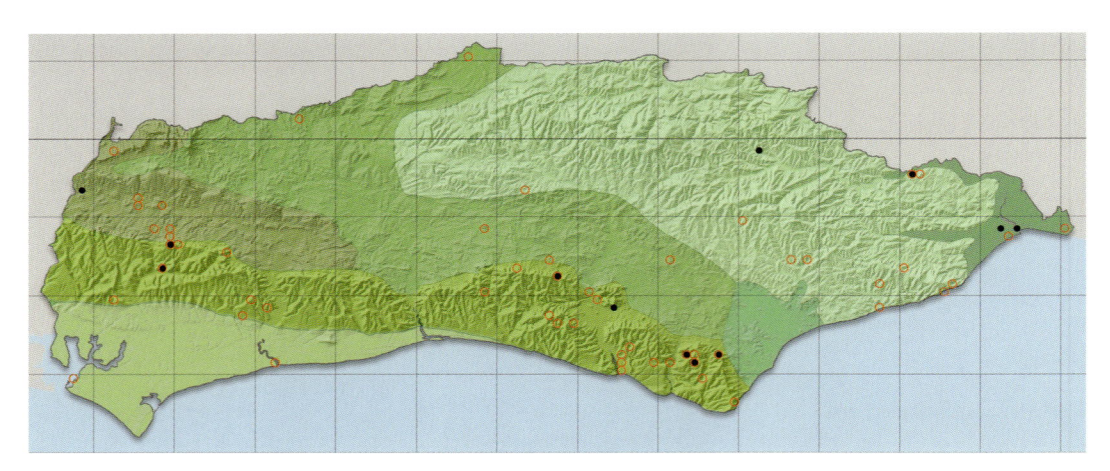

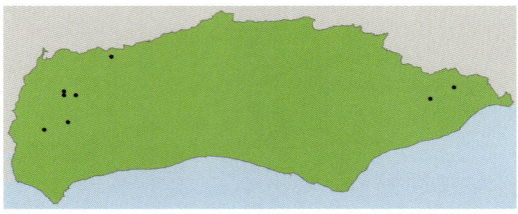

Large map: distribution 2000–2023, with orange circles showing distribution of *Hoplitis claviventris*
Small map: distribution 1844–1999

200 *The Bees of Sussex*

Stelis ornatula ♀ [NO]

Stelis ornatula ♂ [NO]

Out in the field and under the microscope
This bee is best searched for in sheltered spots with abundant bramble or rose stems that are being used by its host, *H. claviventris*, for nesting. In these locations, *S. ornatula* might be found feeding, and, during a survey of sites in the eastern South Downs, Steven Falk reports that he usually found it on ragwort. A different approach is to rear adults from dead bramble or rose stems that have been gathered in late winter.

This black bee can be readily recognised from the pale white spots on the sides of the abdomen. These mirror the white hair bands on the host's abdomen and can be used to distinguish this bee from all other British species of *Stelis*.

Behaviour and interactions
Although *H. claviventris* is its main host in Britain, several other species are given on the continent, including two species recorded in Britain—*H. leucomelana* and *Osmia caerulescens*.

There is an observation from Surrey by David Baldock of a female *S. ornatula* following a female *H. claviventris* "which was flying slowly around a root-plate covered in brambles, presumably searching for a nest site".

Hoplitis claviventris ♀, the host [NO]

On the wing
Sussex records run from late May until late June. Elsewhere, it is reported to be on the wing until late August.

Feeding and foraging
There is little information on plant species visited in Britain, with only Common Ragwort, Common Fleabane, cinquefoil and hawk's-beard cited in the literature.

On the continent, other plant species listed are Autumn Hawkbit, bramble, Cat's-ear, Common Bird's-foot-trefoil, Germander Speedwell, Heath Speedwell, Mouse-ear-hawkweed, Nipplewort and Umbellate Hawkweed.

Stelis phaeoptera Plain Dark Bee
FIRST RECORD 1935, Falmer (Kenneth Guichard) **TOTAL RECORDS** 2

Geography and history
Stelis phaeoptera has been recorded twice in Sussex. The first time was when two females were caught by Kenneth Guichard in Falmer in 1935, and these specimens are in the Natural History Museum. In 2022, 87 years later, a female was swept by Chris Bentley at Rye Harbour Nature Reserve.

Across Britain this is a scarce species, with records from just a handful of counties since 2000. It is seemingly well established in Shropshire and the Welsh Marches, with additional scattered records from across southern England, west Wales, and as far north as Cheshire.

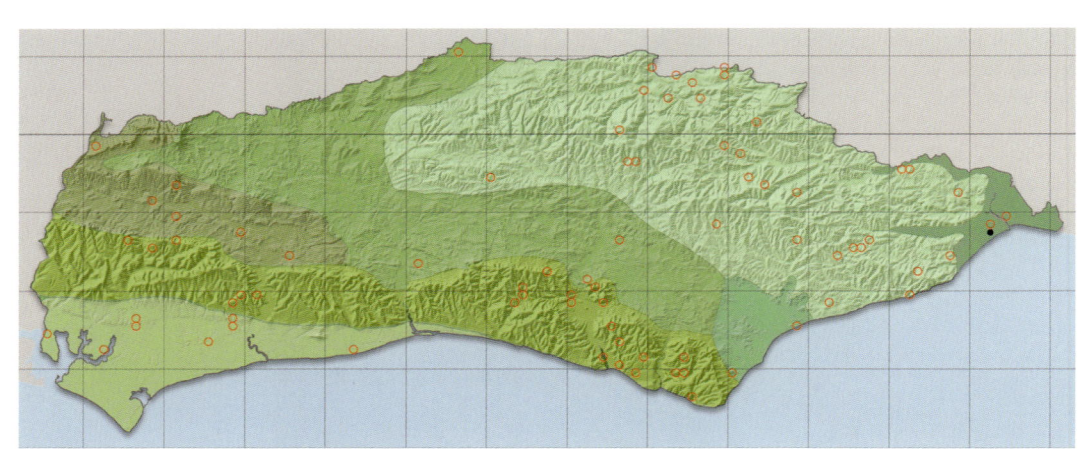

Large map: distribution 2000–2023, with orange circles showing distribution of *Osmia leaiana*
Small map: distribution 1844–1999

202 *The Bees of Sussex*

S. phaeoptera's host in Britain is *Osmia leaiana*, and this is a widely distributed species found in a variety of different locations. These include parks and gardens (where it often uses bee hotels), as well as scrubby grasslands, open woodlands and orchards.

Out in the field and under the microscope
S. phaeoptera is most likely to be found close to the nest sites of its host, *O. leaiana*, where it might be seen investigating or entering nests. In the case of the 2022 record from Rye Harbour Nature Reserve, Chris Bentley reports that this bee "was patrolling a wood pile near to bee hotels (which certainly have *Osmia leaiana*)". It might also be possible to rear adults from stem-nests collected in late winter and then brought into the warmth of a home.

Males and females are completely black other than the presence of very small, weak bands of whitish hairs restricted to the sides of tergites one and two. It most closely resembles another species in the genus, *S. odontopyga*, and the two can only be separated by checking the surface of the propodeal triangle. In *S. phaeoptera* this area is punctate, while *S. odontopyga* lacks punctures here. There is a resemblance to a third species, *S. breviuscula*, but this species is smaller and has white hair bands across the hind margins of the tergites.

Behaviour and interactions
This species targets the brood cells of *O. leaiana*, a species that nests within pre-existing cavities in dead wood, within holes in masonry, and in hollow canes and plant stems. Other hosts are also possible, including *O. caerulescens*, *O. bicornis*, *O. spinulosa* and *Heriades truncorum*.

Stelis phaeoptera ♀ at nest hole [SF]

Osmia leaiana ♀, the host [NO]

On the wing
The flight season for *S. phaeoptera* overlaps with that of its host, which has been recorded in Sussex between mid April and late August.

Feeding and foraging
S. phaeoptera has been recorded visiting plants such as Common Bird's-foot-trefoil, Field Scabious, speedwell, thistle and hawkweed.

Stelis punctulatissima Banded Dark Bee

FIRST RECORD 1879, Hastings (Edward Saunders) **TOTAL RECORDS** 28

Geography and history

Stelis punctulatissima is found in places occupied by its host *Anthidium manicatum*. This last species is a very widely distributed bee found throughout Sussex, most frequently in gardens. It is also found in habitats such as chalk grassland, flower-rich meadow and pasture, heathland, open woodland, soft-rock cliff and vegetated shingle.

Despite the abundance of its host, across Britain *S. punctulatissima* is very scarce and is rarely encountered. It is, however, a widely distributed species and has been recorded as far north as southern Scotland, with most records concentrated in south-east England.

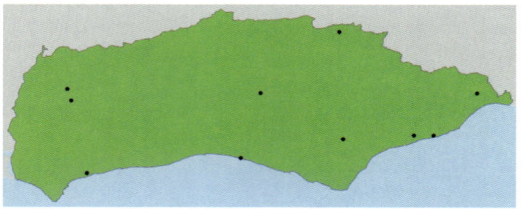

Stelis punctulatissima ♀ [MF]

Stelis punctulatissima ♂ [PW]

Large map: distribution 2000–2023, with orange circles showing distribution of *Anthidium manicatum*
Small map: distribution 1844–1999

Heathland ride, Hargate Forest [IB]

In Sussex, there have been just four records since 2000—from Midhurst in 2007, from Willingdon in 2008, from Pulborough Brooks in 2020, and from Petworth in 2022.

Out in the field and under the microscope

This is the largest of the *Stelis*, and the abdomen can appear quite broad. It is a distinctive species as it is the only *Stelis* where both males and females are black with slim cream-coloured bands along the hind margins of the tergites. The wings are dark while the thorax and abdomen have large punctures.

A considerable amount of good fortune is required to find this species, although Ian Beavis has encountered it once in Hargate Forest near Tunbridge Wells and on several other occasions just over the county boundary in Kent. He describes finding it in a variety of different situations, including flying around bramble, in open ground, at ragwort, around *Calluna* along a heathland ride, flying around a root-plate, and visiting knapweed.

In his gardens in Hailsham and then Willingdon, Martin Jenner reports finding *S. punctulatissima* on Field Scabious. The very first record, by Edward Saunders in 1879, is also from a garden, this time on speedwell.

Behaviour and interactions

S. punctulatissima attacks the brood cells of *A. manicatum*, which nests in pre-existing cavities in dead wood and masonry. It has also been suggested that *S. punctulatissima* targets the nests of different species of *Osmia* and *Megachile*, but this has not been confirmed and is questioned by some.

Anthidium manicatum ♀, the host [NO]

On the wing

Sussex records run from mid June until the beginning of August. Elsewhere, it is reported to fly until late August.

Feeding and foraging

There are records of this bee visiting Black Horehound, bramble, Common Bird's-foot-trefoil, Common Fleabane, Common Knapweed, Common Mallow, Common Ragwort, Field Scabious, hawkweed, Heather, Hooker's Fleabane, Wild Marjoram, speedwell, thistle and Yarrow.

The Bees of Sussex

Andrena – Mining Bees

This is the largest genus of bees in Britain. Nests are excavated in the ground and can be either thinly distributed across the landscape or established close together to form large aggregations. A few species of *Andrena* share a common nest entrance, but all *Andrena* excavate and provision their own brood cells. Females have plumose hairs on the hind legs and on the back of the thorax which they use to carry pollen back to their brood cells. The pollen adheres to the hairs without needing to be wetted and, as the pollen remains dry, *Andrena* are particularly effective pollinators. With the discovery of a species new to Britain, *Andrena ventralis*, in Hampshire in early 2023, the number of species in this genus in Britain has reached 69, of which 58 have been found in Sussex.

Andrena afzeliella a cryptic mining bee
FIRST RECORD 1990, Ambersham Common (Mike Edwards) TOTAL RECORDS 7

Geography and history
In 2022 it was confirmed that *Andrena afzeliella* and *A. ovatula* are distinct species—until this point, they had been treated as one species and referred to as *A. ovatula*. Although *A. afzeliella* is understood to be widely distributed across southern England, it has only been confirmed from Devon, Cornwall, Dorset, Surrey, Essex and Sussex.

Because *A. afzeliella* is such a newly recognised species, few redeterminations of specimens in museum and private collections have been possible, and the true distribution of this species in Sussex is not understood in detail. It is likely, however, to be much more widely distributed than the records suggest. Any records that have not been reviewed are now treated as *A. ovatula* agg.

There are seven confirmed records of *A. afzeliella* from six locations, all previously recorded as *A. ovatula*—Ambersham Common in 1990, Hailsham in 1994, Laughton in 1996, Warbleton in 2015, Chailey Common in 2018, and Norman's Bay in 2021. They encompass heathland, open woodland, vegetated shingle, a garden, and a complex of meadows and pasture. Despite extensive surveying, it has not been recorded in the area south of Tunbridge Wells.

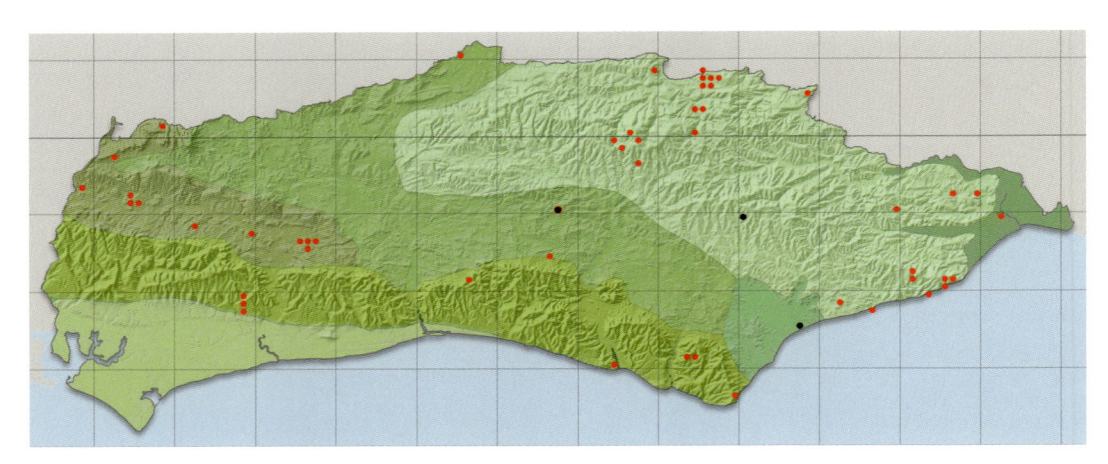

Large map: distribution 2000-2023, with black dots showing distribution of *Andrena afzeliella* and red dots showing distribution of *Andrena ovatula* agg.
Small map: distribution 1844-1999, showing distribution of *Andrena ovatula* agg.

206 *The Bees of Sussex*

A rare photograph of *Andrena afzeliella* ♀, **taken in Spain** [TW]

Out in the field and under the microscope
A. afzeliella most closely resembles *A. ovatula*, with both species having narrow bands of white hairs along the hind margins of tergites two, three and four. In fresh specimens of both species, the hair band on the hind margin of tergite three is unbroken, whereas in the otherwise similar *A. russula* and *A. wilkella* this band on tergite three is interrupted.

The differences between *A. afzeliella* and *A. ovatula* are subtle. For example, female *A. afzeliella* are slightly smaller and have a golden or whitish-grey fringe of hairs on tergite five. In *A. ovatula*, this fringe is dark brown to black in the centre and paler to the sides. The hairs flanking the basitibial plates on the hind tibiae are golden-white in *A. afzeliella* and dark brown in *A. ovatula*. *A. afzeliella* also lacks the short dark hairs found beneath longer yellowish-white hairs on the scutum of *A. ovatula*.

The differences between males of the two species are very slight, and it is often not possible to separate the two. Characters to check include the shape of the genital capsule and the lengths of antennal segments three and four. These are roughly equal in length in *A. afzeliella*, while segment four is slightly longer than segment three in *A. ovatula*.

A. afzeliella is best searched for by checking flowering plants for feeding males and feeding or foraging females. The female swept at Norman's Bay was foraging from Common Bird's-foot-trefoil.

Behaviour and interactions
A. afzeliella has two generations a year, peaking in May and June and then again in July and August, in each case about a month later than *A. ovatula*. It is closely associated with plants in the Pea family, one study showing that just over 90 per cent of the pollen collected by this species came from this family.

On the wing
This species has two flight seasons, the first running from late April until the end of June and the second running from the beginning of July until early August.

Feeding and foraging
As well as collecting pollen from plants in the Pea family, *A. afzeliella* forages from the Rose, Plantain, Borage, Dogwood, Buttercup, Cabbage, Teasel and Daisy families.

Andrena alfkenella Alfken's Mini-miner
FIRST RECORD 1993, Lower Beeding (Peter Hodge) **TOTAL RECORDS** 21

Geography and history
Andrena alfkenella is found in areas with dry, well-drained soils. It is widely distributed across southern counties but is very local.

The few Sussex records are very thinly spread out across the county and are from a variety of different habitats. It has been recorded from heathland, woodland, chalk grassland, vegetated shingle, sand dunes, soft-rock cliffs, and lowland hay meadows and pasture.

All but one of these records have been made since 2000, with the most recent being from Castle Hill in 2022. The other records have been from Great Wood, Telham, Warbleton, Denton Downs, Southerham Farm, Malling Down, Ditchling Common, and farmland near Barnham.

Restored chalk grassland at Southerham Farm [NS]

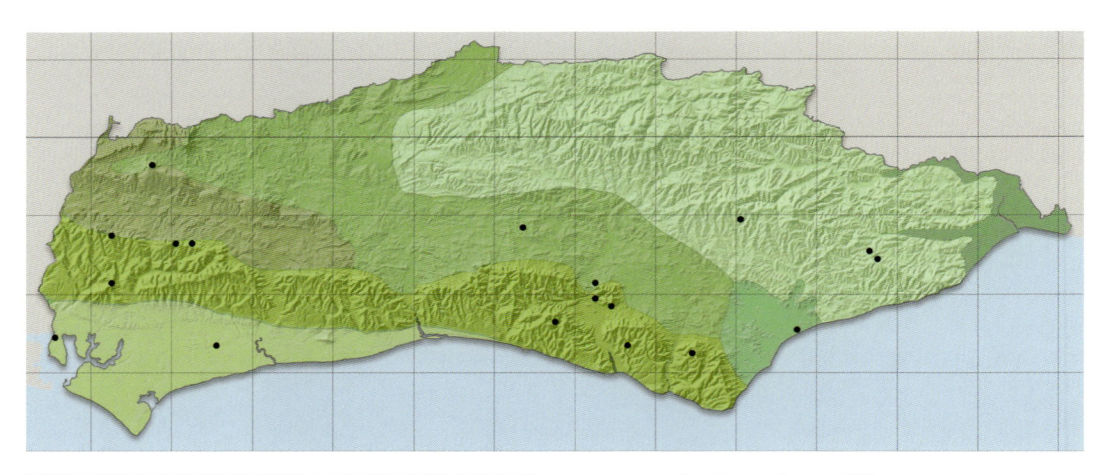

Large map: distribution 2000–2023
Small map: distribution 1844–1999

208 *The Bees of Sussex*

Out in the field and under the microscope
A. alfkenella is one of several small black *Andrena*. Among the key characters to look for on females are the faint punctures on the second tergite, in front of the marginal area. This helps to separate this bee from all the other small black *Andrena* found in Sussex apart from *A. falsifica*.

A. alfkenella can be distinguished from this last species by the dull and roughened marginal area on the first tergite (this area is slightly roughened but shiny in *A. falsifica*) and by the absence of a small step down from the first to the second tergite.

Males of the two generations of *A. alfkenella* are subtly different and for a while were regarded as distinct species. The spring generation males have black hairs on the face and are clearly punctate on tergite one, while the summer generation is white-haired on the face and strongly punctate on the second and third tergites (but not on the marginal areas).

A. alfkenella is most often found feeding or foraging, particularly on plants in the Carrot family. In Norfolk, Nick Owens reports males "swarming around Wild Parsnip", while one Sussex record is of a female on a water-dropwort flower head.

Andrena alfkenella ♀ [JL]

Andrena alfkenella ♂ [JL]

Behaviour and interactions
A. alfkenella has two generations a year. No species of *Nomada* or *Sphecodes* targeting brood cells of this species have been identified.

On the wing
This species has two flight seasons, the first running from late March until early June and the second running from the beginning of July until the beginning of September.

Feeding and foraging
Pollen is collected from the Cabbage, Carrot, Speedwell and Rose families.

Andrena angustior Groove-faced Mining Bee
FIRST RECORD 1896, Hastings (Edward Saunders) **TOTAL RECORDS** 70

Geography and history
Andrena angustior is a species of base-poor open habitats, particularly areas of woodland, adjacent grassland or heathland. In recent years, it has been recorded in Park Corner Heath in the Low Weald, and from Batemans and Nymans in the High Weald.

It is also known from the Wealden Greensand, but there have been few records from here since 2000.

Hay meadow, High Weald [NS]

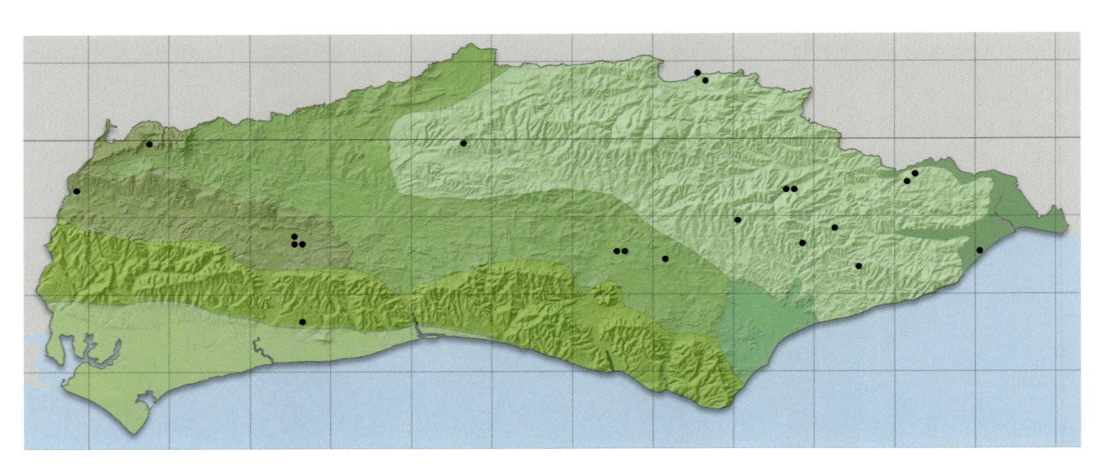

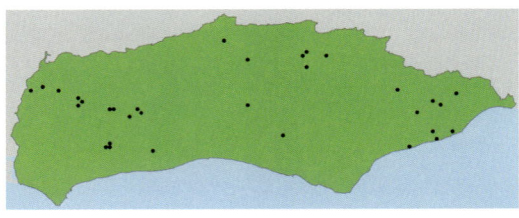

Large map: distribution 2000–2023
Small map: distribution 1844–1999

Andrena angustior ♀ [SF]

Andrena angustior ♂ [PB]

Out in the field and under the microscope
This *Andrena* is best looked for by checking the flower heads of known foodplants in areas of open woodland and heathland. Several observations reference buttercup, while Ian Beavis reports finding males flying close to, and settling on, bare ground along woodland paths.

Superficially, this bee resembles *A. bicolor*. In the field, it is just possible to make out the pale hairs on the clypeus in fresh specimens of both sexes of *A. angustior*—these are dark in *A. bicolor*. Other key characters to look for in *A. angustior* are a very wide and shiny marginal area on the second tergite and very long third segments on the antennae. Females also usually have a shallow, vertical groove on the midline of the clypeus. The wide marginal area found in both sexes helps to distinguish this species from all other *Andrena*.

Most observations have been made in May.

Behaviour and interactions
A. angustior can nest in small, loose aggregations but generally females establish nests as singletons. Nests have been reported alongside paths or on vertical banks.

This is one of the *Andrena* targeted by the cuckoo bee *Nomada fabriciana*, the females laying their eggs within the unsealed brood cells.

On the wing
Sussex records run from mid April until mid June.

Feeding and foraging
Pollen is collected from plants in the Cabbage, Daisy, Carrot, Speedwell and Rose families.

Andrena apicata Large Sallow Mining Bee
FIRST RECORD 1883, Hastings (Francis Morice) TOTAL RECORDS 63

Geography and history
Nationally, *Andrena apicata* is a widely distributed but local bee, its range extending as far north as south Cumbria. It is scarce in counties adjoining Sussex.

In Sussex itself, it is thinly distributed, with a handful of locations generating most records. These include Rewell Wood, the countryside south of Tunbridge Wells, woodland near Haywards Heath, Flatropers Wood, Old Lodge and Hastings Country Park.

Out in the field and under the microscope
A. apicata is one of the first bees to emerge each spring, often appearing on warm days in March. This is when it is possible to see males and females feeding on willow, or, exceptionally, females pushing their way into gorse flowers to collect pollen. Males are often seen flying up and down sunlit tree trunks and fence posts, sometimes in among *A. praecox* males. Both *A. apicata* and *A. praecox* feed predominantly on willow, so they frequently occur in the same locations, although *A. praecox* is usually the more abundant of the two.

Females of the two species are very similar. Female *A. apicata* are generally the larger of the two, and they can be separated by looking at the colour of the hairs on tergite five and at the tip of the abdomen. These are dark brown or black in *A. apicata* and pale in *A. praecox*.

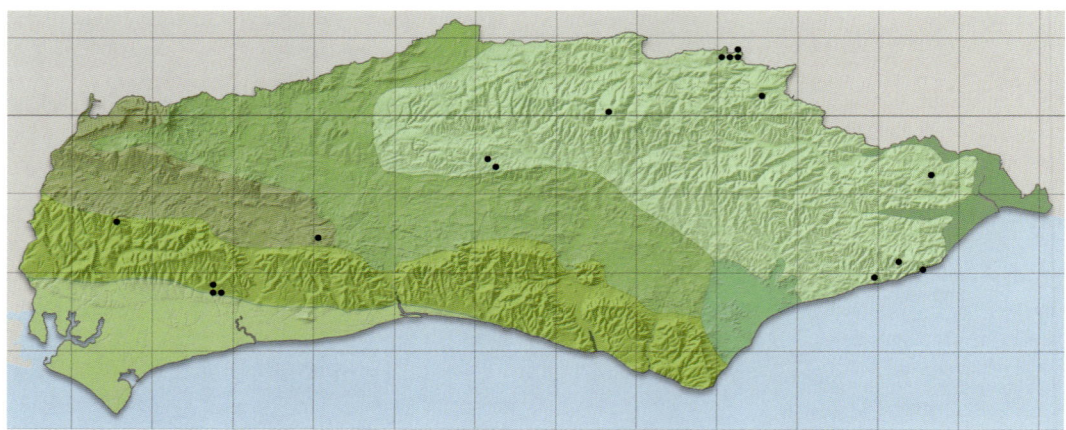

Andrena apicata ♀ [SF]

Large map: distribution 2000–2023
Small map: distribution 1844–1999

212 *The Bees of Sussex*

A. apicata males are similar to *A. praecox* males. They are much smaller than the females and, like *A. praecox* males, have very long mandibles with a broad projection on the underside. The best character to use to separate these two species is the shape of the tip of the eighth sternite. This is notched only in *A. praecox*.

Andrena apicata ♂ [SF]

Behaviour and interactions
A. apicata generally nests in small, scattered aggregations or singly in light soils, but occasionally it can form quite extensive aggregations. These are sited in areas of compacted sand or on banks.

Its brood cells are targeted by *Nomada leucophthalma*.

On the wing
Sussex records run from early March until late April.

Feeding and foraging
This *Andrena* has been recorded visiting alder, Barren Strawberry, Colt's-foot and dandelion. Pollen, however, is largely collected from willow, but also from Blackthorn, Gorse and Wild Cherry.

Willow blossom [PG]

The Bees of Sussex

Andrena argentata Small Sandpit Mining Bee
FIRST RECORD 1974, Ambersham Common (Mike Edwards) **TOTAL RECORDS** 81

Geography and history
Typically, *Andrena argentata* is a species of hot, dry heaths and so is almost entirely restricted to sites on the Lower Greensand. Sites that have generated recent records include Iping Common, Stedham Common and Graffham Common. On three occasions during the 1980s, this bee was also recorded from a sand pit adjoining Rewell Wood.

Iping Common [NS]

Large map: distribution 2000–2023
Small map: distribution 1844–1999

Andrena argentata ♀ [NO]

Andrena argentata mating pair, ♂ above and ♀ below [PB]

Out in the field and under the microscope
To find this *Andrena*, it is best to search the hottest and driest areas of a heathland and look for nest sites or for foraging bees as they move quickly from flower to flower on their preferred foodplants.

Nests are established in areas of bare, loose sand that is free of vegetation, such as along the edges of paths and in sandpits. In these locations, it is possible to find large numbers of males zigzagging above the surface of the sand or females returning to their nests with a pollen load.

Both male and female *A. argentata* have a shiny black abdomen with obvious white hair bands. Both sexes can rapidly lose their colour, the males, especially, acquiring a silvery appearance early on.

Behaviour and interactions
Males seek females as they emerge from their brood cells, trying to mate as soon as the females appear above ground. Sometimes there may be four or five males enveloping a single female.

On flying back to her nest, a female rapidly disappears into the loose sand. She does not leave an obvious entrance because the sand collapses behind her and fills the tunnel as she descends to the brood cells. These are established in firmer sand and might be 14 to 17 cm below the surface.

A. argentata can be abundant on some sites, sometimes appearing in huge numbers. In Surrey, David Baldock estimated 25,000 at one site. It produces just one generation per year, its emergence coinciding with Heather coming into flower.

Nomada baccata (which can be numerous at nest sites) lays its eggs within unsealed cells, while *Sphecodes reticulatus* and *S. ephippius* are also possible brood parasites.

While this bee is understood to be closely associated with hot dry heaths, there are exceptions. Mike Edwards says, "During the warm period between 1996 and 2006 I also found this bee at some 'not so hot' woodlands. These were not in particularly loose sand."

On the wing
Sussex records run from early June until late August.

Feeding and foraging
In Britain, *A. argentata* is normally seen collecting pollen from plants in the Heather family, while on the continent it is reported that pollen is collected from a wider range of plants, including members of the Carrot, Cabbage, Rose and Willow families.

A. argentata has also been recorded visiting Sea-holly, cinquefoil and Wild Parsnip.

Andrena barbilabris Sandpit Mining Bee
FIRST RECORD 1905, Hastings (Edward Saunders) **TOTAL RECORDS** 201

Geography and history
This species is found in places with sandy soils and occurs in a wide range of habitats. These include heathland, open woodland, chalk grassland sites overlain with sand, coastal wetlands and sandpits. Any site with dry sandy soils could potentially support a population.

It is particularly abundant on heathlands on the Lower Greensand such as Iping Common and on the coast at Rye Harbour Nature Reserve. There are additional recent records from scattered locations, including a sandpit near Lewes, the cliffs below Newhaven Fort, Rewell Wood and Seaford Head.

Andrena barbilabris nests in areas of light, sandy soil where it can form enormous aggregations. [JP]

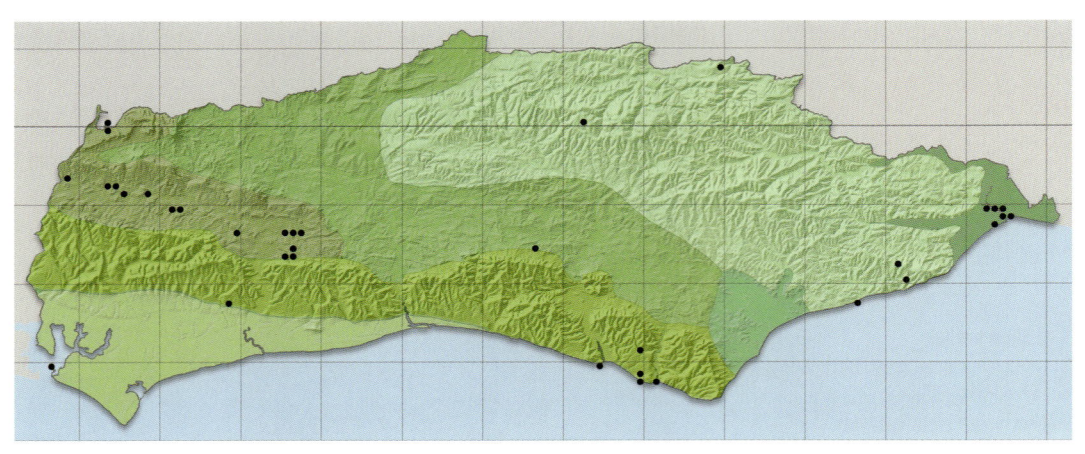

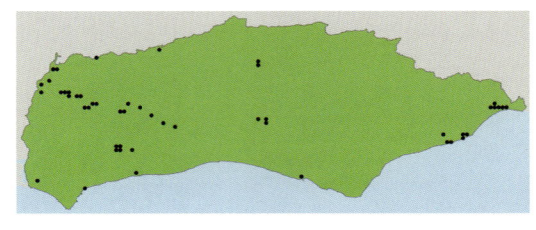

Large map: distribution 2000–2023
Small map: distribution 1844–1999

The Bees of Sussex

Andrena barbilabris ♀ [PC]

Andrena barbilabris ♂ [NO]

Out in the field and under the microscope
In optimum locations, *Andrena barbilabris* can be hard to miss as it nests in huge numbers. The nests tend to be on areas of flat, bare, sandy soil, as on the floor of a sandpit or alongside a path. In these locations, the silvery males can be seen flying rapidly just above the surface, settling briefly or even testing the sand to see if a female is emerging. Females might be seen either returning with pollen loads or excavating a nest.

A. barbilabris females have distinctive reddish-brown hairs on the scutum and a shiny black abdomen with weak white hair bands on the tergites. Males have long white hairs on the sides of the thorax, appear silvery, and have all-black hind legs. Both sexes usually have distinctive longitudinal wrinkles on the first tergite.

Behaviour and interactions
A. barbilabris forms large aggregations, with 32 nests per m^2 reported from one study on the continent. Once a female returns with a pollen load, she pauses briefly to confirm that she is in the correct location and then quickly disappears below ground, the loose sand closing behind her as she works her way down to the brood cells. A report from the continent suggests that it can take two days for each brood cell to be completed and that, in her lifetime, a female will establish two or three nests, each with two or three brood cells.

Sphecodes pellucidus and *Nomada alboguttata* lay their eggs in the brood cells of this species.

On the wing
Sussex records run from late March until the beginning of July.

Feeding and foraging
A. barbilabris forages for pollen from a very wide range of plants, including buttercup, oak, stitchwort, dock, willow, speedwell, dandelion, Blackthorn and Hawthorn.

Andrena bicolor Gwynne's Mining Bee
FIRST RECORD 1877, Worthing (Edward Saunders) TOTAL RECORDS 553

Geography and history
Andrena bicolor is one of the most commonly recorded species of *Andrena* in Sussex and is very widely distributed, occurring in all parts of the county. It is a very versatile bee which can be found in diverse habitats, and it is a frequent visitor to gardens.

Out in the field and under the microscope
As nests are rarely encountered, *A. bicolor* is most likely to be seen feeding. In the early spring, when there is a limited number of plants in flower, Blackthorn and dandelion are always worth a look. As the season progresses, it visits a wider range of plants and could be found on any number of plant species including Round-headed Rampion, which is always worth a look.

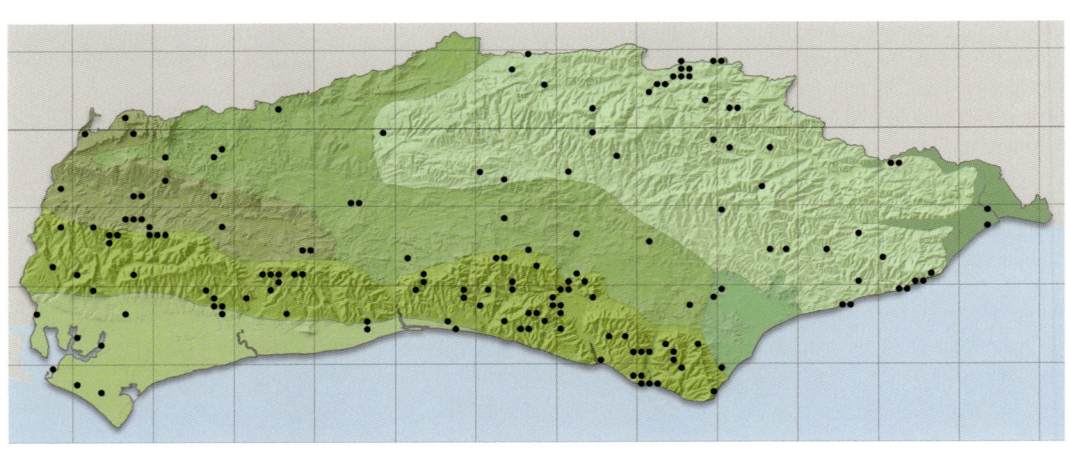

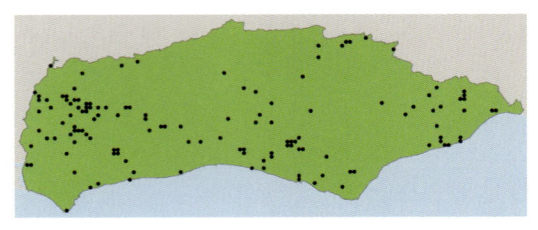

Large map: distribution 2000–2023
Small map: distribution 1844–1999

218 *The Bees of Sussex*

The females, like the males, have large saucer-like punctures on the propodeum. They have black hairs on the clypeus, reddish-brown hairs on the top of the thorax, and a thin covering of brown hairs on the abdomen.

Males in the spring brood have black hairs on the clypeus while summer males often have brown hairs mixed with black here.

Behaviour and interactions

Little is known about this bee's nesting behaviour, although there are reports from Norfolk of it nesting in a sloping bank and from Surrey of a nest in a root plate. It is understood that this is a solitary-nesting species, with a report on the continent indicating that *A. bicolor* nests are excavated to just over 1 m in depth. The cuckoo bee *Nomada fabriciana* targets brood cells of this species.

A study from Spain has shown the benefits that spring bees can obtain by basking within a flower tube on a cool day. The study looked at the effects on *A. bicolor* of the increased temperature within the flower tube, in this case a species of daffodil. It showed that the flower tube was some 8°C above the surrounding air temperature, which raised the temperature of the bee well above that required to sustain flight.

Andrena bicolor ♀ inside a daffodil trumpet [EP]

Round-headed Rampion, Ditchling Beacon [PC]

On the wing

This species has two flight seasons, the first running from late February until the end of May and the second running from the beginning of June until the end of August.

Feeding and foraging

The two generations visit a wide range of plant species. In the spring, pollen is collected from plants such as Bluebell, buttercup, Daisy, dandelion, Germander Speedwell, Hawthorn, Marsh-marigold, Primrose and willow. The summer generation collects pollen from plants such as Round-headed Rampion, bramble, buttercup, Cat's-ear, geranium, harebell, knapweed, Meadowsweet, thistle, cinquefoil and willowherb.

Andrena bimaculata Large Gorse Mining Bee
FIRST RECORD 1900, Bognor Regis (Henry Guermonprez) **TOTAL RECORDS** 97

Geography and history
Andrena bimaculata is closely tied to areas with light sandy soils, and in Sussex most records are from the Lower Greensand. However, there have been just seven records from this area since 2000. In the same period, there have also been six records from five widely separated locations with sandy soils elsewhere in the county—Hastings Country Park, Newhaven, Rewell Wood, Broadwater Warren, and a sandpit near Lewes.

Out in the field and under the microscope
This is a dark bee which resembles a few other species and can be tricky to identify. An additional complication is variability within populations. Some females and males are red-marked on the abdomen, for example, and there are also variations in the hair colour on the face.

Both males and females are among a group of six species that have a small, raised ridge in the middle of the propodeum. Females have pale pollen-collecting hairs on the hind tibiae, and can have red markings on the sides of tergites one and two.

Males are slightly smaller and have dark hairs on the clypeus but otherwise resemble the females. Some have red markings on the marginal areas of tergites one and two.

The most effective way of locating this bee is to search areas of bare sand where males

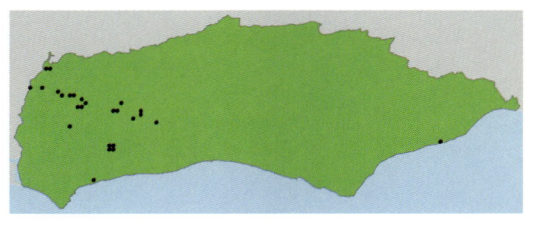

Andrena bimaculata ♀ showing red-marked form [NO]

Large map: distribution 2000–2023
Small map: distribution 1844–1999

220 *The Bees of Sussex*

Andrena bimaculata ♀ showing black form [SF]

Andrena bimaculata ♂ [NO]

might be seen flying fast and low just above the ground. Alternatively, it is worth checking known foodplants. In the spring, *A. bimaculata* is frequently found on gorse flowers, while in the summer it frequently feeds on ragwort and bramble.

Behaviour and interactions
This species is double-brooded, with the first brood more numerous than the second. Nests are often in banks, with individual nests usually scattered over a wide area and with two or three together at most. There may be three brood cells per nest.

Ian Beavis has observed a nest in a boundary ditch and another in the side of an old quarry. He also reports seeing males searching for mates by flying along a Holly hedge and around the branches of a Sycamore tree. In both cases, the males were flying with other species.

A. bimaculata brood cells are attacked by *Nomada fulvicornis*.

On the wing
This species has two flight seasons, the first running from early March until the end of May and the second running from the end of June until mid August.

Feeding and foraging
Females emerging in the spring have been observed collecting pollen from Blackthorn, Gorse and willow, while those emerging in the summer search for pollen from plants such as allium, bramble, Sweet Chestnut, lime, Meadowsweet, privet and Wild Parsnip.

There are also records of *A. bimaculata* visiting Common Ragwort and rape, although it is not known if this was for pollen or nectar.

Andrena bucephala Big-headed Mining Bee

FIRST RECORD 1908, Bersted (Henry Guermonprez) TOTAL RECORDS 80

Geography and history

Andrena bucephala is a scarce bee in Sussex, with most records thinly scattered across the South Downs. It has also been found away from the chalk near Battle, Hastings, Midhurst and Tunbridge Wells, and at Ebernoe Common.

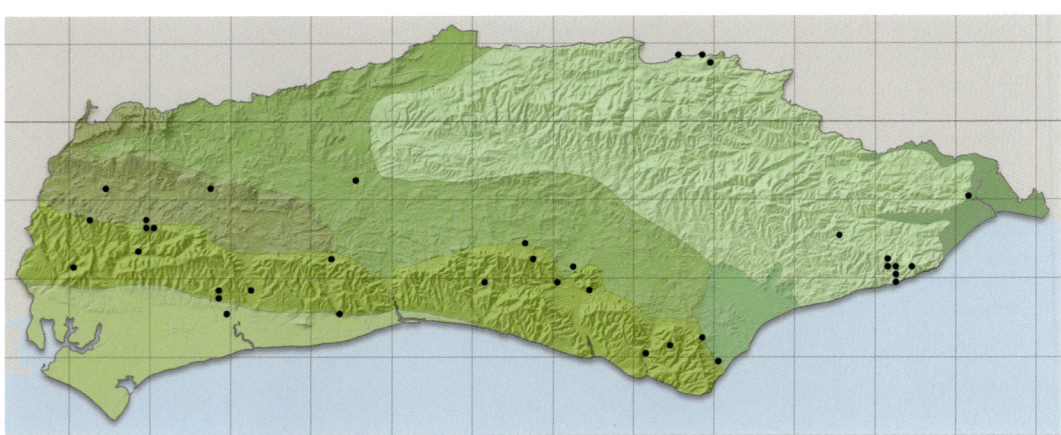

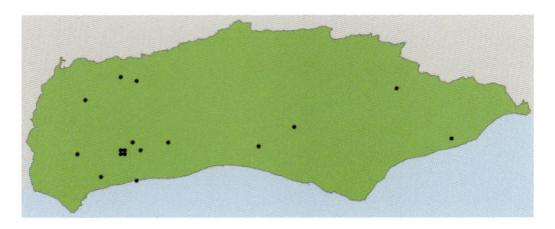

Large map: distribution 2000–2023
Small map: distribution 1844–1999

Between the first record in 1908 and the second in 1974, there was a gap of 66 years. More recently, there has been an increasing number of records, and on the South Downs it has been found in areas of species-rich chalk grassland and open woodlands, including an old chalk quarry near Wiston, Friston Forest and Rewell Wood. It has also been found in Lewes Cemetery and in gardens—including one in Lewes which has provided several records in recent years, most recently in 2022.

Out in the field and under the microscope
Females have distinctive dark, slender abdomens. These have very little hair on them, aside from very thin white hair bands at the back edges of tergites two, three and four. The wings have a very slight yellowish tinge.

Males have very long mandibles that cross, while some also have very large heads (hence both the scientific and common names). The mandibles have a single tooth at the end.

The presence of the cuckoo of this species, *Nomada hirtipes*, can be the first clue that a nest site is nearby, especially if the *Nomada* is present in good numbers. The approach to a nest can be busy with female *A. bucephala* flying to and from the nest. Males are often found visiting flowering trees and shrubs, or careering around the nest site.

Behaviour and interactions
A. bucephala is one of three species of *Andrena*—the others being *A. ferox* and *A. scotica*—in which females often share a nest entrance, each individual establishing and provisioning her own brood cells within the nest. *A. bucephala* nests can support many hundreds of females. Occasionally, two or three nests are established near each other.

Nests are often sited on steep, sunny south-facing banks, deep within a rabbit burrow, or inside an opening within a tree stump. The nests can be perennial, enduring for several years. Males have been reported swarming in large numbers over Holly trees and flowering shrubs.

Brood cells are targeted by *N. hirtipes*.

Andrena bucephala ♀ at a communal nest entrance [PG]

On the wing
Sussex records run from mid April until late June.

Feeding and foraging
This *Andrena* collects pollen from a wide range of plants, but especially from trees and shrubs in the Maple, Rose and Willow families. Hawthorn and Field Maple appear to be particularly important pollen sources.

Andrena chrysosceles Hawthorn Mining Bee
FIRST RECORD 1882, Hastings (Edward Saunders) **TOTAL RECORDS** 343

Geography and history
In Sussex this is an abundant and widely distributed bee which has been recorded in all parts of the county. It is particularly associated with woodland and could be found in any broadleaved wood.

Out in the field and under the microscope
Andrena chrysosceles is a shiny black bee. Females have characteristic orange-yellow tibiae and tarsi on the hind legs with bright golden hairs. They also have white hairs along the back edges of the tergites and orange hairs at the tip of the abdomen. They most closely resemble *A. fulvago* females but are smaller, and the hairs on the hind legs are shorter and less plumose.

Males have orange-yellow marked tarsi on the hind legs and a characteristic white clypeus with a pair of black spots. With practice the males can be identified in the field.

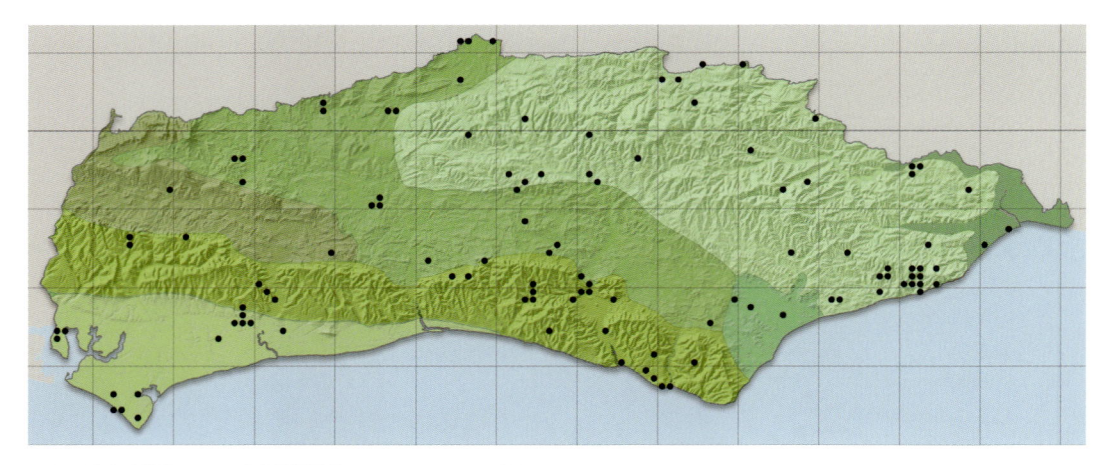

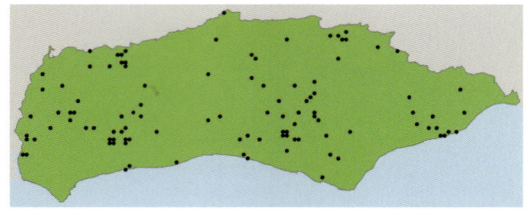

Andrena chrysosceles ♀ [PB]

Andrena chrysosceles ♂ [PB]

Large map: distribution 2000–2023
Small map: distribution 1844–1999

The Bees of Sussex

A. chrysosceles is best searched for by checking flowering plants and shrubs along woodland paths where it might be seen on dandelion, buttercup, Blackthorn or bramble. In 1944, Alfred Brazenor reported that he had collected several males on Blackthorn blossom at Markstakes Common. It is also possible to see males chasing each other over bramble or shrubs.

Andrena chrysosceles mating pair, with ♀ above and ♂ below [PC]

Behaviour and interactions
Little is known about the nesting behaviour of this species, although it has been reported that it nests singly in banks. One observation from Lewes Cemetery is of a solitary nest in a level grass verge supporting abundant flowering dandelion and buttercup.

It is quite common to find individuals that have been attacked by a species of *Stylops*, an internal parasite that is picked up by an adult bee as it visits a flower and is then transported by the bee to its nest. Once inside the nest, the *Stylops* larva enters into a developing bee larva to feed and completes its development when the bee matures. The parasite's head and thorax protrude from the abdomen, and the appearance of the bee changes, with males looking more like females and *vice versa*. In some cases, a very high proportion of a population can be affected.

Nomada fabriciana lays its eggs within brood cells of this species.

Andrena chrysosceles nest site, Lewes Cemetery [JP]

Below: *Andrena chrysosceles* ♀ attacked by a *Stylops* (seen protruding from the bee's abdomen) [EP]

On the wing
Sussex records run from mid March until early August. Typically, *A. chrysosceles* is on the wing until the end of June, with most records from April, May and June.

Feeding and foraging
A. chrysosceles has been observed on many different plant species, including Blackthorn, bramble, Creeping Willow, dandelion, Lesser Stitchwort, Meadowsweet, Wood Avens and Wood Spurge.

A wide range of plants is visited for pollen, including buttercup, Common Rock-rose, Dogwood, Elder, forget-me-not, Hawthorn, Hogweed, Holly, nettle, rose and Rough Chervil.

Andrena cineraria Ashy Mining Bee
FIRST RECORD 1858, Rottingdean, Eastbourne and Seaford (William Unwin) TOTAL RECORDS 207

Geography and history
Throughout the twentieth century, *Andrena cineraria* was an extremely scarce bee in Sussex with records in just 1905, 1918, 1929, 1972, 1993, 1996 and 1997. Since 2000, the number of annual records has risen rapidly.

The first records of *A. cineraria* in Sussex are from 1858 and is by William Unwin who reported that it was "common but local in its distribution. Colonies have been observed on the coast between Brighton and Rottingdean, near Seaford, and also on the coast near Eastbourne."

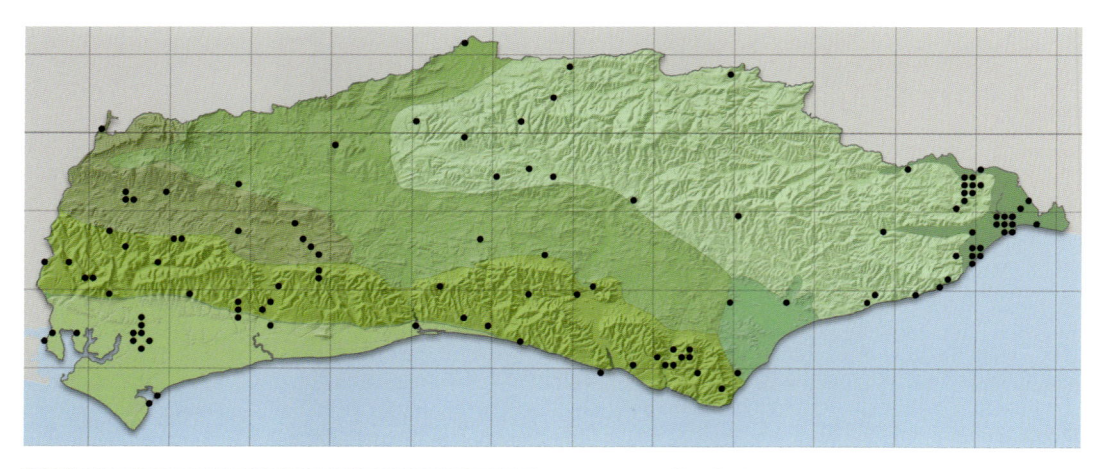

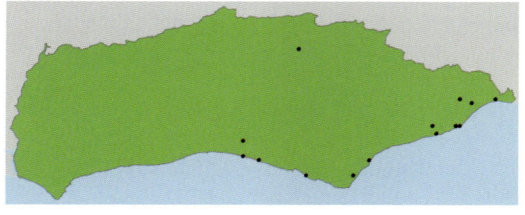

Andrena cineraria ♀ [PC]

Andrena cineraria ♂ [PB]

Large map: distribution 2000–2023
Small map: distribution 1844–1999

Today, *A. cineraria* is widely distributed across much of Sussex but with a particular concentration in the Romney Marsh area. It is also regularly found on chalk grasslands on the South Downs, on the heathlands of the Lower Greensand, in open woodlands, and in gardens where it frequently nests in lawns.

Out in the field and under the microscope

This is a very striking *Andrena* and one that is unlikely to be confused with any other species except for *A. vaga*.

In *A. cineraria* females, the top of the thorax is covered in grey hairs apart from a band of black hairs that runs across the middle—in *A. vaga*, the top is completely covered in grey hairs. In contrast to the thorax, the abdomen is virtually hairless and is an inky-black colour, while the forewings have darkened tips.

Males are similarly marked to the females but are smaller.

A. cineraria is generally found close to nest sites. The nests tend to be sited on areas of flattish, bare or sparsely vegetated ground, and one clue that a nest site might be nearby can be the presence of its parasite *Nomada lathburiana*.

Andrena vaga ♀ [CB]

Behaviour and interactions

A. cineraria often nests in large, dense aggregations, sometimes numbering many thousands of bees. Each nest has just two or three brood cells, with each cell some 10 to 22 cm below the surface. The burrows are left open during foraging trips but are closed during poor weather.

N. lathburiana, its parasite, lays its eggs in unsealed brood cells and can occur in very large numbers. In 2001 David Baldock estimated 5,000 flying with 10,000 of the host at Ham Common in Surrey.

N. goodeniana is known to target *A. cineraria* nests on the continent, and this behaviour is also suspected in Britain.

The handful of August records from Sussex may indicate a second generation.

On the wing

Sussex records run from early March until mid August.

Feeding and foraging

A. cineraria collects pollen from a wide range of plant species, including Blackthorn, bramble, buttercup, hawthorn, Gorse, willow and Wood Spurge.

It has also been reported visiting plants such as Daisy, Common Knapweed, dandelion, Devil's-bit Scabious and Ground-ivy.

Andrena clarkella Clarke's Mining Bee
FIRST RECORD 1882, Hastings (Edward Saunders) **TOTAL RECORDS** 212

Geography and history
Between 1882—when *Andrena clarkella* was first recorded in Sussex—and 1972, there were just eight records for *A. clarkella* in the county. From 1942 onwards, however, it has been recorded with increasing frequency. Over the last ten years, there has been an average of five or six records per year.

This bee is widely distributed across Sussex but with a significant concentration of records from the countryside near Crowborough and Ticehurst. Across the rest of the county, it is also found more locally in places with abundant willow and sandy soils, as at Rye Harbour Nature Reserve.

Andrena clarkella nest site at Rye Harbour Nature Reserve. The emergence of the bee coincides with willow trees producing pollen and nectar. [both PG]

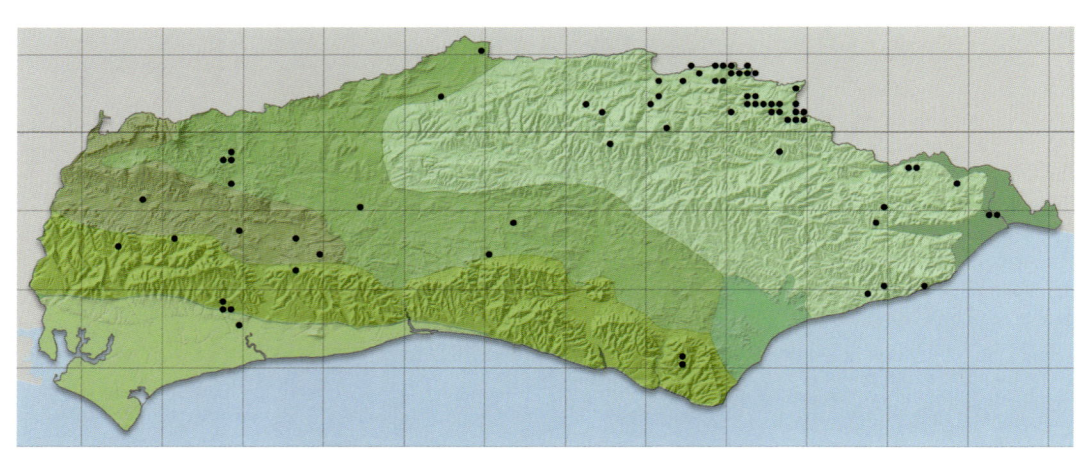

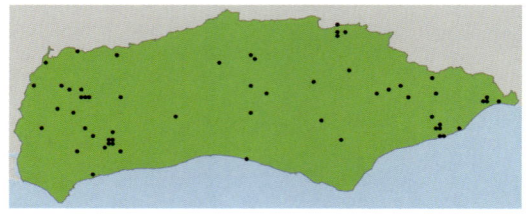

Large map: distribution 2000–2023
Small map: distribution 1844–1999

Out in the field and under the microscope
This is one of the earliest bees to emerge each spring, and the females are unmistakable. The combination of their relatively large size, the black-haired abdomen, and the rusty red covering of hairs on the thorax and orange-red hairs on the hind tibiae means that it cannot be confused with any other species.

Males are quite a bit smaller and are difficult to distinguish from other species of *Andrena* in the field.

A. clarkella is best looked for by searching places where willow is abundant, particularly in open woodlands, heathlands and wetland areas with sandy soil. Males are often seen basking in sunlight on tree trunks, while females are often seen flying to and from their very obvious nests.

Andrena clarkella ♀ [PC]

Andrena clarkella ♂ [PB]

Behaviour and interactions
A. clarkella females excavate their nest burrows into sandy banks (including in a worn section of the moat at Bodiam Castle), in root plates of fallen trees, and on paths or tracks with compacted ground.

It nests in aggregations often encompassing just a handful of nests close together, but these can number many hundreds. In one example from Kent, Ian Beavis reported as many as 2,000 in one location at a density of "50 per square yard in some spots". The burrows are large, and there may be between one and four brood cells per nest, which might be as little as 4 or 5 cm below the surface, although 10 to 30 cm has also been reported.

Males use two mate-seeking strategies, one of which involves flying close to the ground above an aggregation. The other involves flying rapidly upwards from the base of a sunlit tree trunk in a zigzag fashion. The bee then flies quickly to another tree trunk and repeats the process, searching for the females that might alight there.

Nomada leucophthalma is a frequent cuckoo of this species and can be found close to nest sites.

On the wing
Sussex records run from early February until late June, with most from March and April.

Feeding and foraging
This bee's emergence coincides with willow trees—especially Goat Willow and Grey Willow—producing pollen and nectar, which the females use to provision the brood cells.

Other plant species visited include Colt's-foot and dandelion.

Andrena coitana Small Flecked Mining Bee

FIRST RECORD 1873, Littlehampton (Edward Saunders) **TOTAL RECORDS** 84

Geography and history

Andrena coitana is a scarce bee and may be in decline. Historically, it has been largely restricted to scrubby or wooded heathlands on the Wealden Greensand, woodlands on the western South Downs, and scattered locations across the High Weald.

However, since 2000 there have been no records from the Wealden Greensand or the western High Weald and just two from the South Downs. These last two records are from woodland above Cocking and were made in 2000. All other records for this period are from the eastern High Weald. Since 2015 *A. coitana* has been recorded from an even more restricted area, with records from just Battle and Hastings.

This is a species of open woodlands, sandpits, and heathlands with scattered scrub or trees.

Out in the field and under the microscope

This is not an easy bee to find. It is very unobtrusive and has rarely been encountered at a nest site, so most records are of a bee on a flower head or at rest on foliage. At different times during the 1970s it was swept from plants such as Hogweed, Wild Angelica and Cat's-ear. Areas of bare ground in semi-shade should also be checked as the bees could also be active here.

Males have a very distinctive white-marked clypeus which is more extensive than in other similarly marked species. The females are trickier to identify and could easily be confused with other small black *Andrena* such as *A. nitidiuscula* and *A. semilaevis*. A key

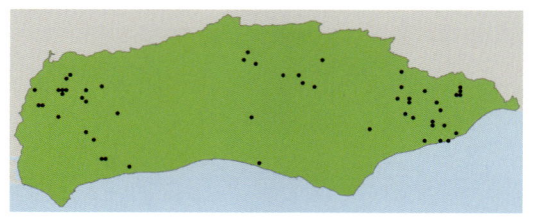

Andrena coitana ♀ [PS]

Large map: distribution 2000–2023
Small map: distribution 1844–1999

Andrena coitana ♂ [PS]

Andrena coitana ♂, showing white clypeus [PS]

difference to look for is the colour of the pollen-collecting hairs on the hind legs, which are pale brown in this species and white in other small black *Andrena*.

Behaviour and interactions

A. coitana nests in areas of semi-shade under trees, usually on well-drained soil such as sand or gravel. One British nest has been described as having a single burrow which was sited in a shady area of woodland. The spoil from this nest formed a pipe that protruded through a covering of moss.

In 2019, Ian Beavis observed males in Fore Wood "coursing along sunlit path-side foliage in company with *A. fucata* and *A. helvola*".

A. coitana brood cells are targeted by the very scarce *Nomada obtusifrons* and probably by *N. roberjeotiana* (a species not known from Sussex).

On the wing

Sussex records run from late June until the end of July. Elsewhere in Britain, it is reported to be on the wing between early June and the end of August.

Feeding and foraging

Pollen is collected from a wide range of plants, including bramble, centaury, hawk's-beard, hawkbit, Harebell, Hogweed, knapweed and thistle. It has also been recorded visiting angelica, Cat's-ear, evening-primrose, Lesser Spearwort, Lesser Stitchwort, mallow, Nipplewort, ragwort, Sheep's-bit and Tormentil, but it is not known if these visits were for pollen and/or nectar.

Andrena confinis Long-fringed Mining Bee
FIRST RECORD 1974, Rewell Wood (Mike Edwards) TOTAL RECORDS 74

Geography and history
Andrena confinis was known as *A. congruens* until 2022. It was first recorded in the county in 1974 from Rewell Wood, and from then until the late 1990s it was regularly recorded in west Sussex, only to become very scarce in recent years. It has been recorded just five times since the year 2000, the last occasion in 2016 when it was recorded from Stedham Common. This pattern of abundance followed by scarcity appears to be a feature of this species.

Most Sussex records are from sandpits and heathlands on the Lower Greensand, with a site near Slindon generating a high proportion of the county's records. Iping Common has also generated a good number of records, although the last of these was in 1990.

There have also been occasional records from sites on the South Downs, including chalk grasslands such as Levin Down, woodlands on the downs near Cocking, and one striking record by Michael Archer from Birling Gap where he found *A. confinis* in 2012. This is the most easterly record in Britain and is some 30 miles from the next nearest known location.

Andrena confinis ♀ [SF]

Out in the field and under the microscope
This is quite a distinctive bee, although the females could be confused with the similar *A. dorsata*. Like females of this last species, fresh *A. confinis* have bright red hairs on the top of the thorax and a

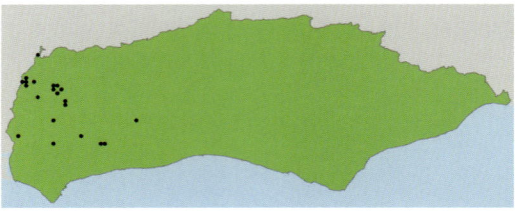

Large map: distribution 2000–2023
Small map: distribution 1844–1999

232 *The Bees of Sussex*

shiny black abdomen with fine white hair bands along the back edges of the tergites. Key differences with *A. dorsata* are the shape of the hind tibiae and their relatively long pollen-collecting hairs. The hind tibiae are slim in *A. confinis*, while in *A. dorsata* they become very broad at the tip.

Males are more difficult and are most likely to be confused with *A. bicolor*. In *A. confinis* antennal segment three is shorter than antennal segment four, while in *A. bicolor* segment three is clearly longer.

A. confinis is often found close to its nest sites, where males and females can be seen flying fast and low just above the ground and occasionally landing on vegetation.

Behaviour and interactions

This *Andrena* has two generations per year and nests in warm, sparsely vegetated banks that are often south-facing. In Germany it is reported that this species nests singly, but in Britain it sometimes establishes very extensive aggregations where the males are often very abundant.

Although no cuckoo bees have been reported for this species in Britain, *Nomada zonata* is given as its cuckoo on the continent. In Britain, this *Nomada* is only known to target the brood cells of *A. dorsata* and was first found in Sussex in 2018.

On the wing

This species has two flight seasons, the first running from mid April until mid May and the second running from the beginning of July until the end of August.

Feeding and foraging

Pollen is collected from several different plant families, including Maple, Carrot, Cabbage, Pea, Beech, Plantain, Buttercup, Rose, Willow and Valerian.

Andrena denticulata Grey-banded Mining Bee
FIRST RECORD 1896, Hastings (Edward Saunders) TOTAL RECORDS 137

Geography and history
Andrena denticulata has been found right across Sussex, with a high proportion of records from the Wealden Greensand and the High Weald.

This species is found in a wide range of habitats which are reflected in the most recent records. These include hay meadows at Marline, chalk heath at Lullington Heath, heathland at Stedham Common, an orchard at Great Dixter, and an area of species-poor grassland near Gatwick Airport. It has also been regularly found in areas of open woodland such as Ebernoe Common and Friston Forest and in species-rich chalk grassland such as Castle Hill, as well as in a garden in Midhurst.

A. denticulata is rarely abundant where it is found.

Lullington Heath [ME]

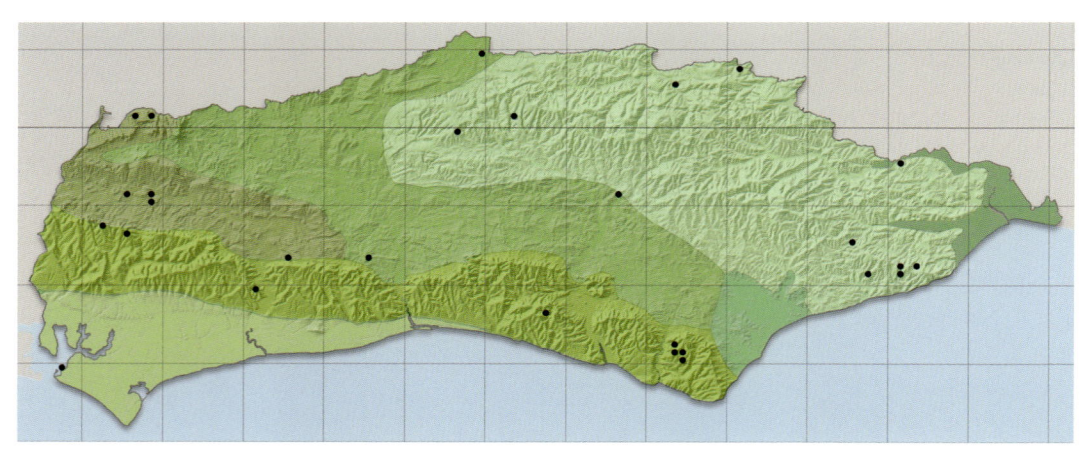

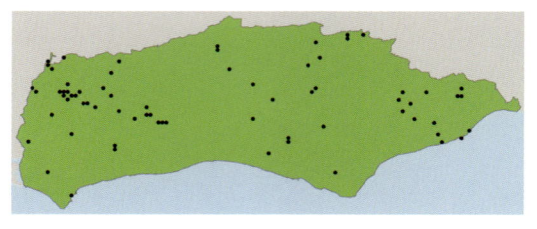

Large map: distribution 2000–2023
Small map: distribution 1844–1999

Andrena denticulata ♀ [NO]

Andrena denticulata ♂ [NO]

Out in the field and under the microscope
At first glance, the females could be confused with the much more abundant *A. flavipes*. However, unlike this last species, the broad, pale hair bands in *A. denticulata* fill the marginal areas of tergites two, three and four, and are preceded by much shorter black hairs. The orange-brown hind tibiae are slightly curved and are exceptionally broad at the tip.

Males are like the females but smaller, and they have a very characteristic ridge at the back of the head which is visible from the side—it gives the head a distinctive appearance. This character is shared with *A. tridentata*, a species that has never been recorded in Sussex and is probably extinct in Britain. The males also have very long curved mandibles.

A. denticulata is often found on flowers, either feeding or gathering pollen. In 1996 David Porter swept a male on ragwort in woodland near Laughton.

Behaviour and interactions
Few *A. denticulata* nests have ever been found, which means that the information available on its nesting behaviour is very limited. Seemingly, it does not form dense aggregations but nests singly or in loose aggregations in areas of sparsely vegetated sandy ground.

Unsealed brood cells are targeted by *Nomada rufipes*.

On the wing
Sussex records run from late June until the end of August.

Feeding and foraging
A. denticulata mainly collects pollen from plants in the Daisy family, especially those with yellow flowers. Pollen sources listed in the literature are Common Ragwort, hawkweed, thistle, Hogweed, knapweed and Shasta Daisy.

Andrena dorsata Short-fringed Mining Bee
FIRST RECORD 1886, Lewes (William Unwin) **TOTAL RECORDS** 749

Geography and history
Andrena dorsata is now one of the most frequently recorded solitary bees in the county, but this reflects a relatively recent change. Until 1972 there were just thirteen Sussex records. This trend is supported by observations by Ian Beavis on both sides of the Kent/Sussex county boundary. He reports finding "0 in 1995, 3 in 1996, 2 in 1997, 1 in 1998, 5 in 1999, and then a huge leap to 34 in 2000".

It is now very widely distributed and occurs throughout Sussex. It can be found in just about any habitat and is often found in gardens.

Out in the field and under the microscope
This bee is usually found in open places, although males can often be seen coursing alongside trees and bushes, especially bramble and gorse. Females are generally found feeding.

Fresh females are very striking. These have reddish-brown hair on the top of the thorax, creamy-white hairs on the sides and at the back of the thorax, and strong white hair bands on the abdomen. The hind tibiae are a distinctive triangular shape and have very short hairs along the hind margin.

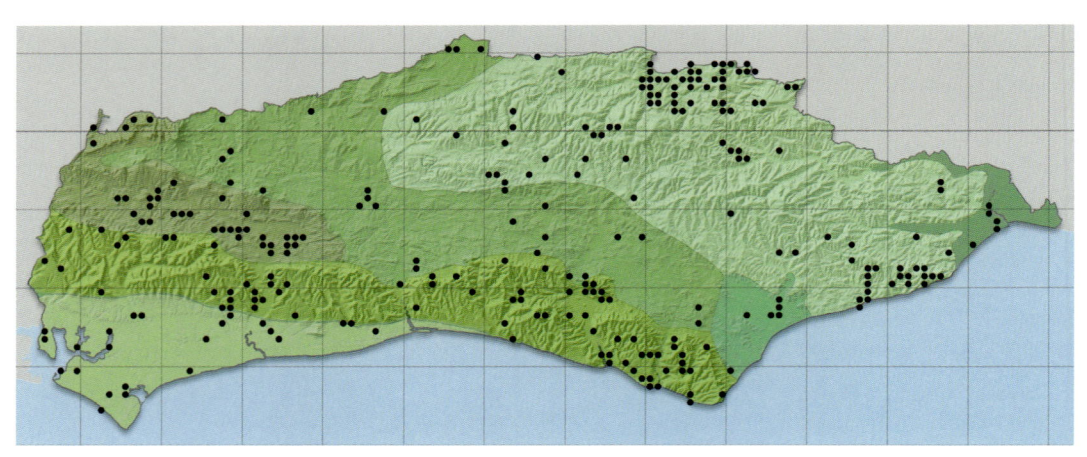

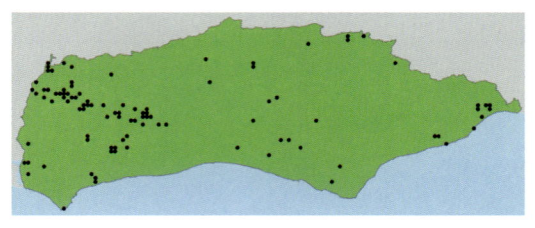

Andrena dorsata ♀ [PB]

Large map: distribution 2000–2023
Small map: distribution 1844–1999

236 *The Bees of Sussex*

Males resemble the females but are much smaller and slimmer. A key character to check is the length of the second tarsal segments on the hind legs—these are elongated in comparison with the otherwise similar *A. russula*, *A. wilkella* and *A. ovatula*.

Behaviour and interactions
Nests are rarely found. Those described tend to be scattered rather than in aggregations, and sited in areas that are either sparsely vegetated or have bare soil. The locations given are south-facing banks and areas of gently sloping or flat ground. Males patrol vigorously, flying along hedges or scrub, often among other species.

A. dorsata brood cells are targeted by *Nomada zonata*.

Andrena dorsata ♂ visiting Blackthorn [NO]

On the wing
This species has two flight seasons, the first running from early March until the end of May and the second running from the beginning of July until late August. There are, however, a handful of Sussex records from September and October (the latest is from 28th October), and these may represent a partial third brood.

Feeding and foraging
In the spring *A. dorsata* will be seen on a range of plants, including Blackthorn, Creeping and Eared Willow, dandelion, Hawthorn, Lesser Celandine and spurge. In the summer, the second generation is seen on plants such as bramble, Burnet-saxifrage, Guelder-rose, Heather, Hogweed, Meadowsweet, mustard, Ox-eye Daisy, Raspberry, Tormentil, White Bryony and Wild Carrot.

Pollen is collected from a wide variety of different plant species.

Andrena falsifica Thick-margined Mini-miner
FIRST RECORD 1974, Ambersham Common (Mike Edwards) **TOTAL RECORDS** 16

Geography and history
Andrena falsifica is a local species that has been recorded just 16 times in the county. Most records are from heathlands and acid grassland, although it has also been found in woodland and from a complex of lowland meadows, pasture and ancient woodland near Hastings.

Modern Sussex records are from Woolbeding Common, Ashdown Forest, Chailey Common, Ditchling Common, Marline and, most recently, Rowland Wood where it was recorded in 2022.

Nationally, this is a very scarce bee that is largely restricted to southern England and to the Welsh coast near Swansea.

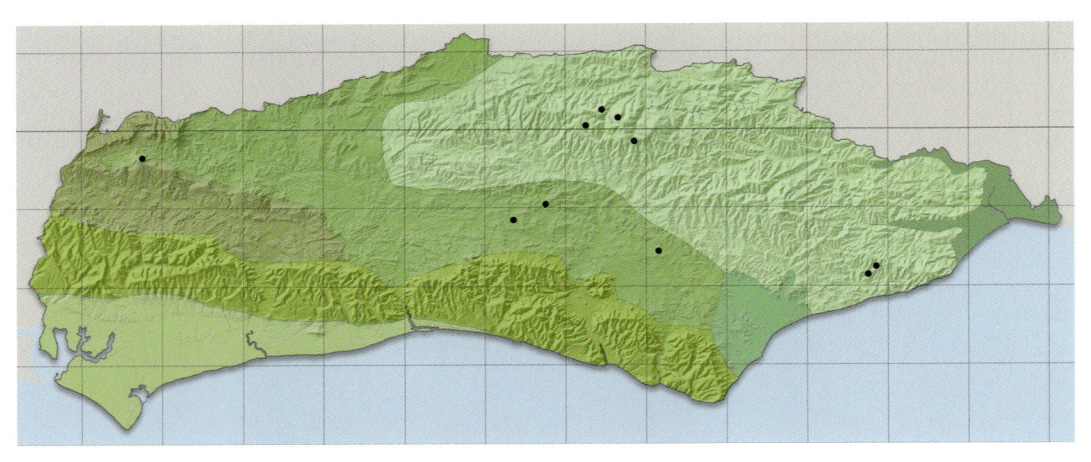

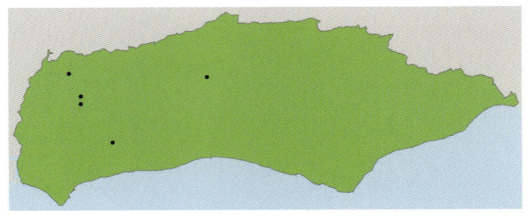

Large map: distribution 2000–2023
Small map: distribution 1844–1999

238 *The Bees of Sussex*

Out in the field and under the microscope
A. falsifica is never abundant and is very easily overlooked, not least as it closely resembles several other small black species of *Andrena*, all of which are particularly challenging to identify.

As well as an all-black body and small white hair bands to the sides of the second tergite, a key character to look for in females is the thickened smooth and shiny area along the hind margin of the first tergite. The thickening creates the effect of a small step down to the second tergite, helping to separate this species from the otherwise similar *A. alfkenella*. The punctures on the top of the scutum are also more widely separated than in this other species.

Males are slimmer and slightly smaller than the females. They also have a thickened smooth and shiny area along the back edge of the first tergite.

Searching known foodplants is likely to be the most successful strategy for finding this bee. The specimen swept at Ditchling Common was feeding on a dandelion flower, while those found in Rowland Wood were foraging from Tormentil on sheltered ride intersections.

Behaviour and interactions
Very little is known about the nesting habits of this species, although it is thought to establish solitary nests in light soils. On the continent, it is understood to nest in sparsely vegetated areas on banks but does not appear to prefer a particular soil type.

Nomada flavoguttata targets this species, laying its eggs within unsealed brood cells.

On the wing
A. falsifica is on the wing from mid March until early July.

Feeding and foraging
A. falsifica has been recorded visiting Tormentil, Daisy, dandelion, Germander Speedwell and Wild Strawberry. It is not known which species are visited for pollen.

Tormentil [PC]

Andrena ferox Oak Mining Bee

FIRST RECORD 1876, Guestling (Frederick Smith) TOTAL RECORDS 7

Geography and history

Andrena ferox is closely associated with mature and open broadleaved woodland that has a significant component of oak in the canopy.

This is an extremely elusive bee. In Sussex, it has only ever been recorded on seven occasions from just five locations, all of which are in east Sussex. The three most recent records are from Hastings in 2010, Flatropers Wood in 2014, and Great Dixter in 2017. It was also recorded in Guestling in 1876 and in Polegate on three separate dates in April 1945. The record from Flatropers Wood is of a female swept by Graeme Lyons within a clearing beneath some powerlines.

Andrena ferox ♀ [PC]

Andrena ferox ♂ [PB]

Large map: distribution 2000–2023
Small map: distribution 1844–1999

240 *The Bees of Sussex*

Outside Sussex, *A. ferox* is now known only from Kent, the New Forest in Hampshire, and near Reigate in Surrey. The record from Flatropers Wood is 12 miles from Puckley, one of the locations in Kent where this bee has been recorded.

Out in the field and under the microscope
There is a significant amount of good fortune involved in finding *A. ferox*. This is because the males and females spend a large amount of time in the canopies of oak trees. Almost as soon as a female emerges from a nest, she is likely to head straight to the tree tops. Several records from the New Forest are, however, of bees feeding closer to ground level on flowering shrubs, and these are always worth checking.

Females have orange-marked tibiae on the hind legs. Males are very slender and often have very large heads. They can also have a sharp spine on each gena, near the base of the mandibles, although this may be reduced or absent, particularly in small specimens.

Behaviour and interactions
A. ferox frequently nests communally, with females sharing an entrance. It does, however, also establish solitary nest burrows, and these can be sited close to a communal nest such that an aggregation might well contain both types of nests. The number of females in an aggregation can potentially run into the hundreds. Nests are made in the ground in areas of rank vegetation or short turf and can be sited deep within old rabbit burrows, mole hills, between the roots of trees, on low banks, or on areas of flat open ground.

Males patrol above the canopies of trees, with mating taking place on leaves several metres above the ground.

Nomada flava, *N. marshamella* and *N. striata* are possible cuckoos of this species, although this has not been confirmed.

Andrena ferox ♀s at a communal nest entrance [PB]

On the wing
Sussex records run from late April until mid June.

Feeding and foraging
While oak flowers are the principal source of pollen, *A. ferox* is thought to collect pollen from other tree species once oak has finished flowering.

A. ferox has also been recorded visiting Blackthorn, Buckthorn, Crab Apple, Field Maple and Hawthorn.

Andrena flavipes Yellow-legged Mining Bee
FIRST RECORD 1879, Hollington (Edward Saunders) TOTAL RECORDS 1,460

Geography and history
Andrena flavipes has increased in abundance in recent years and is now the most frequently recorded solitary bee in the county. It is ubiquitous, occurring in just about every area. It is found on chalk grasslands, meadows and pasture, coastal habitats, heathlands and open woodlands, as well as in gardens and parks.

Out in the field and under the microscope
The females of this species have a broad abdomen, obvious buff hair bands on the tergites, black hairs at the tip of the abdomen, and pale orange pollen-collecting hairs on the hind tibiae. There is also a characteristic indentation on the underside of the hind femur which helps to separate *A. flavipes* from

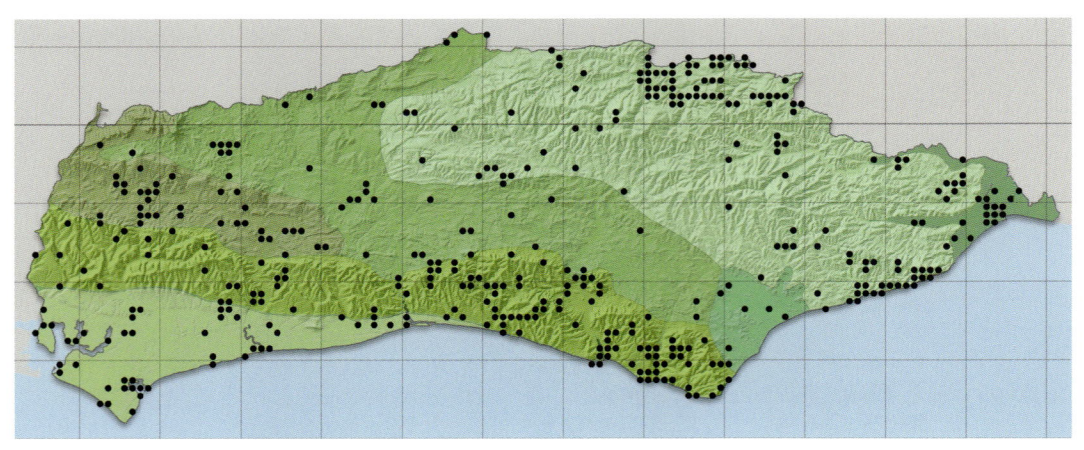

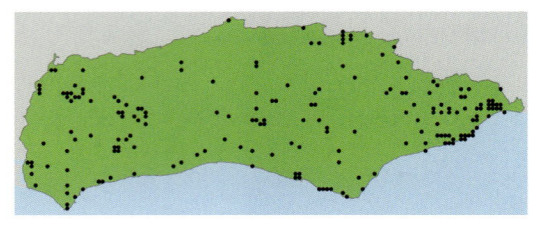

Large map: distribution 2000–2023
Small map: distribution 1844–1999

242 *The Bees of Sussex*

the otherwise very similar but much scarcer *A. gravida* with which it sometimes flies. *A. gravida* also has almost pure white hair bands on the abdomen, in contrast to the buff-coloured hair bands in fresh *A. flavipes*.

Males are like the females but smaller and slimmer, and they lack the orange hairs on the hind tibiae. Each side of the genital capsule has a characteristic notch, a feature that is not shared with any other *Andrena*. Buff-coloured hairs on the face also help to separate male *A. flavipes* from male *A. gravida*, which, like the females, have white hairs on the face.

Males are often found flying rapidly along hedgerows or just above the ground close to a nest site, while females are often found foraging or flying back and forth to their nests.

Behaviour and interactions

A. flavipes often establishes dense aggregations that can cover a wide area, especially on south-facing slopes and banks. Edward Saunders, writing in 1879, reported finding a large colony sited "in a bank by the side of the road near Hollington". The larger aggregations can contain thousands of individual bees, with hundreds of nests sited close together. The brood cells can be as much as 23 cm below ground.

Males often patrol rapidly just above the nest site, zigzagging rapidly as they search for a female before pouncing to mate. Both sexes mate several times during the flight season.

A. flavipes brood cells are targeted by *Nomada fucata*.

This is one of a small number of species targeted by the bee-fly *Bombylius discolor*. Female flies mix their eggs with dust that has been gathered into a special chamber, coating the eggs with the dust before flicking them onto an area of ground being used by nesting bees. On hatching from an egg, a fly larva will attempt to locate an open cell and wait until the bee larva is almost fully developed. At this point the developing fly latches onto its host to feed on its fluids before pupating and emerging as an adult fly.

Andrena flavipes ♀ at a nest entrance [PC]

Bombylius discolor ♀ [PC]

On the wing

This species has two flight seasons, the first running from late February until the end of May and the second running from late June until the beginning of September.

Feeding and foraging

Each generation visits an enormous range of plants, both for pollen to provision brood cells and to feed. The spring generation, for example, has been recorded on spring-flowering trees and shrubs such as Blackthorn, Hawthorn and willow, and on flowering plants such as buttercup, dandelion, Daisy and Hogweed. In the summer, *A. flavipes* might be seen visiting bramble, buttercup, Common Bird's-foot-trefoil, Red and White Clover, and ragwort.

Andrena florea Bryony Mining Bee
FIRST RECORD 1896, Bexhill (Frisby) TOTAL RECORDS 150

Geography and history
Andrena florea is restricted to places where White Bryony grows, which means that it is absent from the northernmost and easternmost areas of Sussex. This plant is its only pollen source in Britain and is restricted to light, base-rich soils, which are mostly found on the chalk of the South Downs and on the sandy soils of the Lower Greensand. It is a plant of hedgerows and scrubby, disturbed ground.

Aside from a solitary record from Bexhill in 1896, *A. florea* was known only from the Lower Greensand until 2007. This is when it was first recorded on the South Downs at Fairmile Bottom, since when it has expanded its range across both this landscape and the Coastal Plain. It was also

Andrena florea ♀ on White Bryony at Seaford Head [PC]

Pollen from White Bryony showing droplets of liquid on the cell floor [PW]

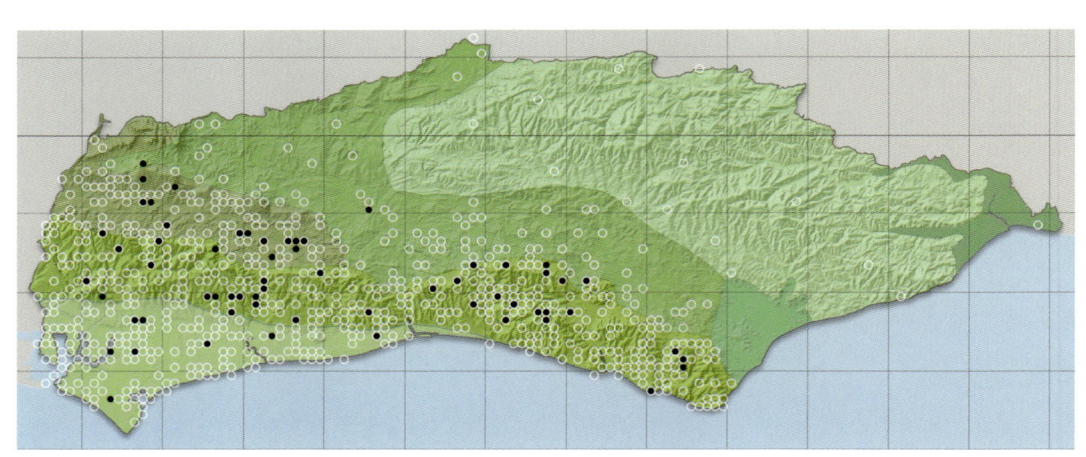

Large map: distribution 2000–2023, with white circles showing distribution of White Bryony
Small map: distribution 1844–1999

recorded from Knepp in 2019. Recent records are from Seaford Head, Wolstonbury Hill, Blackcap and Kingley Vale.

In Britain, *A. florea* is known only from an area bounded by Hampshire, Suffolk and Kent. Within this area, it is very local but can be abundant at nest sites or close to patches of White Bryony.

Andrena florea ♂ [SF]

Out in the field and under the microscope
The best way to find this bee is to look for White Bryony scrambling through scrub. Here, females are often seen foraging, while males are seen patrolling and looking for an opportunity to mate.

It is a distinctive bee, and both males and females have a very shiny abdomen. This usually has reddish-brown bands across the first two tergites, although the amount of reddish brown varies.

While this species has similarities to *A. bucephala*, *A. trimmerana* and *A. rosae*, the close association with White Bryony should help to identify this species.

Behaviour and interactions
A. florea is one of the few British species that collects pollen from just one species of plant—in this case, White Bryony. Elsewhere in Europe, however, it collects pollen from other closely related plant species.

Nests can be established in aggregations of up to 100, with nests sited either close together or scattered. Locations selected for nesting are generally areas of bare or sparsely vegetated ground and are often alongside paths where the ground has been compacted. The brood cells are excavated 5 to 10 cm below the surface, and droplets, possibly of nectar, are deposited against the walls of the cells.

No species of cuckoo bee has been observed targeting this species.

On the wing
Sussex records run from mid May until the beginning of August.

Feeding and foraging
A. florea forages for pollen exclusively from male White Bryony flowers, with nectar taken from both male and female White Bryony flowers. Other plant species are also visited for nectar.

There are also records of this species visiting bramble, hawkweed, hawthorn, Hound's-tongue, Raspberry, Viper's-bugloss and water-dropwort.

Andrena florea ♀ **taking nectar from Hound's-tongue at Seaford Head** [PC]

Andrena fucata Painted Mining Bee
FIRST RECORD 1893, Slindon (Henry Guermonprez) TOTAL RECORDS 168

Geography and history
Andrena fucata is widely distributed across Sussex but is rarely abundant where it occurs. It is particularly associated with open woodlands, but also occurs in scrubby areas on heathlands and along the coast. Eridge Rocks and High and Over have both generated recent records.

Out in the field and under the microscope
This *Andrena* is tricky to identify in the field as it does not have many stand-out characters to look for. Females have brown hairs on the top of the thorax and white hairs on the sides. The abdomen is

Andrena fucata ♀ [SF]

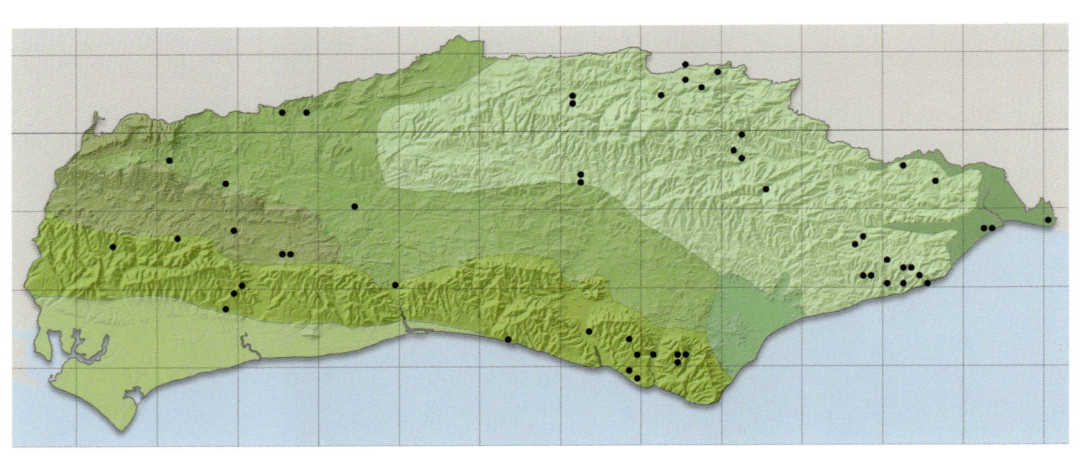

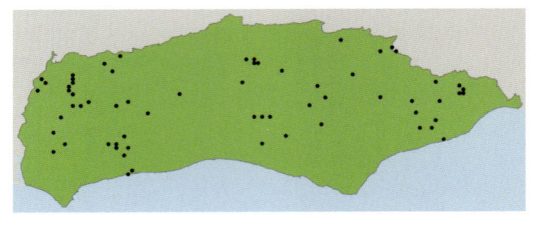

Large map: distribution 2000–2023
Small map: distribution 1844–1999

dark and has no hair bands but does have a covering of hairs. The hairs on the first two tergites are much longer than those over the rest of the abdomen. The pollen-collecting hairs on the hind legs are pale.

The males are dark and slender, and have long curved mandibles with a triangular projection on the underside near the base. The small size of this projection helps to separate *A. fucata* males from all males that share this character except *A. helvola*. This last species has dense bands of pale hairs on the sternites, in contrast to *A. fucata* which has sparse bands of hairs here.

Andrena fucata ♂ [NO]

As well as coming across males and females feeding or foraging—with Wood Spurge flowers particularly worth checking—it is possible to see males zigzagging up tree trunks. Ian Beavis has observed groups of males "spiralling up light-coloured tree trunks in company with *A. helvola*, and coursing over and settling on the sunlit foliage of trees and shrubs beside woodland paths and along woodland edges".

Behaviour and interactions

A. fucata has a relatively long flight season, but most records are from May and June. It usually nests solitarily, although it will sometimes form small aggregations. There is little other information on the nesting habits of this species in Britain.

Nomada panzeri is thought to be a probable parasite of this species here in Britain, while in Scandinavia *N. fusca* also targets this *Andrena*. There is an unconfirmed record for *N. fusca* from Kent in 2018, but this species is otherwise currently unknown from Britain.

On the wing

Sussex records run from the end of March until the beginning of August.

Feeding and foraging

Pollen is collected from a wide range of plants, including buttercup, Common Rock-rose, Creeping-Jenny, hawthorn, mustard, Raspberry, Rough Chervil, Sheep's Sorrel, speedwell, Hemlock Water-dropwort, Winter-cress and Wood Spurge.

Andrena fucata ♀ visiting Raspberry [NO]

Wood Spurge [PC]

Andrena fulva Tawny Mining Bee
FIRST RECORD 1882, Hastings (Edward Saunders) **TOTAL RECORDS** 466

Geography and history
Andrena fulva can be found just about anywhere in the county, taking advantage of open places such as parks and gardens as well as flower-rich grasslands and open woodlands. It can be very abundant.

Out in the field and under the microscope
This is one of the species that heralds the arrival of spring each year. Females are unmistakable—they can readily be recognised in the field as they are unlike any other species. They are about the size of an *Apis mellifera* worker and have a dense covering of red hairs on the top of the thorax and abdomen. In fresh specimens, there are two shades of red—on the thorax the hairs are scarlet, while on the abdomen the hairs are more of an orange-red colour. The head and the sides of the thorax, as well as the underside of the abdomen, are covered in black hairs. The pollen-collecting hairs are also black.

Unlike females, males cannot be identified in the field with confidence. They are much slimmer than the females and have a sparser covering of hairs on the abdomen. The lower face is covered in long white hairs. They are also

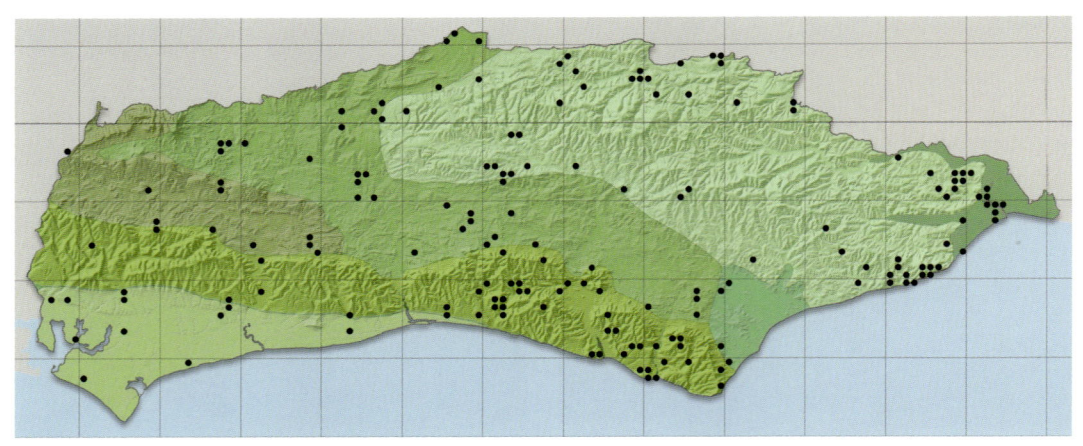

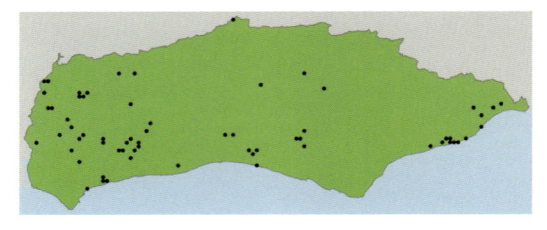

Andrena fulva ♀ visiting Blackthorn [PC]

Large map: distribution 2000–2023
Small map: distribution 1844–1999

248 *The Bees of Sussex*

smaller and less strongly coloured than the females, and, in common with a handful of other species, have long mandibles, each of which has a projection that protrudes downwards from near the base. The key character to look for is the length of the third antennal segment. In A. fulva, this is twice the length of the fourth segment but, in the other species with a projection here, it is roughly 1.5 times the length of this segment

This *Andrena* is best looked for by searching open grassy areas. Most records are from late March and April.

Andrena fulva ♂ [PC]

Andrena fulva ♂ showing the long mandibles and white hairs on the lower face [PC]

Behaviour and interactions
A. fulva often nests in small, loose groups but will also form dense, extensive aggregations. There is one set of observations from Wiltshire by Stuart Roberts who reported 800 nests across an area covering a hectare.

Females excavate their burrows up to 55 cm below ground and, in the process, create a small tumulus of loose soil at the entrance. Favourable locations can often be used for several consecutive years.

Nests are generally sited in areas with level open ground and with a covering of short, sparse vegetation and patches of bare ground. As well as lawns, locations used include flowerbeds, south-facing slopes in grassland, the edges of tracks, arable fields and parks.

Each female might create two or three nests, with each containing four or five brood cells. Males patrol above the nest site, seizing any opportunity to dive onto a female to mate.

In one Welsh study, "18 per cent of *A. fulva* offspring were replaced by a *Nomada panzeri* offspring". *A. fulva* is also the only host of another *Nomada*, the scarce *N. signata*.

It is possible that the form of *N. panzeri* associated with *A. fulva* may prove to be a distinct species.

On the wing
Sussex records run from late February until late July.

Feeding and foraging
Pollen is collected from plants such as beech, Blackthorn, buttercup, campion, Field Maple, Garlic Mustard, Gooseberry, Hawthorn, Holly, oak, Sycamore, Wayfaring-tree, Wild Cherry and willow. *A. fulva* is an important pollinator of fruit trees in the spring.

Andrena fulvago Hawk's-beard Mining Bee

FIRST RECORD 1879, Hastings (Edward Saunders) TOTAL RECORDS 54

Geography and history

In Sussex, *Andrena fulvago* is largely restricted to flower-rich chalk grasslands at the eastern end of the South Downs—most records are from sites between Castle Hill and Beachy Head. There have been no records from the westernmost section of the South Downs since the 1980s.

Since 2012, it has also been found at a handful of other locations, including Lynchmere and Chichester, meadows and pasture near Warbleton, and a small flower-rich pasture at Sheffield Park.

Out in the field and under the microscope

Edward Saunders' 1879 specimen was collected from a hawkweed flower, and checking foodplants is still the best way to find this bee. As many of its foodplants close up by the early afternoon, *A. fulvago* is generally active in the mornings and is more difficult to locate in the afternoons.

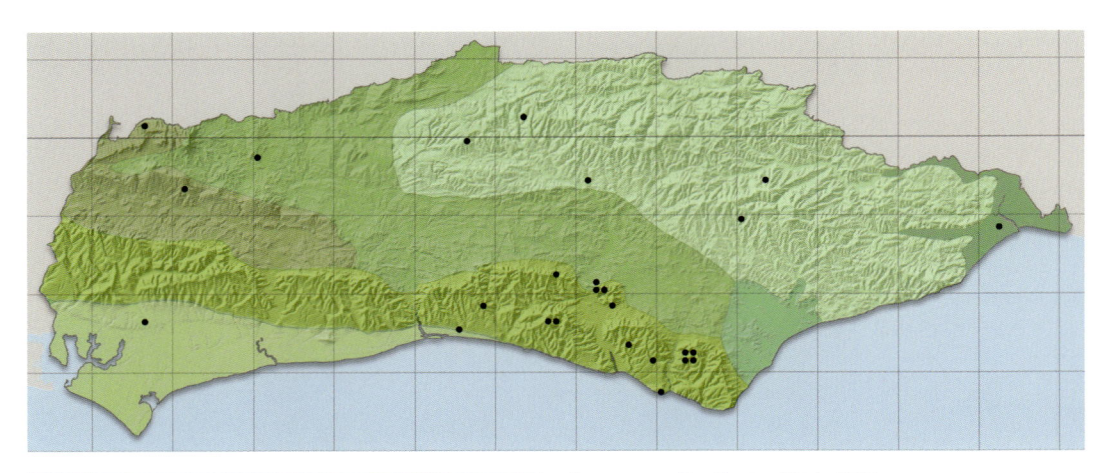

Andrena fulvago ♀ [PB]

Andrena fulvago ♂ [TF]

Large map: distribution 2000–2023
Small map: distribution 1844–1999

250 *The Bees of Sussex*

Females have a slender, shiny abdomen that has weak hair bands and reddish-orange hairs at the tip. They also have bright orange-yellow hind tibiae and bright orange-yellow pollen-collecting hairs. These hairs are often full of yellow pollen.

Males are smaller and slimmer than the females and also have orange-yellow hind tibiae.

Although *A. fulvago* is rarely found in numbers, it can be locally abundant in favourable conditions. In 1981, George Else and Mike Edwards report that "it was unusually abundant on Kingley Vale NNR, West Sussex, where hundreds of individuals were found visiting yellow Asteraceae flowers in areas of open grassland". Although not as abundant, it was also present in good numbers at Malling Down in 2015, again in an area with yellow-flowered plants in the Daisy family.

Behaviour and interactions

A. fulvago generally establishes nests individually or in small aggregations. Little is known of its nesting behaviour in Britain, although reports from the continent suggest it uses sparsely vegetated flat or gently sloping areas.

Until 2017, it was thought that there were no species of cuckoo bee associated with this *Andrena* in Britain. However, *Nomada facilis*—which is known to target *A. fulvago* on the continent—has subsequently been shown to have been present in Britain since at least 1802. This cuckoo bee has not been recorded in Sussex.

On the wing

Sussex records run from early May until the end of July.

Feeding and foraging

This *Andrena* specialises in collecting pollen from yellow flowers in the Daisy family, including Cat's-ear, dandelion, hawk's-beard and Mouse-ear-hawkweed.

Beaked Hawk's-beard [PC]

Mouse-ear-hawkweed [PC]

Andrena fuscipes Heather Mining Bee
FIRST RECORD 1896, Hastings (Edward Saunders) **TOTAL RECORDS** 206

Geography and history
Andrena fuscipes is restricted to areas with Heather, Bell Heather and Cross-leaved Heath, but it is not as widely distributed as these plant species. Its principal populations are on the heathlands of the Lower Greensand and the High Weald. Sites that have generated recent records include Iping Common, Ashdown Forest and Broadwater Warren. Away from these areas, it is also known from locations such as Selwyns Wood, Lullington Heath and Chailey Common.

Broadwater Warren [IB]

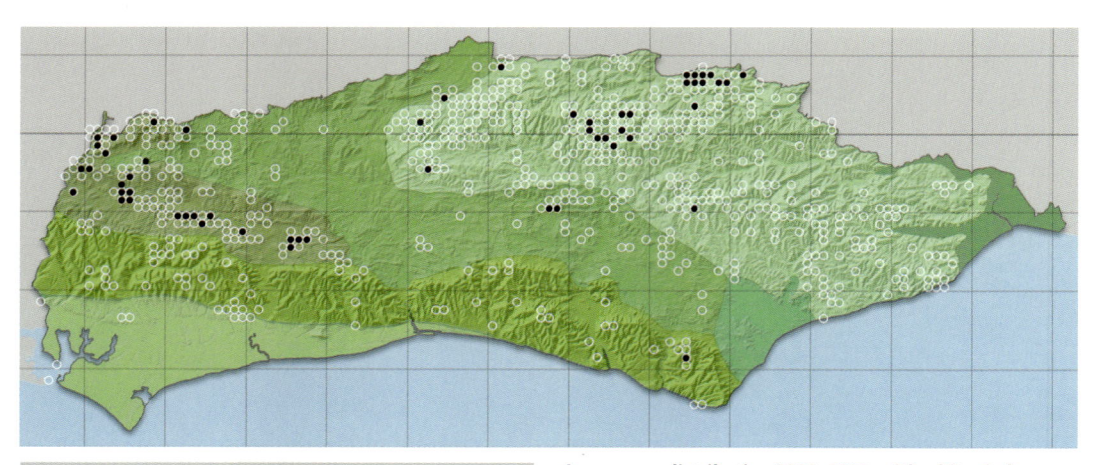

Large map: distribution 2000–2023, with white circles showing distribution of Bell Heather, Cross-leaved Heath and Heather
Small map: distribution 1844–1999

252 *The Bees of Sussex*

Out in the field and under the microscope
This species of *Andrena* prefers areas with exposed sand and Heather, and therefore any such areas are worth investigating. Its parasite *Nomada rufipes* can be at least as abundant as its host and is often the first sign that *A. fuscipes* is present.

Males might be seen zigzagging rapidly back and forth above flowering Heather, while females might be seen collecting pollen from these same flowers.

A female *A. fuscipes* is a dark bee with dense, pale hair bands on the abdomen and slightly darker brown hairs on the thorax. The surfaces of the tergites are covered with pale brown hairs.

Males are smaller than the females and can appear to be a silvery colour. Both males and females have very polished galea, a characteristic that is not shared with any other species of *Andrena*.

Most records are from August, coinciding with Heather coming into flower.

Andrena fuscipes ♀ [PC]

Andrena fuscipes ♂ [PB]

Behaviour and interactions
A. fuscipes usually establishes solitary nests in among Heather stems in sandy ground, but very occasionally it will form large aggregations.

Brood cells of this species are attacked by *N. rufipes*.

On the wing
Sussex records run from the end of July until early September. The vast majority of records are from August.

Feeding and foraging
Pollen is almost exclusively collected from plants in the Heather family, although *A. fuscipes* does also gather pollen from other plants once these have finished flowering, including yellow-flowered plants in the Daisy family.

There are also records of this species visiting Common Bird's-foot-trefoil, willowherb, Sheep's-bit and thistle.

Andrena gravida White-bellied Mining Bee
FIRST RECORD 1881, Hastings (F. Collett) TOTAL RECORDS 31

Geography and history
Aside from early records from Hampshire (where it was last recorded in 1849) and Essex (where it was last recorded in 1850), this species has historically been restricted to Kent and east Sussex. It now appears to be increasing in abundance, but it is still very local.

The earliest Sussex record is from 1881 when Edward Saunders was able to identify a specimen in a collection belonging to Mr F. Collett. Seven years later, males were described by William Bennett, a local butcher and entomologist, as "extremely common on the hills immediately surrounding Hastings". It is likely that six undated specimens of *Andrena gravida* in the Booth Museum, under the name Bennett and collected in Hastings, relate to this observation.

After a flurry of records at the turn of the century, there was then just one record between 1924 and 1991. This is when *A. gravida* was found near Udimore by Alfred Jones. Since then, it has been recorded with increasing frequency, including at Friston Forest, Great Dixter and south of Tunbridge Wells, all since 2015. These records are from locations that include an orchard, a lawn adjoining a flowerbed and allotment, and an area of Blackthorn scrub. It has also been found crawling along a pavement in Rye.

The most recent records—from 2021, 2022 and 2023—are from the downs near Iford, Rye Harbour Nature Reserve, Rye, and lastly the cemetery in Lewes.

Out in the field and under the microscope
A. gravida is similar to the very abundant *A. flavipes* with which it sometimes flies. However, *A. gravida* is a little larger and has white hair bands on the abdomen as well as snowy white hairs on the face and on the sides of the thorax. This colouring is in contrast to the buff-coloured hairs found on *A. flavipes*. Females of this last species also have an indentation on the underside of the hind femur, a character that is missing in *A. gravida*.

Males are very like the females but smaller and slimmer.

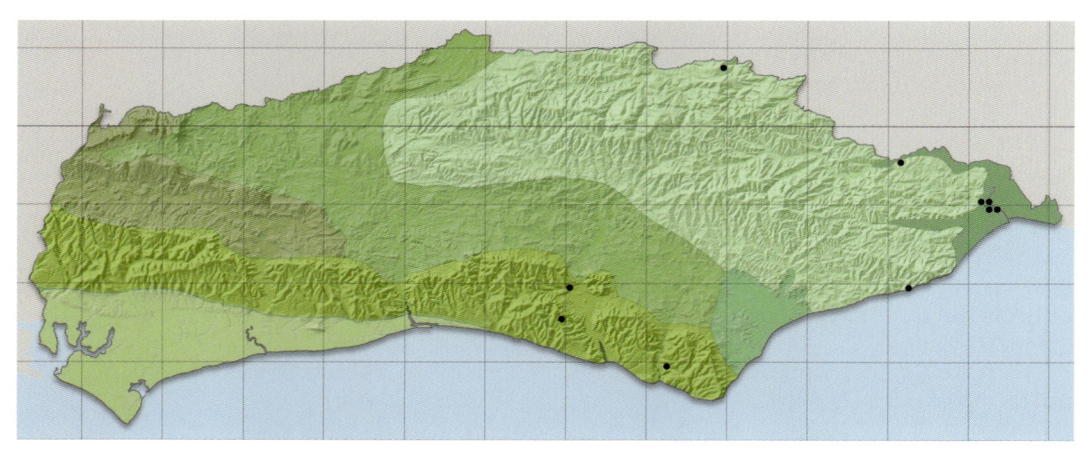

Large map: distribution 2000–2023
Small map: distribution 1844–1999

Andrena gravida ♀ [CG]

Andrena gravida ♂ [SF]

Behaviour and interactions
In 2021 Chris Bentley discovered a nest site in Rye, and this was still present in 2022. This was sited below the edge of a sunny, sparsely vegetated and tightly mown lawn adjoining a flowerbed and an allotment with fruit trees. On the first visit, males were flying rapidly above the flowerbed and fast and low above the sward, with *A. flavipes* males seen flying in amongst them. There were some 40 *A. gravida* males altogether and just a handful of females. A male was observed pouncing on a female, with mating taking place on the ground.

A nest site was also found in Lewes Cemetery in 2022 and was still active in 2023. This is sited within an area of closely mown grassland, with *A. flavipes* and *A. dorsata* nesting in the same area. Three nests were sited within 0.5 m of each other with one of the nest entrances blocked with loose soil as the female left.

Nomada bifasciata lays its eggs in the brood cells of this species. This bee was first recorded in Britain in 2018 in Kent but has not yet been recorded in Sussex—even though it can now be found just over the county boundary in Tunbridge Wells.

Nomada bifasciata ♀ [SF]

On the wing
Sussex records run from mid March until mid May.

Feeding and foraging
A. gravida collects pollen from a wide range of plants, including plants in the Daisy, Willow, Cabbage and Rose families. Individual bees in Lewes Cemetery were observed foraging for pollen from dandelion flowers located about 20 m from the nest site.

There are also records of this species visiting apple, cherry, stitchwort and White Clover.

Andrena haemorrhoa Orange-tailed Mining Bee

FIRST RECORD 1858, Firle, Lewes and Ringmer (William Unwin) TOTAL RECORDS 977

Geography and history

The very first reference to *Andrena haemorrhoa* in Sussex was made by William Unwin in 1858. He described finding it near Lewes, Firle and Ringmer, adding that it was "abundant".

It is still one of the commonest spring *Andrena* across the county and is found in all sorts of habitats. These include various types of grassland, open woodland and heathland. It is also known from parks and gardens. It has been most frequently recorded on the South Downs in areas of chalk grassland with scrub, and on the coast between Hastings and Camber Sands.

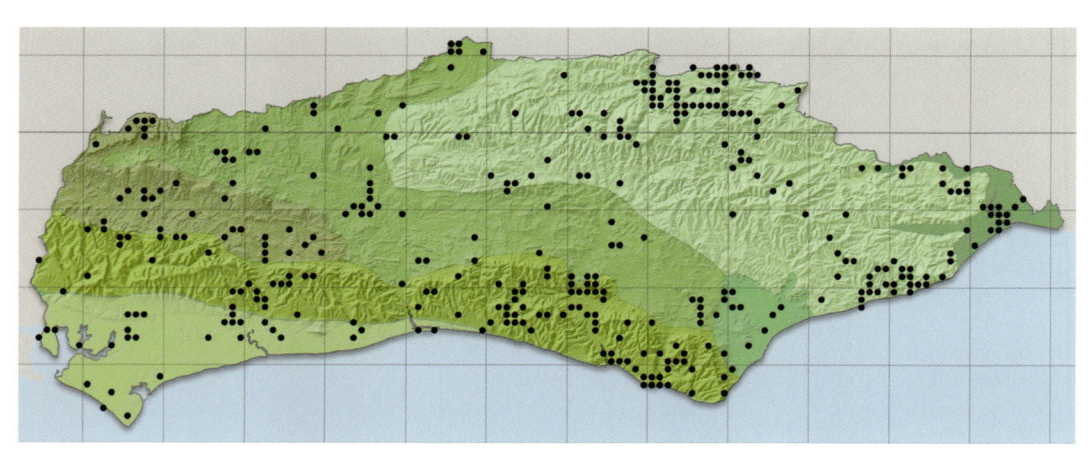

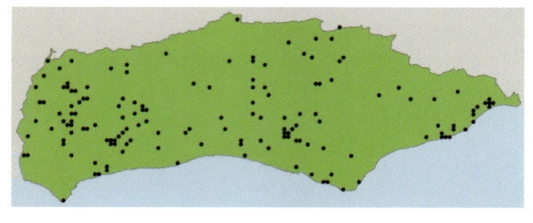

Large map: distribution 2000–2023
Small map: distribution 1844–1999

256 *The Bees of Sussex*

Out in the field and under the microscope

Hawthorn bushes are a good place to search for this bee, and here it is possible to find females foraging on the flowers as well as males feeding. Females can also often be seen foraging from willow catkins and from fruit trees. In addition, Steven Falk reports seeing worn females clustered on Wild Mignonette on the South Downs in June.

Females are quite distinctive, and with practice can be identified in the field. These have dark, fox-red hairs on the top of the thorax and white hairs on the sides. They also have a shiny, virtually hairless abdomen that has reddish-orange hairs at the tip, as well as pale orange pollen-collecting hairs on the hind legs.

Males are trickier to identify, especially once their colour fades. Fresh examples resemble the females but are smaller and slimmer and have long antennae. Like the females, the males have a small, upstanding ridge in the centre of the propodeum, a feature shared with just three other species found in Sussex—*A. pilipes*, *A. tibialis* and *A. bimaculata*. *A. haemorrhoa* can be identified by the pale hairs on the clypeus (these other species all have black hairs here).

A. haemorrhoa has a long flight season, although most observations are from April, May and early June.

Andrena haemorrhoa mating pair, ♀ on the left and ♂ on the right [PC]

Behaviour and interactions

There is little information about the nesting behaviour of *A. haemorrhoa*. Nests generally seem to be scattered rather than in aggregations, a view supported by an observation made by Peter Greenhalf in 2022 in Rye. Here, a solitary nest was located at the base of the edge of a lawn. Small aggregations have, however, been reported from the continent. It overwinters as an adult in the brood cell.

Brood cells are targeted by *Nomada ruficornis*.

Andrena haemorrhoa nest [ME]

On the wing

Sussex records run from early March until late July.

Feeding and foraging

A. haemorrhoa collects pollen from a very wide range of plants in a variety of families. Among these are species of apple, buttercup, cherry, daisy, dandelion, dogwood, maple, stitchwort, hawthorn and willow.

The Bees of Sussex

Andrena hattorfiana Large Scabious Mining Bee
FIRST RECORD 1904, Slindon (Henry Guermonprez) **TOTAL RECORDS** 49

Geography and history
Andrena hattorfiana is restricted to southern England and south Wales, its range currently extending as far north as Norfolk and as far west as Pembrokeshire. Within this area, its distribution is closely linked to that of Field Scabious, but also Small Scabious and Greater Knapweed. In Sussex, Field Scabious is largely confined to the South Downs where it occurs on areas of species-rich chalk grassland, including some road verges.

Within the county, most records for *A. hattorfiana* are from the eastern South Downs. It has been recorded from near Litlington, from a bank dominated by Field Scabious adjacent to the car park at Lewes Golf Course, and from chalk grasslands within Brighton. There is also a handful of records from other locations away from the downs, including a 2020 record from a garden in Midhurst where Field Scabious had been grown specifically to attract this bee.

Out in the field and under the microscope
On the eastern South Downs, the most reliable place to find *A. hattorfiana* is Castle Hill where ten females were counted on a single visit in

Field Scabious, Castle Hill [PG]

Large map: distribution 2000–2023, with white circles showing distribution of Field Scabious
Small map: distribution 1844–1999

258 *The Bees of Sussex*

2023. However, *A. hattorfiana* could be found anywhere where Field Scabious is present within this landscape, so it is always worth checking the flower heads. It is also worth searching Greater Knapweed flowers—in 1943 Alfred Brazenor swept 13 specimens at Ovingdean, including one that was "on Greater Knapweed".

The females can be readily identified in the field and are slender bees that come in two colour forms. In Sussex, most are largely black and have a shiny, slightly elongated abdomen that has weak, white hair bands and red hairs at the tip. A small proportion, however, are red-marked on the front few tergites.

Andrena hattorfiana ♀ showing black form [PC]

Andrena hattorfiana ♀ showing red-marked form [PC]

Andrena hattorfiana ♂ [SF]

Males are typically black and, in common with males of a very few species, have a white-marked clypeus.

Both males and females could potentially be confused with *A. marginata*, which also forages from scabious flowers and has similar markings. *A. marginata*, however, is a much smaller bee.

Any female that is fully laden with Field Scabious pollen is a very striking insect. This is because this plant produces pollen that is a vivid pink, and fully laden females have fluorescent pink pollen grains on their hind legs and at the back of the thorax.

Behaviour and interactions

A. hattorfiana nests either solitarily or in small aggregations, each nest containing up to six brood cells. The hairs on the hind legs are specially adapted to collect the relatively large scabious pollen grains.

Brood cells are targeted by the very scarce cuckoo bee *Nomada armata*, a species that is not known from Sussex.

On the wing

Sussex records run from early May until the end of July.

Feeding and foraging

A. hattorfiana is strongly associated with a small number of plant species in the Teasel family, especially Field Scabious but also Small Scabious. *A. hattorfiana* is, however, seemingly a poor pollinator of Field Scabious.

In addition to collecting pollen from these plants, *A. hattorfiana* will also collect pollen from Greater Knapweed, and some Dorset populations are entirely dependent on this plant alone.

Andrena helvola Coppice Mining Bee
FIRST RECORD 1902, Linch Down (Henry Guermonprez) **TOTAL RECORDS** 116

Geography and history
Andrena helvola is essentially a bee of open, broadleaved woodland, where regular management creates opportunities for plants such as Wood Spurge to flower and set seed. The bee particularly thrives in coppiced woodland. Places with recent records include Ashcombe Bottom, Friston Forest, Ebernoe Common, Rewell Wood, West Dean Woods, Fore Wood, and woodland at Great Dixter.

There are also modern records from chalk grasslands on the downs such as Cradle Hill, Mount Caburn and Wolstonbury Hill, while elsewhere in Sussex there are also recent records from Chailey Common, Knepp and Hastings.

Fore Wood [IB]

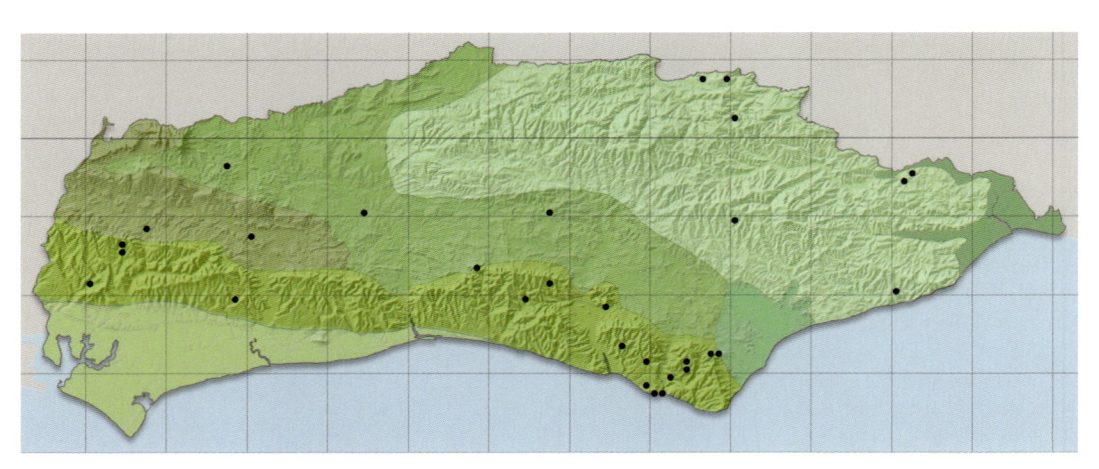

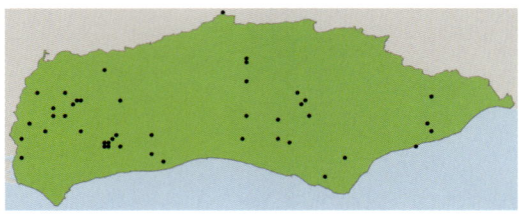

Large map: distribution 2000–2023
Small map: distribution 1844–1999

260 *The Bees of Sussex*

Andrena helvola ♀ visiting Wood Spurge [PB]

Andrena helvola ♂ visiting Wood Spurge [PB]

Out in the field and under the microscope
While this *Andrena* is frequently found on Wood Spurge, it is often also seen flying close to trees and shrubs. It is generally found in small numbers, although exceptionally it can be more abundant. In 1938 Kenneth Guichard visited a deserted cottage garden near Worthing and describes how "isolated bushes of Hawthorn were swarming with common *Andrenas*, amongst which were a few ♀s and many ♂s of *Andrena helvola*".

More recently, in Kent, Ian Beavis has found females settled on Hawthorn blossom, on sallow catkins and on sunlit foliage. He also reports that in 1996 he found "a group of males swarming in sunshine around a row of young sycamores […] and settling briefly on their foliage". In Sussex, Steven Falk has recorded *A. helvola* on Hawthorn and rose.

Females have white hairs on the face, reddish-brown hairs on the top of the thorax and long, pale yellowish hairs on tergites one and two. These and the hairs on the top of the thorax do, however, fade to white over time.

Males have very long mandibles with a small projection on the underside at the base. The males also have dense fringes of long white hairs along the hind margins of the sternites, which separates them from the otherwise similar *A. fucata* and *A. lapponica*.

This species has a long flight period, with a peak in May.

Behaviour and interactions
A. helvola nests solitarily or in scattered aggregations in woodland clearings and along woodland rides. Mating has been observed taking place on flowers.

Brood cells are targeted by *Nomada panzeri*.

On the wing
Sussex records run from late March until the end of June.

Feeding and foraging
Pollen is collected from a wide range of plants in the Buckthorn, Gooseberry, Daisy and Rose families.

A. helvola is often seen visiting spring-flowering trees and shrubs such as Blackthorn, Raspberry, Guelder-rose, hawthorn and Field Maple, plus flowering plants such as Wood Spurge and dandelion. It is particularly associated with Wood Spurge.

Andrena humilis Buff-tailed Mining Bee
FIRST RECORD 1879, Hastings (Edward Saunders) TOTAL RECORDS 62

Geography and history
Andrena humilis is an elusive bee in Sussex and is largely confined to areas with sandy soils, with most records from sites across the Wealden Greensand and the High Weald. It has also been recorded on a handful of occasions on the South Downs, usually in places where the chalk is overlain by sand.

Amongst the diverse locations with recent records are a garden in Midhurst, an area of landscaped gardens and parkland at Sheffield Park, a meadow at Batemans, and a complex of meadows and pasture near Warbleton, as well as Birling Gap, Rye Harbour Nature Reserve, and the coast near Hastings.

Hay meadow, High Weald [NS]

Large map: distribution 2000–2023
Small map: distribution 1844–1999

Out in the field and under the microscope
A. humilis is best searched for by checking yellow-flowered plants in the Daisy family.

The females are dark and have a shiny abdomen with pale, reddish hairs towards the tip (although the colour of these does fade). It has a resemblance to *A. fulvago* but is larger, duller, and lacks the yellow on the hind legs of this last species.

Males are of similar build to the females but slightly smaller. They are one of a handful of species with a white or yellow marking over much of the clypeus.

Andrena humilis ♀ [PB]

Andrena humilis ♂ [SF]

Behaviour and interactions
In 1937 Kenneth Guichard discovered a nesting aggregation near the coast between Worthing and Goring-by-Sea. This *A. humilis* colony was sited along a south-facing gravel bank some 20 yards long, with its parasite *Nomada integra* more numerous than the host. *A. flavipes* was also nesting in the same location.

On this occasion, *A. humilis* was seen visiting only Hawkweed Oxtongue flowers, and the bees were described as "working with extraordinary rapidity, gathering pollen and nectar with feverish haste and then dashing onto the next flower almost with the speed of an *Anthophora*". The flowers had closed by 1 o'clock in the afternoon, by which time *A. humilis* had largely disappeared.

Nests are usually in small, compact aggregations with four and five brood cells per nest. Each nest might be excavated to a depth of between 12 and 25 cm and are closed overnight. A study in Sweden showed that foraging flights lasted on average ten minutes.

Brood cells are only attacked by *N. integra*.

On the wing
Sussex records run from early May until the beginning of August, with most from May and June.

Feeding and foraging
Pollen is collected from plants in the Daisy family, especially those with yellow flowers. Those reported in the literature include Beaked Hawk's-beard, Hawkweed Oxtongue, Mouse-ear-hawkweed and Rough Hawkbit.

Andrena labialis Large Meadow Mining Bee

FIRST RECORD 1876, Eastbourne (Frederick Smith) **TOTAL RECORDS** 195

Geography and history

Andrena labialis is found on flower-rich grasslands and along woodland rides where there are plentiful clovers, vetches and trefoils. It is widely distributed across the county but local. It can be present in good numbers in favourable locations.

There are recent records from a woodland, scrub and grassland mosaic adjacent to Ebernoe Common, from Rewell Wood, from meadows and pasture in the High Weald (for example, at Nymans and near Willingdon), from grasslands on the Pevensey Levels, from Rye Harbour Nature Reserve, and from chalk grassland at Beachy Head.

Clovers, vetches and trefoils in flower at Rye Harbour Nature Reserve [PG]

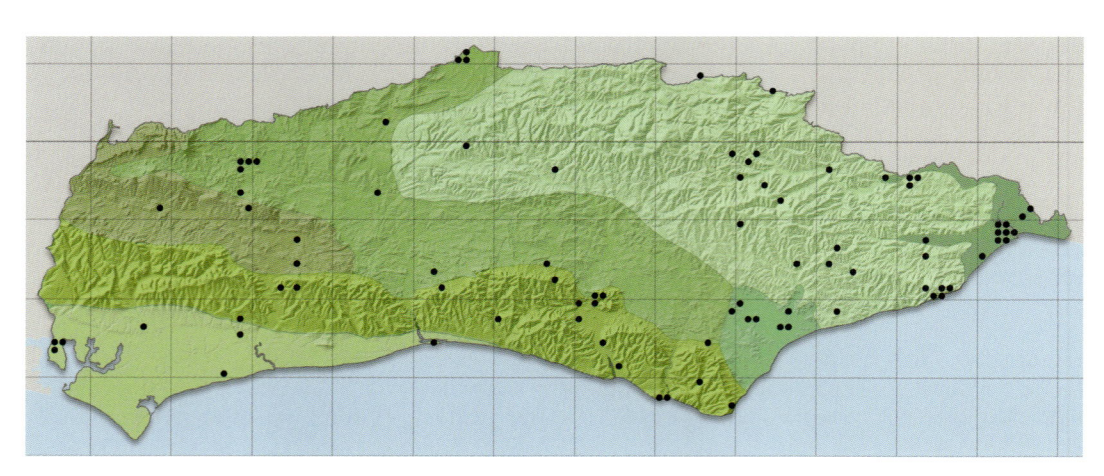

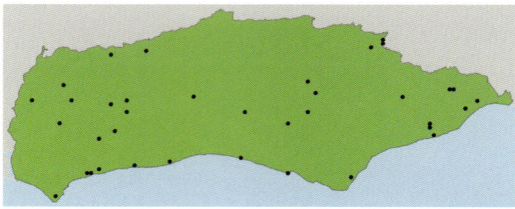

Large map: distribution 2000–2023
Small map: distribution 1844–1999

Out in the field and under the microscope

This is one of the larger *Andrena*, and it is dark with a shiny and densely punctate abdomen. Both sexes have slim, white hair bands on the hind edges of the second, third and fourth tergites. These are interrupted on the second and third tergites.

In common with nine other species of *Andrena*, the males have a white-marked area on the clypeus. However, this marking extends onto the lower face and is much more extensive than in the other white-marked species. Males are slightly smaller than the females and can be more readily identified in the field.

This species is usually found foraging from plants such as Hawthorn, bird's-foot-trefoil and clover, but males might also be seen flying low over areas of grassland or along sunny woodland rides.

Andrena labialis ♀ [DG]

Andrena labialis ♂ [PC]

Behaviour and interactions

One of the earliest records of *A. labialis* in Sussex was made by Edward Saunders in 1891. He writes that the bees "swarmed along the low cliffy banks to the east of Rustington" and adds that they were flying with male and female *Sphecodes rubicundus*, *A. nigroaenea* and *Eucera longicornis*. The *S. rubicundus* females were entering the *A. labialis* nest burrows. *A. labialis* outnumbered these other species 200 to 1.

On the continent it is suggested that *A. labialis* will establish aggregations with up to 600 nests, but it will also nest solitarily. As well as the type of nesting situation described by Edward Saunders, this *Andrena* uses vertical cuttings and the surface of hedge banks. There is also a single report of three females using the same entrance burrow, indicating some level of communal nesting.

Unusually for an *Andrena*, in Britain this species is targeted by a *Sphecodes*, *S. rubicundus*, and not by a species of *Nomada*. On the continent, however, several species of *Nomada* are given as parasites of *A. labialis*, including *Nomada succincta*, a species that is restricted to the Channel islands in the British Isles (where it targets *A. nigroaenea* nests).

On the wing

Sussex records run from the beginning of May until mid August.

Feeding and foraging

Pollen is collected from a range of plants, including members of the Pea, Buttercup, Carrot and Daisy families. Specific plant species listed as sources of pollen include Common Bird's-foot-trefoil, Red and White Clover, Sainfoin and Gorse, plus Common Rock-rose, Meadowsweet and Viper's-bugloss.

Andrena labiata Red-girdled Mining Bee

FIRST RECORD 1858, Brighton and Portslade (William Unwin) **TOTAL RECORDS** 137

Geography and history

Andrena labiata was a more widespread species in the 1930s when it could take advantage of extensive areas of species-rich grassland and managed woodland. However, in subsequent years the extensive application of artificial fertiliser to increase the productivity of grasslands and the reduction in woodland management triggered a decline.

Most modern Sussex records are from the Wealden Greensand and the High Weald, where it can still be locally abundant in areas of flower-rich grassland and areas of open woodland with sandy soils. In the High Weald it has been recorded from Great Dixter, Fore Wood, an area of species-rich grassland near Warbleton, and near Ashurst. On the Wealden Greensand, it has been recorded from areas of acid grassland such as Lord's Piece and Weavers Down, and also from gardens in Midhurst.

It is not often found on the South Downs, despite the abundance of its preferred foodplant, Germander Speedwell. It has, though, occurred at Fairmile Bottom, Deep Dean and Seaford—all areas with sandy substrates nearby. Elsewhere in the county, there are recent records from Chichester, Ebernoe Common and Knepp.

Andrena labiata ♀ visiting Germander Speedwell [PB]

Out in the field and under the microscope

This is a striking bee with red and black markings. It is unlike any other species of *Andrena*, although it could at first glance be confused with a species of *Sphecodes*. In females, however, the dark pollen-collecting hairs on the hind legs will help to confirm an identification—

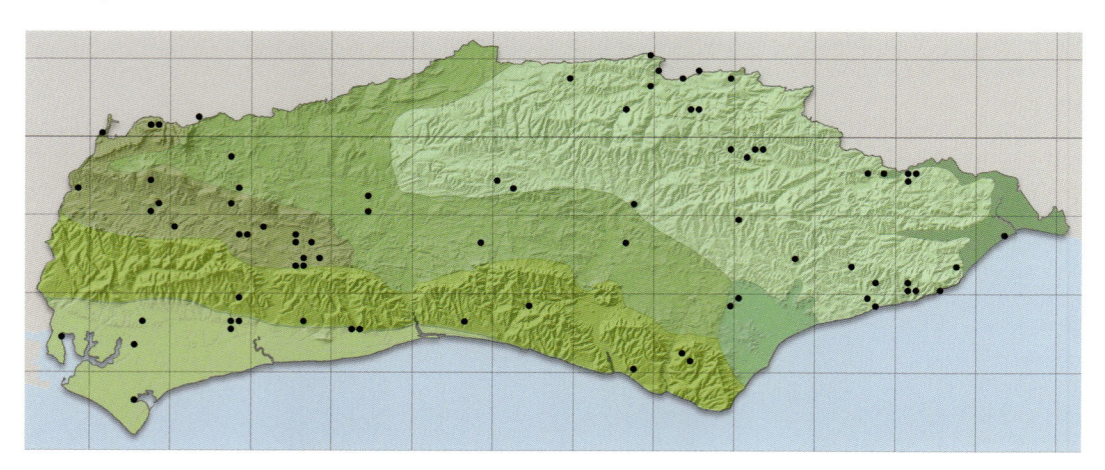

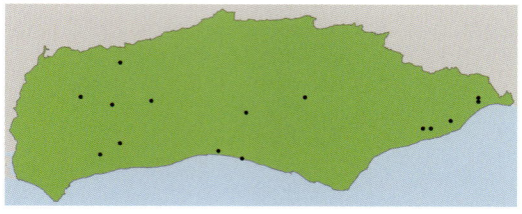

Large map: distribution 2000–2023
Small map: distribution 1844–1999

Andrena labiata ♂ visiting Greater Stitchwort [PB]

Andrena labiata ♂ showing white-marked clypeus [NO]

these are always absent in *Sphecodes*. *A. labiata* females are red on tergites two and three, and black on tergites one, four, five and six.

Males are similarly marked with black and red abdomens, and are unique among British bees in having this abdominal colour pattern in combination with a white-marked clypeus and lower face.

Searching patches of Germander Speedwell and Greater Stitchwort, or nearby areas of flower-rich vegetation, is a good way to find this bee.

Behaviour and interactions
Females nest in large aggregations—sometimes with several hundred nests—or solitarily, in sunny locations with short or sparse vegetation. In 1921, Rosse Butterfield reported finding a colony on a sunny bank near Church-in-the-Wood, Hollington, where he also found its parasite, the very scarce *Nomada guttulata*. More recently, Ian Beavis found a sizable colony nesting on a flower-rich bank, and elsewhere a female nesting solitarily.

On the wing
Sussex records run from the beginning of April until the end of June.

Feeding and foraging
This bee is closely associated with plants in the Speedwell family, especially Germander Speedwell. It is also often seen on Greater Stitchwort.

As well as speedwell, it will collect pollen from plant families such as Buttercup, Cabbage, Daisy, Pink and Rose.

Germander Speedwell [PC]

Andrena lapponica Bilberry Mining Bee
FIRST RECORD 1976, Marley Common (Mike Edwards) TOTAL RECORDS 29

Geography and history
Andrena lapponica is largely confined to northern and western areas of Britain—it can often be found in places such as the west Pennines and Exmoor—but does also occur more locally in other areas, including Sussex. This bee largely forages for pollen from Bilberry, a species that is typically found in dry woodlands here.

Historically, *A. lapponica* was known from two areas of the county—the western Wealden Greensand and Ashdown Forest. Of these, it is now known only from the Wealden Greensand, where it was recorded as recently as 2022 on Midhurst Common. On Ashdown Forest, it has not been recorded since 1993.

There is the possibility that this species was also known from the Hastings area, although the evidence is weak—a specimen labelled "Hastings" survives in the Cliffe Castle Museum, Keighley,

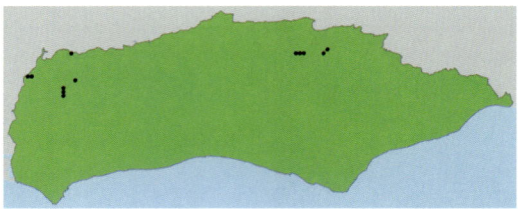

Large map: distribution 2000–2023, with white circles showing distribution of Bilberry
Small map: distribution 1844–1999

but there is no date on the label, the name of the collector is not given, and there are no records for Bilberry in this part of the county.

Out in the field and under the microscope
This bee should be searched for by checking stands of Bilberry.

Females most resemble *A. varians*. They have reddish-brown hairs on the top of the thorax and long pale hairs on the first two tergites. The remaining tergites are covered in shorter dark hairs. The face is entirely black-haired.

Males are smaller and slimmer, and have extremely long mandibles that comfortably cross over when closed, forming the shape of an 'X'. They belong to a small group of *Andrena* which has a downward-pointing projection on the underside of each mandible, close to the base. In this species, the projections are relatively short, while the central area of the clypeus is very polished.

Behaviour and interactions
Nests are thinly distributed in suitable areas of bare ground within Bilberry-rich habitat. These can include footpaths and unvegetated slopes and banks. Small aggregations have been reported on the continent.

It had long been understood that this *Andrena*'s brood cells were targeted by *Nomada panzeri*. However, in 2022 it was recognised that *N. panzeri* is in fact two species and that it is the second of these, *N. glabella*, that attacks *A. lapponica*. This *Nomada* has been recorded once in Sussex.

On the wing
Sussex records run from the beginning of April until late June.

Feeding and foraging
In Sussex, Bilberry is the principal source of pollen, although elsewhere in Britain Cowberry is also used.

A. lapponica has also been reported visiting plants such as Blackthorn, Common Dog-violet, dandelion and willow.

Bilberry [PC]

Andrena marginata Small Scabious Mining Bee
FIRST RECORD 1854, Lewes (William Unwin) **TOTAL RECORDS** 34

Geography and history
Andrena marginata is an extremely scarce bee which in recent years has been recorded sporadically from just two widely separated locations on the South Downs, both of which have abundant scabious-rich chalk grassland. These locations are the Harting Down area and Castle Hill.

Across Britain it has an unusual distribution. There are records from the Highlands and a scattering of sites in northern England, but otherwise records are concentrated in southern England and south Wales. In some of these locations, *A. marginata* can be frequent. In Surrey, for example, David Baldock reported that it was "widely distributed but very local". There, as elsewhere, *A. marginata* can be found on acid grassland sites as well as chalk grassland.

Andrena marginata was recorded from the same patch of Field Scabious at Castle Hill in 2019, 2020 and 2021. [JP]

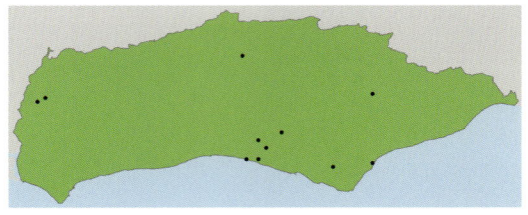

Large map: distribution 2000–2023, with white circles showing distribution of Small Scabious, Field Scabious and Devil's-bit Scabious
Small map: distribution 1844–1999

270 *The Bees of Sussex*

In Sussex, it was previously known from a wider area than just Harting Down and Castle Hill, including Dallington, Barcombe and Eastbourne. The earliest reference to *A. marginata* in the literature is by Frederick Smith, Senior Assistant in the Zoological Department at the Natural History Museum. He reported being sent a specimen from Sussex by S. Smith in 1847, but with no details of where it had been found in the county. Seven years later, William Unwin reported that one had been found near Lewes, the first confirmed Sussex record. It was clearly a locally common species on the downs near Ovingdean in the 1940s—the Booth Museum holds a series of 40 specimens collected by Alfred Brazenor from this area between 1943 and 1945, 33 of which were collected in July and August 1944.

A. marginata was thought to have disappeared from the downs near Lewes until it was found by James Power again at Castle Hill in 2019, after an absence of 32 years. It has since been found again in the same location in 2020 and 2021. The population centred on Harting Down was discovered by Mike Edwards in 1982, and his most recent record for this area is from 2022.

Andrena marginata ♀ [PC]

Andrena marginata ♂ [SF]

Out in the field and under the microscope

The optimum way to find *A. marginata* is to search scabious flower heads where females might be foraging for pollen. Patience will be required, however, as it now appears to be present at its Sussex locations in very low numbers—just singletons were observed at Castle Hill in 2019, 2020 and 2021. Each of these specimens was foraging from, or flying among, Field Scabious flowers.

It is a very distinctive bee. Females have red or black abdomens and black hairs on the hind tibiae. Superficially, they resemble *A. hattorfiana* females but are much smaller.

The males are slightly smaller than the females and always have dark abdomens. They can be distinguished by the white marking, with two black spots, that covers much of the clypeus. Below the white marking, the clypeus is distinctive in that it curves outwards to a pair of characteristic points.

Behaviour and interactions

Populations on areas of acid grassland tend to fly later than those on chalk grassland, the later flight period coinciding with the flowering period of its principal pollen source here, Devil's-bit Scabious. In time, these populations may prove to be a new 'hidden' species.

Brood cells are targeted by the very scarce cuckoo bee *Nomada argentata*.

On the wing

Sussex records run from early July until mid August.

Feeding and foraging

A. marginata collects pollen almost entirely from different species of scabious. On chalk grassland it principally targets Small Scabious but also frequently visits Field Scabious, while Devil's-bit Scabious is the principal pollen source on acid grassland.

Andrena minutula Common Mini-miner
FIRST RECORD 1872, Linch (Henry Guermonprez) **TOTAL RECORDS** 680

Geography and history
Andrena minutula is the most widely distributed and most abundant of the small black *Andrena*. It is found throughout Sussex, occurring in a wide range of different habitats. These include grasslands, woodlands and heathlands. It is also often seen in gardens.

Out in the field and under the microscope
Because of its small size, *A. minutula* is very easy to overlook. It is also very challenging to identify. This can only be done with the aid of a microscope, but, even then, this species is readily confused with other members of a sizeable group of very similar-looking small black *Andrena*. Of these it most closely resembles *A. minutuloides* and *A. subopaca*.

Andrena minutula ♀ visiting Hogweed [NO]

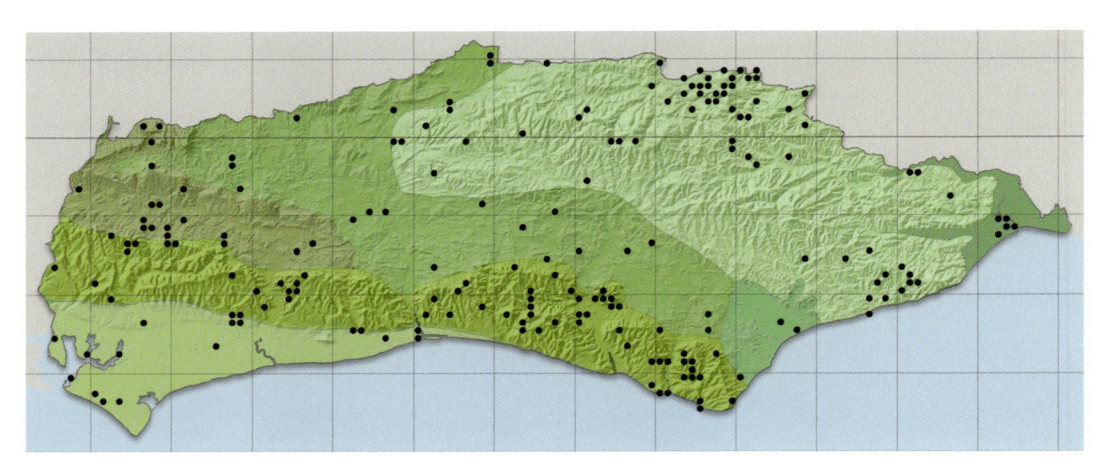

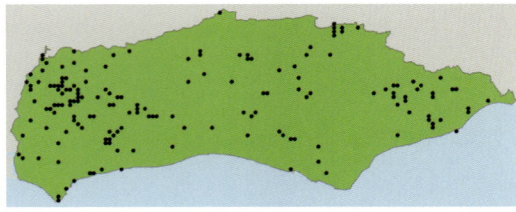

Large map: distribution 2000–2023
Small map: distribution 1844–1999

In fresh specimens of female *A. minutula*, the white hairs to the sides of the second and third tergites are very dense and obscure the hind margin of the tergites (these abrade with time). Other characters to look for in this species are the relatively dense punctation on the scutum and the dull surface of the scutellum.

Males that emerge in the spring differ from those in the summer generation. The spring males have a dull thorax and either black hairs on the face or a mix of black and white hairs here. In contrast, summer males have a shinier scutum and just white hairs on the face.

A. minutula is best searched for by checking flower heads, including spring-blossoming shrubs.

Andrena minutula ♂ [SF]

Andrena minutula mating pair [PB]

Behaviour and interactions
A. minutula has two flight periods each year, and until 1919 the two generations were treated as distinct species, *A. parvula* for the spring generation and *A. minutula* for the summer generation.

There is little known about its nesting habits, although it is understood to nest solitarily. On the continent, it is reported that it nests on sparsely vegetated banks and along path edges. Ian Beavis has observed nesting in south-facing banks in the Tunbridge Wells area.

Brood cells are targeted by *Nomada flavoguttata*.

On the wing
This species has two flight seasons, the first running from mid March until early June and the second running from late June until late August.

Feeding and foraging
A very wide range of plant families are visited for pollen. These include plants in the Buttercup, Deadnettle, Rose, Carrot, Daisy, Speedwell and Willow families.

Andrena minutuloides Plain Mini-miner
FIRST RECORD 1910, Midhurst (Henry Guermonprez) **TOTAL RECORDS** 115

Geography and history
Andrena minutuloides is a scarce bee largely restricted to areas of flower-rich chalk grassland on the South Downs. It has also been found at a handful of sites in the High Weald and the Wealden Greensand, including Broadwater Warren and heathland near West Chiltington.

The South Downs, where it can be locally abundant, is one of its national strongholds. Here, it is widely distributed and has recently been found at Devil's Jumps, Fairmile Bottom, Saddlescombe, Malling Down and Beachy Head. It has also been recorded from a garden in Lewes where Wild Carrot is established in the lawn.

Out in the field and under the microscope
This is a small black *Andrena*. Unlike the very similar *A. minutula*, the spring and summer generations of *A. minutuloides* are identical.

Even fresh specimens of *A. minutuloides* and *A. minutula* are notoriously difficult to separate, a situation that is exacerbated by their natural variability and the degree of wear in each individual. In fresh females of *A. minutuloides*, the surface of the abdomen is visible through the small patches of hairs on the sides of the second

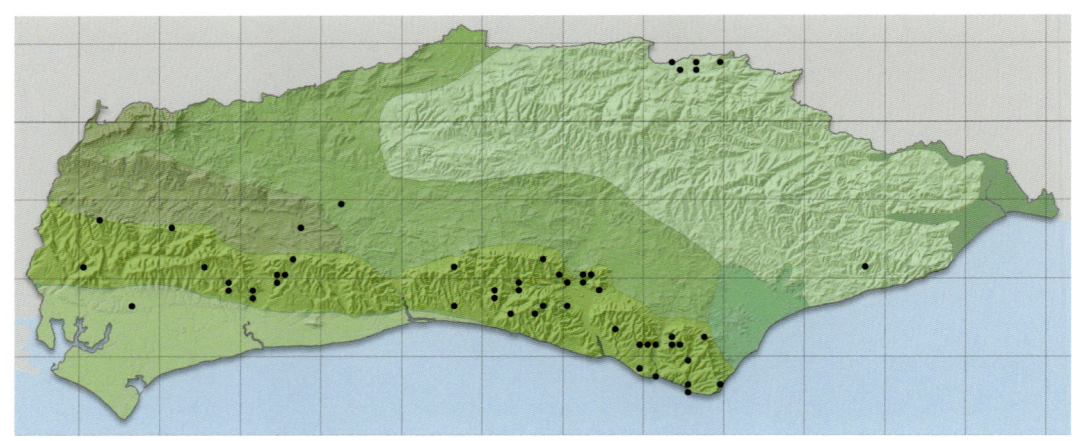

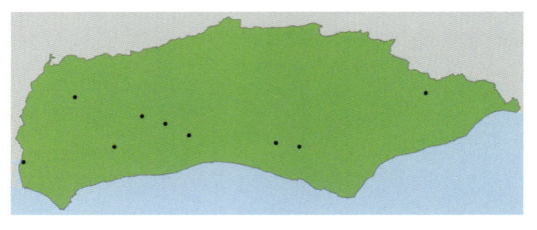

Wild Carrot growing in the sward at Beachy Head [SF]

Large map: distribution 2000–2023
Small map: distribution 1844–1999

and third tergites (the surfaces are completely obscured in fresh *A. minutula*), while the surface of the scutellum is shiny (it is dulled in *A. minutula*). Males also have a shinier scutellum than male *A. minutula*.

Worn specimens of *A. minutuloides* are all but impossible to reliably separate from *A. minutula*, and it is likely that many records for both species are based on misidentifications.

A good way to find this bee is to search flower heads of plants in the Carrot family such as Wild Carrot, Angelica, Burnet-saxifrage and Upright Hedge-parsley.

Andrena minutuloides ♀ [SF]

Wild Carrot [PC]

Behaviour and interactions

Although no nests have been found in Britain, it is understood that *A. minutuloides* nests solitarily and does not form aggregations. It has two flight periods, and the second generation is generally more numerous than the first. Each generation appears around a month later than the equivalent generation of *A. minutula*.

On the continent, *A. minutuloides* is targeted by *Nomada flavoguttata*, and this is likely to be the case in Britain too, although this behaviour has not been confirmed here.

On the wing

This species has two flight seasons, the first running from the end of April until mid June and the second running from early July until the end of August.

Feeding and foraging

A. minutuloides forages for pollen from a wide range of plants, including members of the Carrot, Maple, Daisy, Cabbage, Rose, Willow and Speedwell families.

Andrena nigriceps Black-headed Mining Bee
FIRST RECORD 1919, Fairlight (William Butterfield) TOTAL RECORDS 8

Geography and history
Andrena nigriceps is a very scarce and local bee that has been recorded just eight times in Sussex, most recently at Seaford Head in 2007, close to where it was found in 1968. Prior to the 2007 record, it had not been seen in the county since 1977. This is when it was recorded twice during August at Fairlight, the same location as the first record in 1919.

A. nigriceps was also found on the Wealden Greensand in 1975, when it was recorded twice, once near Upperton and once from Sullington Warren.

Across Britain, *A. nigriceps* is patchily distributed as far north as the Scottish borders and as far west as Cornwall.

Andrena nigriceps ♀ visiting Hogweed [NO]

Large map: distribution 2000–2023
Small map: distribution 1844–1999

Out in the field and under the microscope

Andrena nigriceps ♂ [HW]

This is a late-season bee and is most likely to be encountered foraging on plants such as ragwort, thistle, Tansy and knapweed.

Females have a covering of black hairs on the face and reddish-brown hairs on the top of the head and thorax. The first four tergites are covered in long pale hairs, with the fifth covered in long black hairs. The front corners of tergites three and four have patches of much shorter black hairs. The tibiae on the hind legs are very broad at the tip.

Males are smaller and slimmer than the females. They generally have white hairs on the face, reddish-brown hairs on the top of the thorax, and a covering of long pale hairs on the abdomen.

Male *A. nigriceps* can only be separated from male *A. simillima* through careful observation—key features are slight differences in the length of the third and fourth antennal segments, and the colour of the tarsal segments on each of the legs. In *A. nigriceps*, the third antennal segment is only slightly longer than the fourth—it is much longer in *A. simillima*—while the tarsi are a dark brown in *A. nigriceps* and yellow in *A. simillima*.

Behaviour and interactions

Although now very scarce, there are indications that *A. nigriceps* could have been a locally abundant species in the past. There is a report from Norfolk in 1901 by Colbran Wainwright which described *A. nigriceps* as "very abundant on ragwort".

It has also been thought to nest solitarily, although a small aggregation of some 20 nests was reported from Lancashire. One further nesting observation, from Norfolk by Nick Owens in 2016, is of a female nesting in a sandy bank.

Brood cells are targeted by *Nomada rufipes*.

A. nigriceps is very closely related to *A. simillima*, and the two species may yet prove to be forms of a single species.

On the wing

This species is on the wing from early July until late September.

Feeding and foraging

A. nigriceps collects pollen from a wide range of plants including members of the Beech, Rose, Bellflower, Bedstraw and Daisy families.

Andrena nigroaenea Buffish Mining Bee
FIRST RECORD 1878, Bognor Regis (Henry Guermonprez) **TOTAL RECORDS** 610

Geography and history
Andrena nigroaenea is very abundant and widely distributed in Sussex. It has most frequently been recorded on the South Downs, but is also regularly found in the area south of Tunbridge Wells, on the Wealden Greensand, and along the coast between Bexhill and Lydd Ranges. It can be found in just about any habitat and is regularly found in gardens.

Out in the field and under the microscope
Females of this species are about the size of *Apis mellifera* and are quite distinctive in the field. Key features to look for are the covering of brown hairs on the thorax and abdomen, the orange pollen-collecting hairs on the hind legs, and the presence of hairs on the clypeus that are either just black or

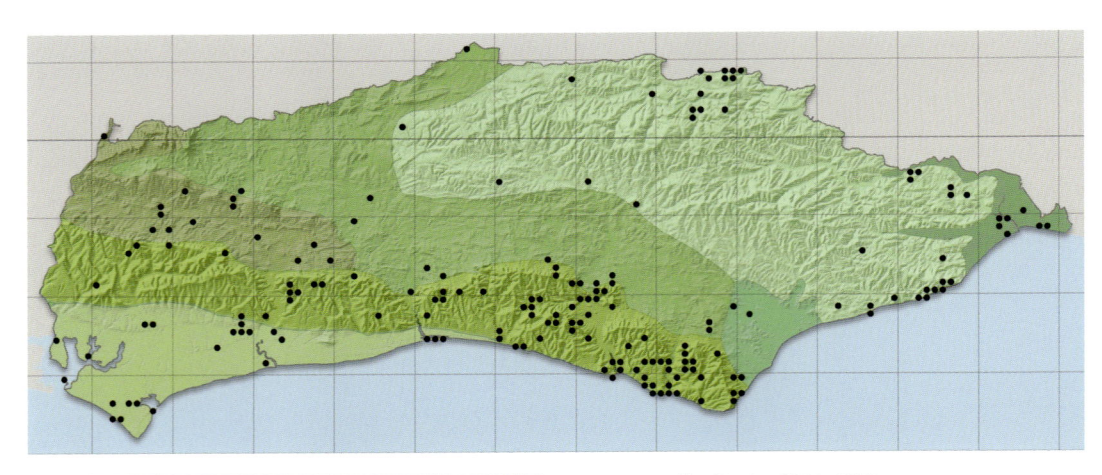

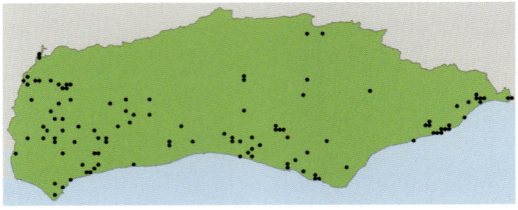

Andrena nigroaenea ♀ [PB]

Andrena nigroaenea ♂ [PC]

Large map: distribution 2000–2023
Small map: distribution 1844–1999

278 *The Bees of Sussex*

a mix of black and ginger. They most resemble *A. bicolor* but are generally larger and lack the large punctures on the propodeum found in this other species.

A. nigroaenea males resemble the females but are slightly smaller and slimmer.

This bee is best searched for by checking flowering plants during the flight season, although it is also possible to come across males flying rapidly around gorse bushes or in small numbers coursing along hedgerows or along the edge of woodland.

Behaviour and interactions

It had been assumed that *A. nigroaenea* nests singly, but observations from elsewhere in the country have shown that this species does nest in aggregations, with 30 nests reported by Nick Owens from one Norfolk garden.

Andrena nigroaenea nest [NO]

The chemical attractant that is emitted by females is mimicked by the floral bouquet of unpollinated Early Spider-orchids. This attracts male *A. nigroaenea*, and during 'mating' the orchid pollinia become attached to the male's head. In Britain, *A. nigroaenea* is largely responsible for pollination of this scarce orchid, which in Sussex is found at Castle Hill and on chalk grasslands near Eastbourne.

There are a handful of records for *A. nigroaenea* from late July and August, and these support the view that this species may have a small second generation in Sussex.

In Sussex, *A. nigroaenea* brood cells are targeted by two species, *Nomada goodeniana* and *N. fabriciana*. Elsewhere, cells are also targeted by *N. succincta*, a species that is not found in Sussex.

Early Spider-orchid [PC]

On the wing

Sussex records run from late February until late August, with most records from April, May and June.

Feeding and foraging

Pollen is collected from a very wide range of flowering plants and shrubs, including plants in the Buttercup, Rock-rose, Cabbage and Pea families. Specific species listed include White Bryony, bramble, Hawthorn, White Clover and dandelion.

Andrena nitida Grey-patched Mining Bee
FIRST RECORD 1858, Lewes (William Unwin) TOTAL RECORDS 710

Geography and history
Andrena nitida is one of the most frequently recorded solitary bees in Sussex. It is found throughout the county in many different habitats, including chalk grassland, meadows and pasture, heathland, woodland and soft-rock cliffs. It is also known from cultivated farmland and gardens.

Out in the field and under the microscope
This is a striking bee that can be identified in the field. Fresh specimens have a reddish-brown covering of hairs on the top of the thorax, and this contrasts strongly with the black, shiny abdomen.

Andrena nitida ♀ [PC]

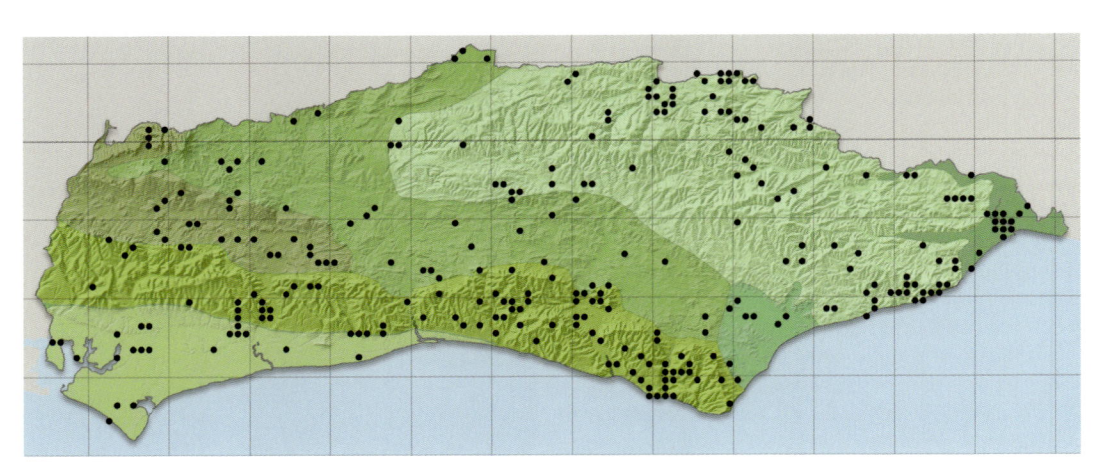

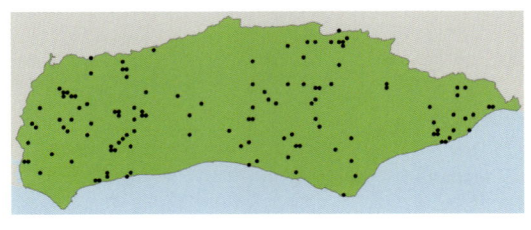

Large map: distribution 2000–2023
Small map: distribution 1844–1999

The Bees of Sussex

Andrena nitida ♀ showing pale hairs on the clypeus [PB]

Andrena nitida ♂ [PB]

There are patches of white hairs at the sides of tergites one, two and three, a character that is not shared with any other species of *Andrena*. These hairs generally persist and are therefore a good field character to look for. The pollen-collecting hairs on the hind tibiae are black.

Males resemble the females but are slightly smaller and slimmer. They also lack the white hair patches on the abdomen, instead having weak white hair bands to the side and back edge of the second and third tergites. The clypeus is covered in white hairs, aside from a narrow band of black hairs adjacent to each of the eyes.

A. nitida is most likely to be found foraging or feeding on spring-flowering plants such as dandelion, Lesser Celandine, Blackthorn, Hawthorn and willow. Males might also be found chasing along hedgerows or patches of scrub, often in the company of other species.

Lesser Celandine [PC]

Behaviour and interactions
This species generally nests singly but can form loose aggregations. Nests can be sited in lawns and other types of grassland, on woodland banks and in woodland clearings.

In Britain, brood cells are targeted by *Nomada goodeniana*. Other species also attack this bee on the continent, including *N. flava*, *N. fulvicornis*, *N. marshamella* and *N. succincta*.

On the wing
Sussex records run from mid March until mid July.

Feeding and foraging
Flowers in a very wide range of plant families are visited for pollen. Among these are Buttercup, Daisy, Pink, Rock-rose and Willow, as well as members of the Rose family such as Blackthorn, Wild Cherry and Hawthorn.

Andrena nitidiuscula Carrot Mining Bee
FIRST RECORD 1927, Pett (John Le Brockton Tomlin) **TOTAL RECORDS** 47

Geography and history
Andrena nitidiuscula has been found in small numbers in scattered locations across the county, with the number of records rising appreciably since the late 1990s. The Hastings, Seaford Head and Chichester Harbour areas have generated the most records, and it has recently also been found in the gardens at Great Dixter, and at Southerham Farm, Tide Mills, Woods Mill and Ditchling Common.

In Britain, its range extends as far west as Devon and as far north as the south Midlands, with east Sussex its most easterly location. A key criterion is the presence of plants in the Carrot family, especially Wild Carrot.

Wild Carrot, Tide Mills [WP]

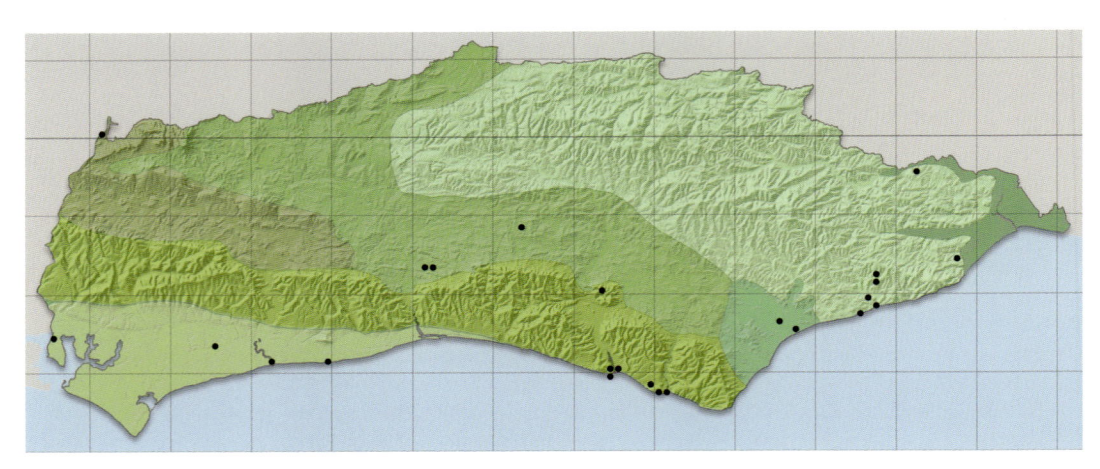

Large map: distribution 2000–2023
Small map: distribution 1844–1999

282 *The Bees of Sussex*

Out in the field and under the microscope
A lot of good fortune is required to find *A. nitidiuscula*—it has been found sporadically over the years, generally only as singletons. It is most often seen as it forages, particularly on Wild Carrot. Steven Falk reports "small numbers associated with Wild Carrot on cliff-top grassland near Hope Gap". It has also been found basking on areas of bare ground.

Females are superficially very similar in appearance to other species of small black *Andrena* but are usually slightly larger. The head, thorax and abdomen are entirely black, and there are small patches of white hairs to the sides of the second, third and fourth tergites. A key character is the presence of punctures on the marginal area of the third tergite.

Males are similar in size to other small black *Andrena* males but, among a few differences, have a distinctly thickened hind margin to the head.

Andrena nitidiuscula ♀ [LO]

Andrena nitidiuscula ♂ [TF]

Behaviour and interactions
It is thought that *A. nitidiuscula* generally either nests solitarily or in small aggregations, although large aggregations have also been reported. Nests have been observed in areas of either bare or sparsely vegetated ground.

Nests are targeted by *Nomada errans* (known only from Dorset) and possibly *N. rufipes*.

On the wing
Sussex records run from late June until early August. Elsewhere, it is reported to be on the wing until early September.

Feeding and foraging
Pollen is collected exclusively from plants in the Carrot family, especially Wild Carrot but also Upright Hedge-parsley and Wild Parsnip.

Andrena niveata Long-fringed Mini-miner
FIRST RECORD 1874, Worthing (Edward Saunders) TOTAL RECORDS 28

Geography and history
Andrena niveata was first reported as a British species from specimens that were swept by Edward Saunders: "Worthing June 1874, 1 ♀, and again in June 1888, when I caught 3 ♂ and 3 ♀ near the sea coast to the west of the town."

It is very scarce in Britain, and since 2000 has only been recorded in Glamorgan, Kent, Surrey and Sussex. Nationally, it is found in a variety of habitats, including heathland, fens, chalk grassland, and soft-rock cliffs along the coast. It has also been recorded from cultivated farmland and an allotment.

Arable field margin below Cradle Hill [SF]

Large map: distribution 2000–2023
Small map: distribution 1844–1999

284 *The Bees of Sussex*

All of the modern Sussex records are from locations on the South Downs between Burpham and Beachy Head, and these include chalk grassland at Seaford Head, a road verge at Birling Gap, and a vineyard and the edges of arable fields below Cradle Hill. Many of these records are from Steven Falk's survey of sites in the eastern South Downs where he found that *A. niveata* was always associated with Charlock, Hedge Mustard and Rape.

It was most recently recorded during a survey of Brighton's 'Bee Banks' in 2020 and on arable margins near Burpham in 2022. In the Brighton survey, Graeme Lyons reports that it was feeding on Hoary Mustard. Near Burpham, it was locally abundant and occurred along the margins of four of the eight fields surveyed.

This area between Burpham and Beachy Head supports a nationally significant population—of the 40 or so records from across Britain since 2000, just over half have been from here.

Out in the field and under the microscope

Fresh specimens of female *A. niveata* are readily distinguished from the other small black *Andrena* by the covering of long white hairs on the marginal area of the fourth tergite. These mask the surface of the tergite beneath. Females also have dense patches of long white hairs to the sides of the second and third tergites. Overall, the white hairs can give this bee a silvery appearance.

Males resemble the females but are smaller and slimmer.

A. niveata is best searched for by checking the flowers of plants in the Cabbage family, especially in areas of disturbed ground, for example along the margins of arable fields. It may be flying around suitable plants with other small black bees such as *A. minutula*, *A. minutuloides* and *Lasioglossum minutissimum*.

Andrena niveata ♀ [SF]

Andrena niveata ♂ [SF]

Behaviour and interactions

There is little information on the nesting behaviour of *A. niveata*, although during one survey near Alfriston Mike Edwards found that it was "nesting commonly in an area of sparsely vegetated south-facing soil" within a vineyard.

There are no known species of cuckoo bee that target this bee.

On the wing

Sussex records run from the beginning of May until mid July.

Feeding and foraging

Pollen is collected just from plants in the Cabbage family, including Hoary Mustard, Charlock, Hedge Mustard, rape, rocket and Cabbage.

A. niveata has also been seen visiting plants such as Wild Carrot, speedwell, dandelion, hedge-parsley and water-dropwort.

Andrena ovatula Small Gorse Mining Bee
FIRST RECORD 2013, Old Lodge (Chris Bentley) TOTAL RECORDS 7

Geography and history
In 2022 it was confirmed that two forms of *Andrena ovatula* represent two distinct species, *A. afzeliella* and *A. ovatula*. This means that while *A. ovatula* is understood to be widely distributed in Sussex, there are few confirmed records, with pre-2022 records now treated as an aggregate, *A. ovatula* agg., unless it has been possible to re-examine specimens. There are 178 records for the aggregate, with the earliest dating from Hastings in 1888.

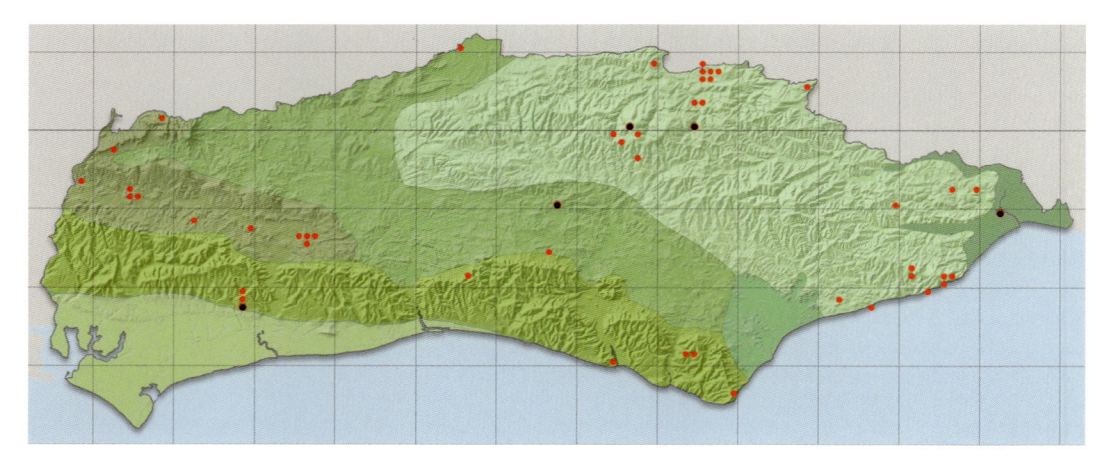

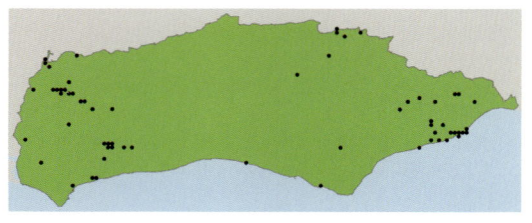

Old Lodge [NS]

Large map: distribution 2000–2023, with black dots showing distribution of *Andrena ovatula* and red dots showing distribution of *Andrena ovatula* agg.

Small map: distribution 1844–1999, showing distribution of *Andrena ovatula* agg.

Meanwhile, seven specimens in private collections have been re-examined and confirmed as the species *A. ovatula*. These are from Chailey Common (where *A. afzeliella* is also found), Old Lodge and Rewell Wood. These locations encompass heathland and open woodland. *A. ovatula* is believed to be distributed across the county in areas with sandy or gravelly soils, and is understood to exploit a range of habitats.

Out in the field and under the microscope
The identification of *A. ovatula* is not straightforward. It most closely resembles *A. afzeliella* but could also be confused with *A. russula* and *A. wilkella*. Like these other species, female *A. ovatula* have narrow bands of white hairs along the hind margins of tergites two, three and four. However, fresh specimens can be separated from *A. russula* and *A. wilkella* by the unbroken hair band on the hind margin of tergite three.

The differences between *A. ovatula* and *A. afzeliella* are more subtle. Female *A. ovatula* are slightly larger and have a dark brown to black fringe of hairs on tergite five. In *A. afzeliella*, the fringe here is golden to whitish grey and, although it can be brown in the centre, it is never dark brown. The hairs flanking the basitibial plate on the hind tibia are golden white in *A. afzeliella* and dark brown in *A. ovatula*. *A. ovatula* also has short dark hairs on the scutum beneath longer yellowish-white hairs. In *A. afzeliella*, all of the hairs here are yellowish white.

The differences between males of the two species are very slight and they cannot always be separated. A character to check is the length of antennal segments three and four. These segments are roughly equal in length in *A. afzeliella*, but segment four is slightly longer than segment three in *A. ovatula*. There are also subtle differences in the shape of the genital capsule.

A. ovatula is best searched for by checking flowering plants in suitable habitats.

Andrena ovatula agg. ♀ [SF]

Andrena ovatula agg. ♂ [SF]

Behaviour and interactions
A. ovatula is double-brooded, flying approximately a month earlier than *A. afzeliella*. It is reported to be very dependent on plants in the Pea family for pollen, with one unpublished set of data suggesting 99.6 per cent of pollen used to provision brood cells comes from this family. Second generation females have also been observed by Ian Beavis collecting Heather pollen on heathland sites near Tunbridge Wells.

On the wing
This species has two flight seasons, the first running from mid March until early June and the second running from early June until mid September.

Feeding and foraging
A. ovatula appears to have a particular association with shrubby plants in the Pea family such as Broom, greenweed and gorse.

Andrena pilipes Black Mining Bee
FIRST RECORD 1878 (Ovingdean and Brighton, Brazenor) **TOTAL RECORDS** 36

Geography and history
Andrena pilipes is a coastal species associated with sandy soils and soft-rock cliffs. In recent years, it has been regularly recorded in small numbers from Seaford Head and from the cliffs and scrubby clifftop grassland to the west of Newhaven Harbour. It has also been found on the cliffs near Bulverhythe.

A. pilipes is almost indistinguishable from the very similar *A. nigrospina*, which was only recognised as resident in Britain in 1994. This last species is very scarce and has not been recorded in Sussex.

Out in the field and under the microscope
Both *A. pilipes* and *A. nigrospina* are unlike any other species of *Andrena*. Not only are they among the largest species in the genus, but they are also almost entirely clothed in black hairs.

Newhaven cliffs [WP]

Large map: distribution 2000–2023
Small map: distribution 1844–1999

In females, key differences between the two species are that the white pollen-collecting hairs on the hind legs are marginally more extensive in *A. pilipes*, and also that in this species there are denser punctures on the surfaces of the first three tergites.

Males of the two species are a little more straightforward to separate. Male *A. pilipes* have black or brown hairs on the top of the head and thorax, as opposed to the grey hairs found in *A. nigrospina*.

A. pilipes is often found at rest on the rocky surface of a cliff or feeding from a plant such as dandelion.

Andrena pilipes ♀ [SF]

Andrena pilipes ♂ [PB]

Behaviour and interactions
Although *A. pilipes* establishes significant nesting aggregations in other parts of the country—for example, at Prawle Point in Devon—the colonies in Sussex appear to be more thinly distributed. Nests are usually sited in vertical cliffs, with brood cells targeted by *Nomada fulvicornis*.

On the wing
This species has two flight seasons, the first running from mid April until the end of June and the second running from the beginning of July until the end of August.

Feeding and foraging
Pollen is collected from a wide range of range of plant species including Charlock and White Mustard.

Plant species visited by the spring generation also include willow, Blackthorn, spurge, dandelion and Alexanders, while those visited by the summer generation include Thrift, bramble, wall-rocket, thyme, Creeping thistle and water-dropwort.

Andrena pilipes visiting Thrift [WP]

Andrena praecox Small Sallow Mining Bee

FIRST RECORD 1882, Hastings (Edward Saunders) **TOTAL RECORDS** 147

Geography and history

Andrena praecox is restricted to damp areas supporting willow on sandy soils. It is almost entirely absent from the South Downs and low-lying areas between Chichester Harbour and Brighton, and it has not been recorded from the Pevensey Levels. Instead, it is predominantly found on the Wealden Greensand as well as the Low and High Wealds.

 A. praecox is principally a species of open woodlands, heathlands, and coastal, scrubby habitats. Recent records are from places such as Etchingham, Hartfield, Ebernoe Common and Nymans.

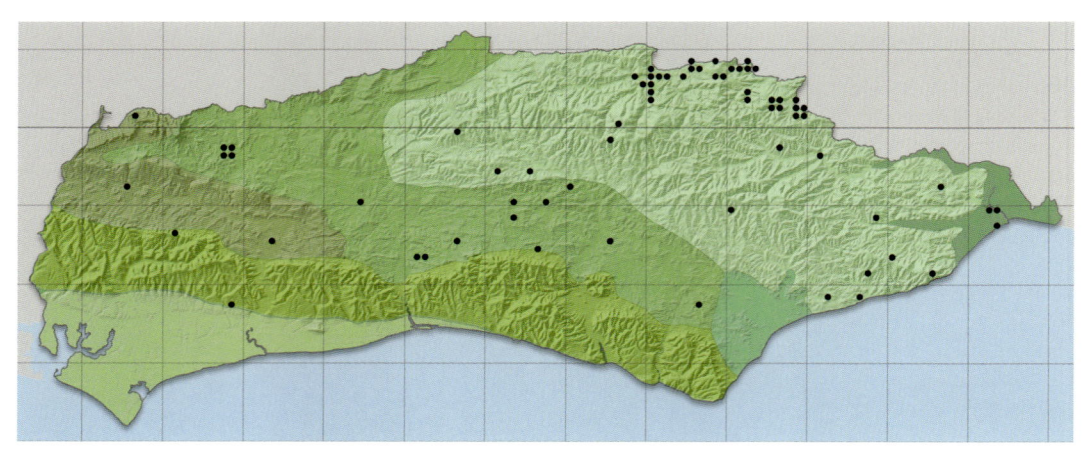

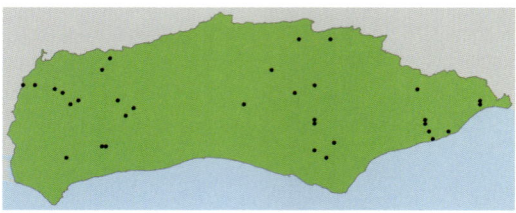

Large map: distribution 2000–2023
Small map: distribution 1844–1999

Out in the field and under the microscope
Females are most often found on willow flowers, while the males might be seen flying low over patches of sunlit bare ground or close to sunlit tree trunks. Here, they might be either at rest or zigzagging up the tree trunk. Males and females can sometimes be present around a single tree in good numbers, flying with other species. In the High Weald, Ian Beavis has also found it "along hedgerows, field boundaries and stream-sides in more open country". Although locally distributed, it can be abundant in some locations.

A. praecox is very similar in appearance to its close relative *A. apicata*, and the two species are not easy to separate from each other. The potential for confusion is further compounded by their foraging preferences—they both collect pollen from willow trees. *A. praecox* is generally the smaller of the two species.

Female *A. praecox* have pale hairs on the clypeus and a band of dark hairs along the inner eye margins (these are best viewed from the side). The top of the thorax is brown-haired, with long pale hairs on the abdomen. The hairs on the first two tergites are long in comparison with those on other tergites.

As well as its smaller size, a key difference between females of this species and *A. apicata* is the presence of light brown hairs, as opposed to dark brown or black hairs, on the fifth tergite and the tip of the abdomen.

Males are slightly smaller than the females but are otherwise similarly coloured. The mandibles are extremely long, with the tips overlapping. They also have a pronounced projection at the base of the mandible on the underside, and they can be separated from *A. apicata* by the notched margin to sternite eight.

Behaviour and interactions
This is one of the first solitary bees to emerge each year, and it has a relatively short flight period. It can form small, compact nesting aggregations in areas of light, sandy soil.

Brood cells are targeted by *Nomada ferruginata*.

On the wing
Sussex records run from mid March until late June, although more typically this species is on the wing until the end of April or early May.

Feeding and foraging
Pollen is collected from several different willow species, especially Grey and Goat Willow.

Willow at Rye Harbour Nature Reserve [PG]

Andrena proxima Broad-faced Mining Bee

FIRST RECORD 1882, Hastings (Edward Saunders) **TOTAL RECORDS** 41

Geography and history

In 2019 it was confirmed that *Andrena proxima* is in fact two closely related species, *A. proxima* and *A. ampla*. It is understood that *A. ampla* has a south-west bias, and it is not believed to be present in Sussex—the examination of specimens in museum and private collections has yielded records for just *A. proxima*.

A. proxima has been recorded in the county on a handful of occasions but may be increasing in abundance. These records are thinly distributed across Sussex, especially in the east of the county. It has been found in a variety of different habitats, including a wetland pocket within a sandpit, chalk

Andrena proxima has been recorded foraging from Hogweed on the Pevensey Levels. [SF]

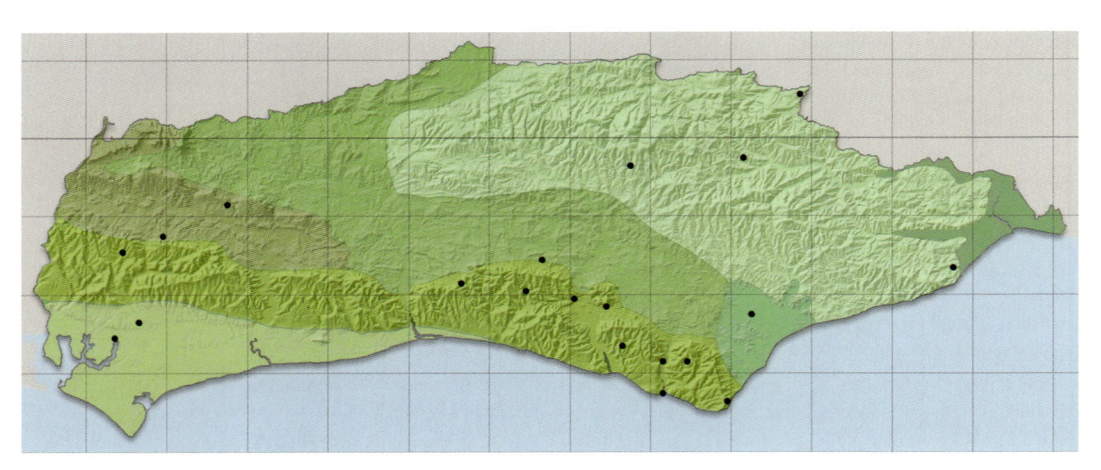

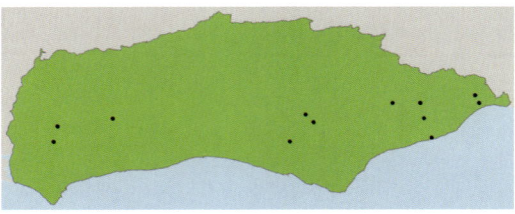

Large map: distribution 2000–2023
Small map: distribution 1844–1999

292 *The Bees of Sussex*

grassland, lowland wet grassland, and a hay meadow. Most recently, it was found in Lewes Cemetery in both 2022 and 2023.

Out in the field and under the microscope

A. proxima is best searched for by checking the flowers of plants in the Carrot family. Here, females might be seen scrabbling across the tiny flowers gathering pollen. They might also be found close by, as in Lewes Cemetery in 2022. Here, males and females were flying together in an area of coarse, shaded vegetation in among the gravestones, not far from extensive stands of Alexanders and Cow Parsley.

A. proxima is a dark bee. Females have a wide face and lack punctures on the clypeus. They also have widely broken bands of white hairs on tergites two, three and four. These small hair bands slant slightly upwards, away from the surface of the abdomen, and have a rectangular appearance.

Females closely resemble *A. ampla* females. The differences between the two species are subtle, but, amongst other characters, *A. proxima* can be identified by looking for a slightly roughened scutum and for the presence here of just a few black hairs beneath the brown hairs.

Males are like the females but smaller and slimmer. They are more challenging to separate from *A. ampla* males, but among the characters to look for are the form of the punctures on the scutum. In *A. proxima* these almost touch and are large, shallow and crescent-like, while in *A. ampla* they are denser, finer and clearly separated from each other.

Andrena proxima ♀ visiting Alexanders [PC]

Andrena proxima ♂ [TF]

Behaviour and interactions

A. proxima is reported to establish its nests in areas that are sparsely vegetated or with short turf. It was believed to only nest solitarily or in small aggregations, but observations by George Else in Devon suggest that it may also nest communally.

Brood cells are targeted by *Nomada conjungens*, which can be abundant around nest sites.

On the wing

Sussex records run from late April until mid July.

Feeding and foraging

This bee collects pollen just from plants in the Carrot family such as Cow Parsley, water-dropwort, Pignut, Hogweed, Wild Carrot and Alexanders.

Unmanaged roadside verge with Alexanders in Lewes [JP]

Andrena russula Red-backed Mining Bee
FIRST RECORD 1976, Beachy Head (Alfred Jones) **TOTAL RECORDS** 17

Geography and history
This species was known as *Andrena similis* until 2022. It is a very scarce bee, recorded just 17 times in Sussex. These records are from scattered locations across the county, including Ashdown Forest, Crowborough, Hastings, Ranscombe Lane below Mount Caburn, Falmer, and a clay pit near Uckfield. The two most recent records are from West Dean Woods and Sheffield Park in 2019.

These locations are very varied and encompass heathlands, a lowland meadow and a flowery road verge. It has also been found on chalk grasslands, in open woodlands and on coastal cliffs.

In the early part of the twentieth century, it was regarded as a local species nationally that could be abundant in some places. Since then, however, it appears to have declined and may be extinct in Kent and Surrey.

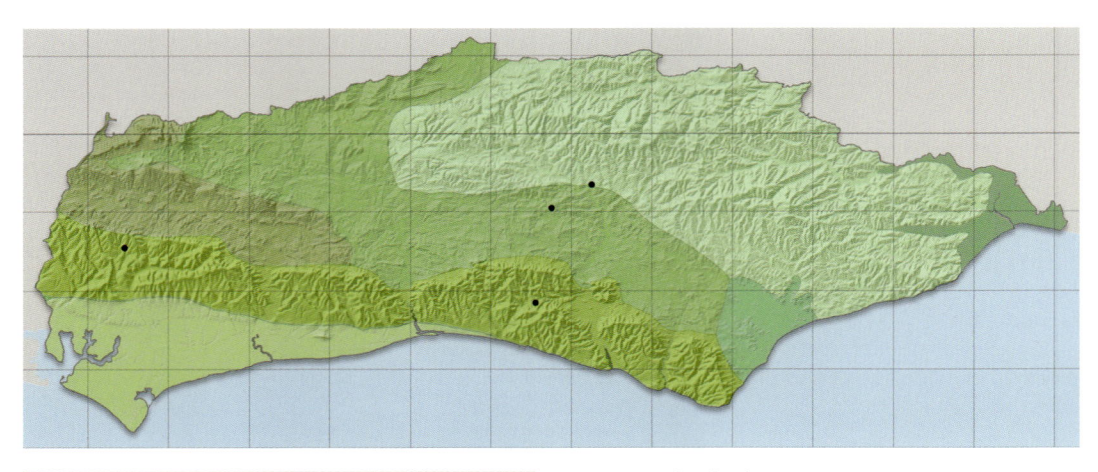

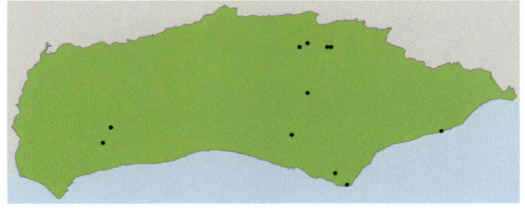

Large map: distribution 2000–2023
Small map: distribution 1844–1999

Out in the field and under the microscope
A. russula superficially resembles *A. ovatula*, *A. afzeliella* and *A. wilkella*.

Female *A. russula*, like the females of these other species, have narrow bands of whitish hairs on the hind margins of tergites two, three and four. In the case of *A. russula*, however, the hair band on the third tergite is very widely interrupted. In *A. wilkella* this hair band is less widely interrupted, while in *A. ovatula* and *A. afzeliella* the hair band is continuous. In fresh specimens of *A. russula*, the hairs on the top of the thorax are a bright reddish orange.

Males have very small hair bands to the sides of tergites two, three and four, and reddish-orange hairs on the thorax, although as with females these fade with time.

A. russula is most likely to be found feeding, and the 2019 record from Sheffield Park was of a female on Common Bird's-foot-trefoil.

Behaviour and interactions
Information on the nesting behaviour of *A. russula* is very patchy. It has been reported that it sometimes forms large compact aggregations but may also nest solitarily.

No species of *Nomada* have been reported as targeting the brood cells of this bee in Britain, although on the continent *N. striata* is suggested.

On the wing
Sussex records run from late March until the end of June.

Feeding and foraging
Pollen is gathered from plants in the Pea family, including bird's-foot-trefoil, Red Clover and gorse.

Common Bird's-foot-trefoil [PC]

Red Clover [JP]

Andrena scotica Chocolate Mining Bee
FIRST RECORD 1941, Brighton (A. Brazenor) **TOTAL RECORDS** 5

Geography and history
In 2022 it was confirmed that, apart from spring-flying males, *Andrena scotica* cannot be separated with absolute confidence from *A. trimmerana*. As a result of this, most records for these two species have been assigned to an aggregate, *A. scotica* agg., encompassing 849 records.

As things stand, there are currently just five confirmed records for *A. scotica*, even though it is very widely distributed and can also be locally abundant. It is understood to be present in just about all parts of the county and occurs in a wide range of habitats including grassland, woodland and heathland, as well as in parks and gardens.

Out in the field and under the microscope
Until 2022, it had been thought that the presence or absence of reddish markings on the sides of the first and second tergites was a reliable character to separate the two species, *A. scotica* and

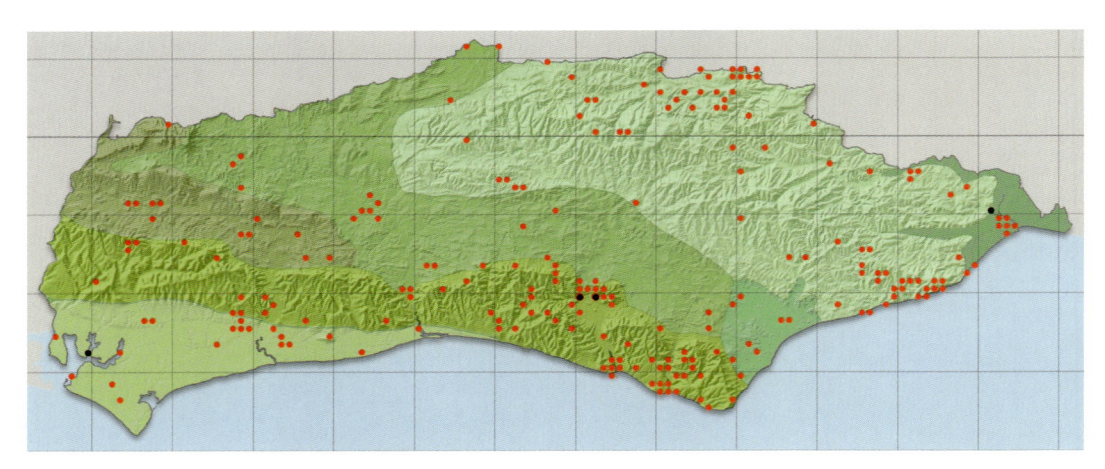

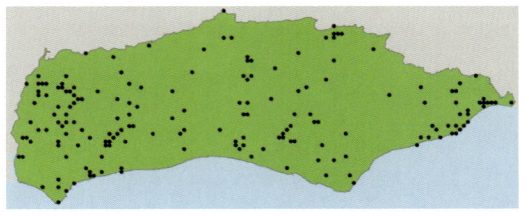

Andrena scotica agg. ♀ [PB]

Andrena scotica ♂ showing the two teeth at the tip of each mandible [IT]

Large map: distribution 2000–2023, with black dots showing distribution of *Andrena scotica* and red dots showing distribution of *Andrena scotica* agg.

Small map: distribution 1844-1999, showing distribution of *Andrena scotica* agg.

A. trimmerana. While *A. scotica* usually does lack red on these tergites, however, it can be red-marked here like *A. trimmerana*. Equally, *A. trimmerana* can display entirely dark terga.

Spring-flying *A. scotica* males can, however, be separated from spring-flying *A. trimmerana* males by the presence of two teeth towards the tip of the mandibles, and by the absence of a long spine on the gena, near the base of the mandibles. Some examples do have a spine here, but it is never as long as in *A. trimmerana*, which in any event has just one tooth on the mandibles.

A. scotica males are often seen patrolling near trees, hedges or banks, often among other species. They might also be seen feeding. Females are usually encountered foraging, especially on trees and shrubs such as Hawthorn, Blackthorn and willow, and their large nesting aggregations can be obvious. The presence of *Nomada marshamella* or *N. flava*, both of which are very distinctive black and yellow bees, can also be a good sign that a nest site is nearby.

Behaviour and interactions

A. scotica often nests communally but also establishes solitary nests. These can be within very extensive aggregations. Communal nests represent a form of social behaviour and, while individual bees may share a nest entrance, there is no evidence of other co-operative behaviour within the nest. Many hundreds of bees may share the same nest entrance and main tunnel, with each female establishing and provisioning her own brood cells. Nests are established in a variety of situations, including in lawns and banks, within fields and beneath hedges.

In common with other species in the genus, *A. scotica* is parasitised by a conopid fly *Myopa buccata*. This fly might be seen at rest on a flowerhead waiting for an approaching bee. As the bee lands on a nearby flower to feed or forage, a female *M. buccata* will take off and land momentarily on the abdomen, in the process placing an egg inside the abdomen. This is achieved by inserting the ovipositor between two of the tergites. On hatching, the fly larva feeds internally on the bee's body tissue. The bee eventually dies, having first buried itself in the ground. *A. scotica* is also frequently attacked by a species of *Stylops*, an internal parasite.

Brood cells are targeted by two species, *N. marshamella* and *N. flava*.

Andrena scotica agg. ♀ with a *Stylops* parasite [PC]

Myopa buccata ♀ [PC]

On the wing

A. scotica is typically on the wing between early March until mid June. Any specimens swept during July are likely to be *A. trimmerana*.

Feeding and foraging

A very wide range of flowering plants is visited for pollen and/or nectar, including buttercup, stitchwort, White Bryony, willow, mustard, birch, apple, pear, Blackthorn, Raspberry, Hawthorn, Field Maple, Hogweed, Wood Spurge and dandelion.

Andrena semilaevis Shiny-margined Mini-miner
FIRST RECORD 1904, Lavant (Henry Guermonprez) **TOTAL RECORDS** 242

Geography and history
Andrena semilaevis is a widely distributed bee in Sussex and is found in a range of open habitats. These include heathland, species-rich chalk grassland and broadleaved woodland. It is also known from gardens.

Locations that have generated recent records include Great Dixter, Ashburnham Place, Waltham Brooks, Malling Down and Ditchling Common.

Out in the field and under the microscope
This is a common small black bee. It is tricky to separate both males and females from other similar species, but with the aid of a microscope it is possible to see that there are faint punctures on the

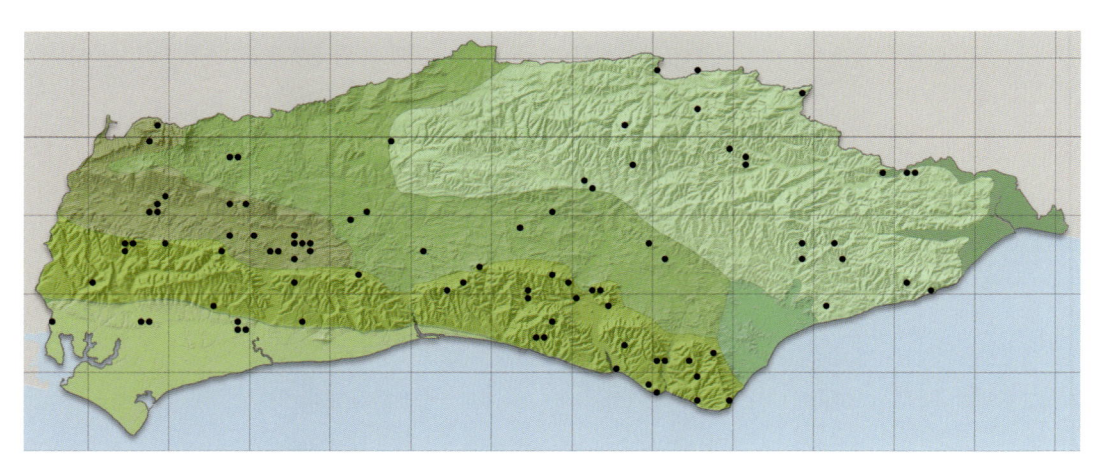

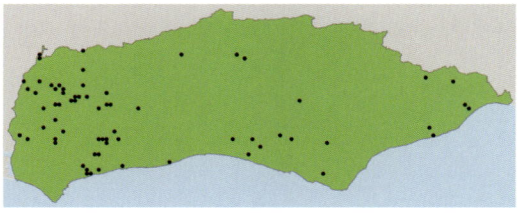

Large map: distribution 2000–2023
Small map: distribution 1844–1999

298 *The Bees of Sussex*

surfaces of tergites two and three, and that the marginal areas of tergites two, three and four are shiny and very smooth. These characteristics are not shared with any other small black *Andrena*. The tergites have a noticeably undulating appearance when viewed from the side.

This *Andrena* is often found visiting the umbels of plants in the Carrot family with white flowers—Steven Falk in his survey of the eastern South Downs reports finding it frequently on Cow Parsley. It has also been seen visiting plants such as Germander Speedwell and buttercup.

Cow Parsley [PC]

Behaviour and interactions
A. semilaevis has a particularly long flight period in some years, records extending from late April until late August. As elsewhere in southern England, the handful of Sussex records from July and August are indicative of a second generation which overlaps with the first.

Nesting has rarely been observed, although it is thought to nest solitarily.

Brood cells are targeted by *Nomada flavoguttata*.

On the wing
Sussex records run from late April until the beginning of August. In other parts of the country, it is reported to be on the wing until late August.

Feeding and foraging
Pollen is gathered from a wide range of plant species, including the Speedwell, Daisy and Carrot families.

Andrena simillima Buff-banded Mining Bee

FIRST RECORD 2018, West Sussex (unknown) TOTAL RECORDS 1

Geography and history

The status of *Andrena simillima* in Sussex is something of an enigma—there is a single reference to this bee in the county with "West Sussex" given as the location. This is in the *Handbook of the Bees of the British Isles*, published in 2018. It has not been possible to trace a specimen in any of the likeliest museums or private collections.

A. simillima is largely a coastal species, although a population was discovered on Salisbury Plain in Wiltshire in the 1990s. In counties adjoining Sussex, it is known from the east Kent coast (where it was last recorded in 1989) and from Hampshire. Elsewhere, it is restricted to scattered locations around the coast of south-west England, especially Cornwall.

Andrena simillima ♀ [PB]

Andrena simillima ♂ [SF]

Large map: distribution 2000–2023, left blank as the one record has no fixed location
Small map: distribution 1844–1999

A further potential twist is that this species may yet prove to be a form of *A. nigriceps*, a species that has been recorded on a handful of occasions in the county.

A. simillima is a species of coastal grasslands, soft-rock cliffs, chalk grasslands and, possibly, heathlands.

Out in the field and under the microscope

This *Andrena* is most frequently found visiting bramble, Greater Knapweed and Black Knapweed, but also plants such as Goldenrod, thistle and Wild Carrot.

Females most resemble *A. nigriceps*, although they are slightly smaller. Unlike this other species, *A. simillima* females have pale hairs on the face, and they usually lack the short black hairs on the corners of tergites three and four. But, as with *A. nigriceps*, the top of the thorax is covered in reddish-brown hairs.

Males of the two species are even closer in appearance. One character to look for is the length of the third antennal segment. This is clearly longer than the fourth segment in *A. simillima*, and only slightly longer than the fourth in *A. nigriceps*.

Behaviour and interactions

Patrick Saunders, working in Cornwall, has shown that bramble and Black Knapweed are important sources of pollen for *A. simillima*. He also described this species as establishing solitary nests and small aggregations with between 10 to 15 females nesting close together in one location. Nests were established in areas of "sloping or vertical bare ground", such as along the edge of a compacted path and in a west-facing hedge-bank.

Its brood cells are believed to be targeted by *Nomada rufipes*, although this relationship has not been confirmed here or on the continent.

On the wing

A. simillima is on the wing during July and August.

Feeding and foraging

This bee is mainly associated with plants in the Daisy family, particularly Greater and Black Knapweed which are important sources of pollen. Bramble is also an important source of pollen.

Greater Knapweed [JP]

Andrena subopaca Impunctate Mini-miner
FIRST RECORD 1902, Woolbeding (Henry Guermonprez) **TOTAL RECORDS** 362

Geography and history
Andrena subopaca is a very widely distributed bee in Sussex. It is often found in woodlands—Park Corner Heath, Powdermill Wood, Ebernoe Common and West Dean Woods have all generated recent records—but it is also known from other habitats. These include species-rich chalk grasslands across the South Downs, heathlands on the Lower Greensand and in the Low and High Weald, as well as areas of lowland meadow and pasture, also in the High Weald.

It is often considered to be a woodland bee, but further north in Britain it commonly occurs in habitats which lack trees, including open moorlands and acid grasslands.

Park Corner Heath [NS]

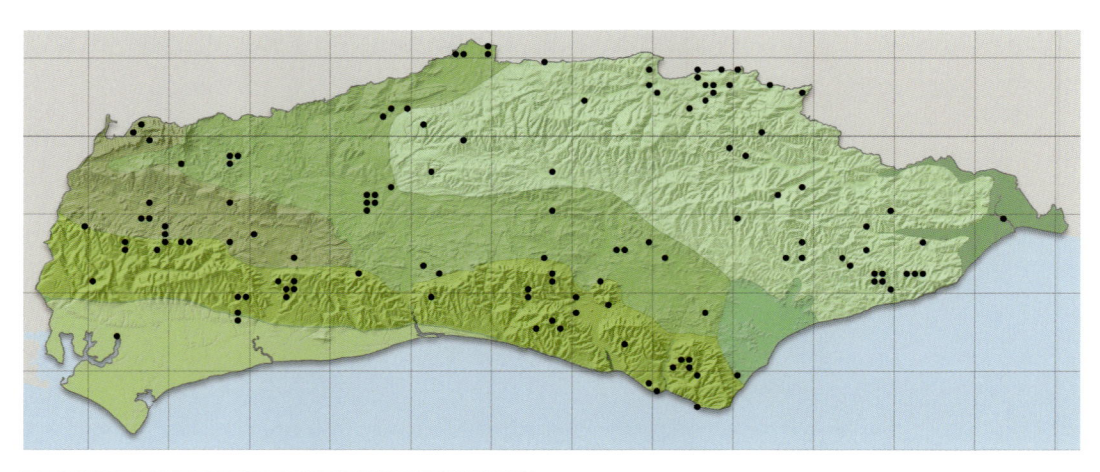

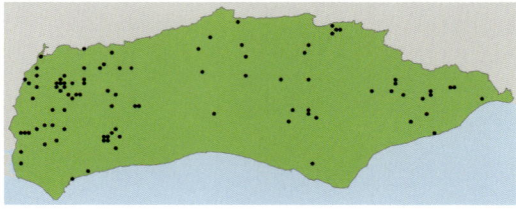

Large map: distribution 2000–2023
Small map: distribution 1844–1999

Out in the field and under the microscope
This small black *Andrena* is a challenging species to identify. It most resembles *A. minutula* and *A. minutuloides*, but females can be distinguished from these other species by the fewer, finer punctures on the scutum—the punctures resemble very fine pinpricks. Also, *A. subopaca* females have a duller scutum and more widely separated punctures on the clypeus.

Males are trickier still to identify. As with the females, they have relatively few pinprick-like punctures on the scutum. Two key characters to look for are the length of the antennal segments—the fourth and fifth combined are longer than the third—and the marginal area of the first abdominal segment which is quite dull, aside from a narrow, very shiny rim along the hind edge. Unlike the first generation of *A. minutula*, the hairs on the clypeus are white.

A. subopaca can be abundant in sunny, sheltered areas (especially within woodland), and searching the flowers of known foodplants is a good way of locating this bee.

Andrena subopaca ♀ [SF]

Andrena subopaca ♂ [SF]

Behaviour and interactions
This species nests in small aggregations which are generally sited on bare or sparsely vegetated banks. South-facing banks in woodland have been known to be used.

Brood cells are targeted by *Nomada flavoguttata*.

On the wing
In Sussex, most records run from the end of March until the end of June. As elsewhere in the country, there are a small number of records between July and September, and these support the view that *A. subopaca* possibly has two generations a year. The first is clearly much more abundant, however, as most records are from April, May and June.

Feeding and foraging
Pollen is collected from a range of plants, including members of the Cabbage, Pink, Speedwell and Rose families.

Andrena synadelpha Broad-margined Mining Bee
FIRST RECORD 1888, Bognor Regis (Henry Guermonprez) TOTAL RECORDS 68

Geography and history
Andrena synadelpha is a local bee that has been largely recorded from the south and west of the county. It is found in a range of habitats, including woodland, chalk grassland and hedgerows. In other parts of the country, it has also been recorded in gardens.

Most modern records in Sussex are from the South Downs, with recent records from places such as Ditchling Beacon, Seaford Head, Southerham Farm and Castle Hill. Woodlands that have provided recent records include Fore Wood and Ebernoe Common.

Fore Wood [IB]

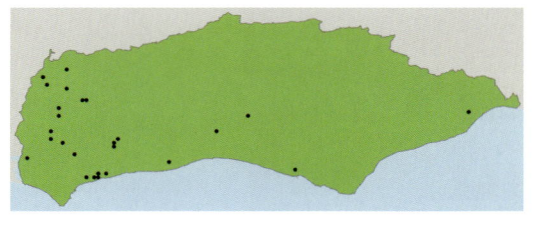

Large map: distribution 2000–2023
Small map: distribution 1844–1999

304 *The Bees of Sussex*

Out in the field and under the microscope
Females have brown hairs on the top of the thorax and long hairs on the first two tergites. A key character to look for is the exceptional length of the marginal areas on tergites three and four—these occupy most of the tergite, a character not found in any other species of *Andrena*.

Males are much smaller and have extremely long mandibles. From the side, the back of the head appears right-angled. They most resemble *A. varians* males, but they can be separated from that species by the presence of black hairs along the inner margins of the eyes.

It is not an easy bee to find, but searching flowering shrubs is one good strategy. George Else has reported finding it visiting Hawthorn and Field Maple flowers in sunlit clearings in the New Forest, and Steven Falk has also highlighted this association on the South Downs.

Andrena synadelpha ♀ [PC]

Andrena synadelpha ♂ [NO]

Behaviour and interactions
On the continent it has been suggested that *A. synadelpha* nests in large aggregations. However, in Britain it has only ever been observed nesting solitarily or in small groups.

Brood cells are targeted by *Nomada panzeri*.

Field Maple [PC]

On the wing
Sussex records run from early April until late July.

Feeding and foraging
A. synadelpha collects pollen from a very wide range of plant families, including Beech, Cabbage, Primrose, Maple and Rose.

The Bees of Sussex

Andrena tarsata Tormentil Mining Bee
FIRST RECORD 1896 (Hastings, Edward Saunders) TOTAL RECORDS 20

Geography and history
Andrena tarsata is a very scarce bee that has not been seen in Sussex since 2009. The very first reference to it in the county is from Hastings in 1896, but there were then no further records until it was found near Crowborough in both 1931 and 1932. It was then recorded sporadically throughout the 1970s, 1980s and 1990s from St Leonards Forest, Hastings, Ashdown Forest and Ambersham Common.

Between 1996 and 2009, Ian Beavis made a series of records from Broadwater Warren and the area south of Tunbridge Wells. These, plus a solitary record from Lynchmere Common in 2000 and a solitary record from Ashdown Forest in 2004, are the only recent records for *A. tarsata* in Sussex.

Nationally, this bee has a northerly distribution and is associated with moorlands, heathlands, acid grasslands and, less frequently, open woodlands. It may be in decline in southern England (it was last recorded in Kent in 1986), but appears to be stable in the northern part of its range.

Broadwater Warren [IB]

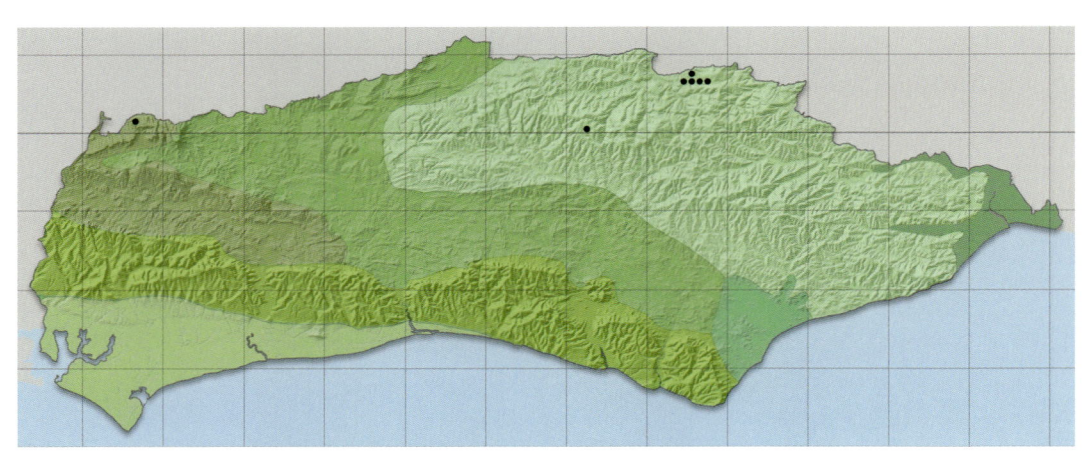

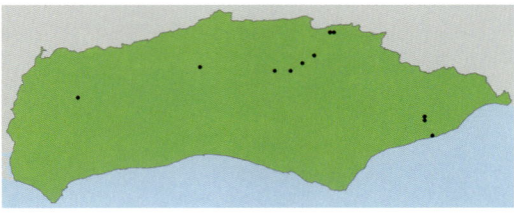

Large map: distribution 2000–2023
Small map: distribution 1844–1999

Out in the field and under the microscope

This is a small dark bee, and females could easily be confused with several other species of small black *Andrena*. However, both sexes have pale orange-brown hind tarsi whereas these other species are completely dark here. *A. tarsata* is the only species of *Andrena* where the females have three teeth on the mandibles.

Males have a white marking on the clypeus and could only be confused with *A. coitana*, although the white marking is slightly more extensive in this last species.

The main source of pollen for *A. tarsata* is Tormentil, and therefore any patches of this plant should be checked for this bee. Tormentil grows in places with short turf and acidic soils, and these might include heathlands, bogs and woodlands. In Broadwater Warren, Ian Beavis recorded females "gathering pollen from [Tormentil] growing at the edge of heathy rides". He also noted several males "flying around a patch of sunlit bracken".

Andrena tarsata ♀ [PS]

Andrena tarsata ♂ [PS]

Behaviour and interactions

A. tarsata nests in sparsely vegetated areas, usually solitarily but occasionally in aggregations, often in south-facing banks or areas of sloping ground. In one report from Scotland by Andrew Jarman and Brian Little, an aggregation of 50 to 70 burrows had been established along a path for about 40 m. Nearby, a further 40 to 50 burrows were also discovered in an area of bare ground covering 2 m². Also in Scotland, some 100 males were found close to a nest site, either patrolling slowly or at rest on vegetation. Females returning to the nest were nearly always followed by males.

Brood cells are targeted by *Nomada roberjeotiana* (not known from Sussex) and *N. obtusifrons*.

On the wing

A. tarsata is on the wing from mid June until early September. Modern Sussex records are from July and August.

Feeding and foraging

Pollen is collected almost exclusively from cinquefoil, particularly Tormentil.

Andrena thoracica Cliff Mining Bee

FIRST RECORD 1858, coast between Brighton and Eastbourne (William Unwin) **TOTAL RECORDS** 77

Geography and history

The earliest reference to *Andrena thoracica* in Sussex is from 1858, when William Unwin wrote, "This bee appears to usually effect [sic] the coast and has been taken at Newhaven, Seaford, Brighton and Eastbourne, in May and June, not uncommonly." Thirty years later, William Bennett described *A. thoracica* as "common at Hastings, Ecclesbourne and Fairlight". This last stretch of coast is now the only known location for this bee in Sussex.

Hastings Country Park, Covehurst undercliff [AP]

Large map: distribution 2000–2023
Small map: distribution 1844–1999

Between 1896 and 1922, it was regularly recorded in Felpham by Henry Guermonprez, but it has not been recorded here since then.

Nationally, it is known from heathlands in inland areas, and in Sussex it was found at Ambersham Common between 1974 and 1982, and at Heyshott Common in 1979.

Out in the field and under the microscope

It is hard to miss this bee. The females are among the largest of the *Andrena* and have a covering of dense reddish-brown hairs on the top of the thorax, black pollen-collecting hairs on the hind legs, and a shiny black abdomen that has a covering of dense, very short black hairs on the third, fourth and fifth tergites.

The males resemble the females but are much smaller.

The best way to find this bee is to search along cliff faces where the bee is known to nest. Here, it is possible to find females returning to their nests with a pollen load or males either flying rapidly along the cliff face or at rest on the surface of the rock.

The undercliffs at Hastings Country Park and the low cliffs between Galley Hill and Glyne Gap are reliable locations to search.

Andrena thoracica ♀ [AP]

Andrena thoracica ♂ [NO]

Behaviour and interactions

A. thoracica nests both in cliffs and on areas of flat ground, including on worn pathways. It can form enormous aggregations—there is a report from the continent of 1,000 nests in one location.

Brood cells are targeted by *Nomada goodeniana*. On the continent, *N. fulvicornis* is also given as a parasite.

On the wing

This species has two flight seasons, the first running from late February until mid May and the second running from late June until the end of August.

Feeding and foraging

In the spring, brood cells are provisioned with pollen from a variety of plants, including members of the Buttercup, Elm, Willow, Maple and Daisy families. In addition, females collect pollen from shrubs in the Rose family such as Blackthorn.

In the summer, females collect pollen from a wide range of other plants, including members of the Beech, Knotweed, Lime, Rose, Pea and Daisy families. The 1979 record from Heyshott Common is of a bee swept off a Creeping Thistle flower.

Andrena tibialis Grey-gastered Mining Bee

FIRST RECORD 1882 (Hastings, Edward Saunders) TOTAL RECORDS 49

Geography and history

Andrena tibialis is a scarce bee which has been recorded four or five times a year at most. It has largely been known from four main areas of the county—the South Downs (especially the easternmost section), the Wealden Greensand, the coast between Hastings and Camber Sands, and sites to the south of Tunbridge Wells.

It is found in a variety of different habitats, including heathland, chalk grassland and sandpits. It is also found in gardens—throughout the 1990s it was regularly recorded in a garden in Midhurst, and in 2020 it was found in a garden in Lewes.

Out in the field and under the microscope

This is one of only a few *Andrena* which have a small, raised ridge in the middle of the propodeum.

Females have a shiny black abdomen that has greyish hairs on the upper surface, and pale orange hind tibiae with pale orange pollen-collecting hairs Freshly emerged females have a covering of brown hairs on the top of the thorax (which fade to white as the bee ages) and white hairs on the sides and rear of the thorax as well as on the face.

Andrena tibialis ♀ [SF]

Large map: distribution 2000–2023
Small map: distribution 1844–1999

310 *The Bees of Sussex*

The males are smaller than the females and have pale orange tips to the otherwise dark hind tibiae.

This *Andrena* is frequently found on plants such as dandelion, buttercup and willow. Males might also be found patrolling around hedgerows or isolated trees and shrubs, sometimes in the company of other species such as *A. bimaculata*.

Behaviour and interactions
A. tibialis establishes solitary nests in sunny areas with short or sparse vegetation. It has a single flight period in Britain, but in some areas on the continent it has two generations a year.

In Britain brood cells are attacked by *Nomada fulvicornis*. *N. goodeniana* is also cited as a parasite on the continent.

Andrena tibialis ♂ [SF]

On the wing
Sussex records run from mid March until late June.

Feeding and foraging
The females collect pollen from a wide range of plants, including members of the Buttercup, Willow, Maple and Rose families.

Wild Cherry, a member of the Rose family and a confirmed source of pollen for *Andrena tibialis* [PC]

Andrena trimmerana Trimmer's Mining Bee
FIRST RECORD 2011, Iping Common (James Power) TOTAL RECORDS 7

Geography and history
A paper published in 2022 has shown that modern keys for this species are not reliable and that many forms of *Andrena trimmerana* are in fact indistinguishable from *A. scotica*. As a result, almost all of the 195 pre-2022 records for *A. trimmerana* have been assigned to the aggregate *A. scotica* agg., and only those where it has been possible to check a specimen have been confirmed as *A. trimmerana*.

Nonetheless, *A. trimmerana* is believed to be widely distributed in scrub-rich habitats dominated by Blackthorn, particularly along the coast, on the South Downs, on the Wealden Greensand and on the High Weald.

Confirmed locations are Iping Common, Chichester Harbour, the cliffs to the west of Newhaven, Abbot's Wood, the soft-rock cliffs at Bulverhythe, Hastings Country Park and Rye Harbour Nature Reserve.

Andrena trimmerana ♂s might be seen patrolling above areas of coarse vegetation, as here at Hastings Country Park. [JP]

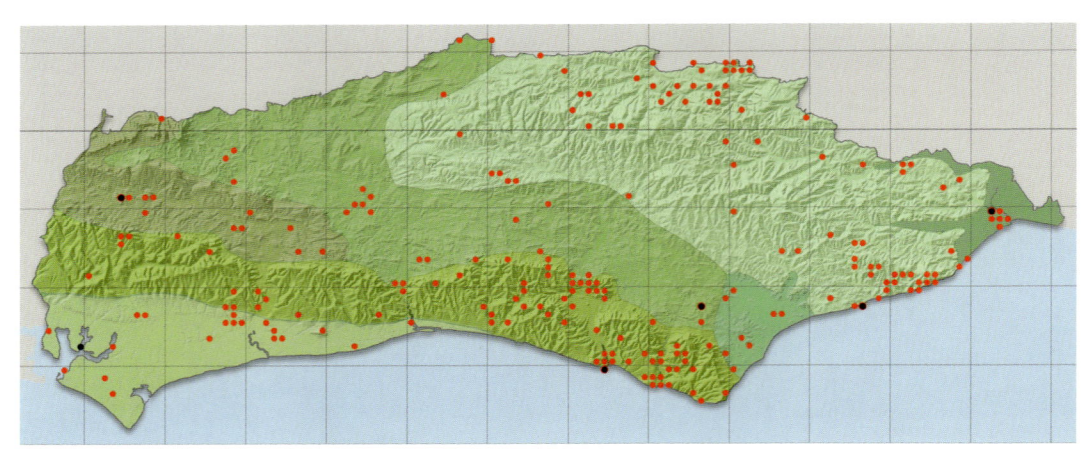

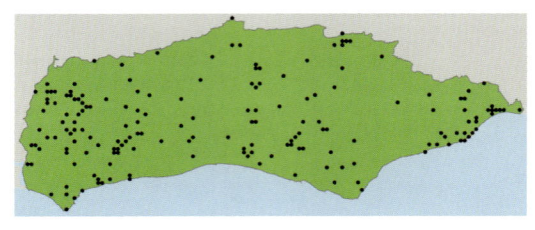

Large map: distribution 2000–2023, with black dots showing distribution of *Andrena trimmerana* and red dots showing distribution of *Andrena scotica* agg.

Small map: distribution 1844–1999, showing distribution of *Andrena scotica* agg.

312 *The Bees of Sussex*

Andrena scotica agg. ♀ [PB]

Andrena trimmerana ♂ [SF]

Andrena scotica ♂ with two teeth at the tip of the mandibles [IT]

Andrena trimmerana ♂ with one tooth at the tip of the mandibles [IT]

Out in the field and under the microscope
Only spring-flying males can be confidently identified as *A. trimmerana*. These are dark bees with very long mandibles that lack the second tooth found in *A. scotica* males. They also have a long spine on the gena, near the base of the mandibles—in *A. scotica* this spine is either reduced or absent.

In all other cases, the flight period is a better guide to identification, with any males or females flying from mid-June onwards likely to be *A. trimmerana*. Even this is not certain, however, as individuals of *A. scotica* are very occasionally found outside the normal flight period.

Searching flowering Blackthorn scrub and willow in spring, plus bramble and thistle in the summer, is a good strategy to locate *A. trimmerana*.

Behaviour and interactions
A. trimmerana nests solitarily in banks or in areas of sloping ground. Ian Beavis reports nests "in a south-facing bank [. . .] in bare vertical ground at the edge of an arable field (bordered by woodland)". In 2023, a series of nests were observed within a shallow wheel rut within Ashcombe Bottom.

Nomada marshamella attacks the brood cells of this species.

On the wing
Sussex records typically run from early March until mid May and again from mid June until mid September.

Feeding and foraging
Pollen is collected from Blackthorn, willow and gorse in the spring, and from bramble and thistle in the summer.

Andrena vaga Grey-backed Mining Bee
FIRST RECORD 1945, Bignor (Christopher Andrewes) TOTAL RECORDS 28

Geography and history
In Sussex, *Andrena vaga* was first recorded in Bignor in 1945, some 72 years before it was found in the county again, this time at Rye Harbour Nature Reserve. The very first record was made by Christopher Andrewes in a garden where a female was investigating a bank being used by large numbers of nesting *A. flavipes*.

This Sussex record was only the third for Britain, and Christopher Andrewes speculated "whether this fine species is truly resident, or in a marvellous year for butterfly-immigration, it came from across the Channel". This specimen is in the Natural History Museum.

After its discovery in Bignor in 1945, *A. vaga* was recorded in Kent the following year, but it then disappeared from the country until its rediscovery on the coast, again in Kent, in 2008. Since then, it has also been found on the coast in Hampshire and Essex, and inland in Kent.

In 2017 Peter Greenhalf made the discovery that *A. vaga* had found its way back to Sussex when he netted a female covered in yellow pollen at Rye Harbour Nature Reserve. This was on a small sandy bank close to extensive stands of willow. *A. vaga* was seen here again in 2018 and 2019. It is also now known from Dungeness in Kent, just along the coast from here.

Cliffs and beach at Glyne Gap [AP]

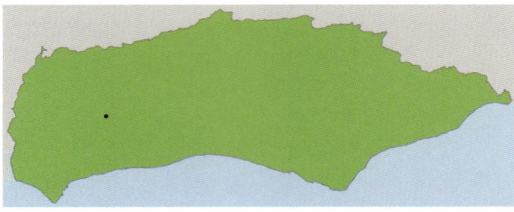

Large map: distribution 2000–2023
Small map: distribution 1844–1999

In 2018 Andy Phillips discovered *A. vaga* nesting in the cliffs at Glyne Gap. Again, stands of willow were nearby. The colony here is small—there was a maximum count of ten individuals in 2022—with the nests in a small area of crumbling, sandy cliff measuring 2 by 1 m.

Out in the field and under the microscope

As this *Andrena* may be expanding its range, any location with areas of exposed sandy cliffs or level sandy ground close to abundant willow is worth investigating in the spring.

This bee could only be confused with the similar *A. cineraria*. In male and female *A. vaga*, however, the top of the thorax is entirely grey-haired whereas *A. cineraria* has a band of black hairs here.

Andrena vaga ♀ [SF]

Andrena vaga ♂ [SF]

Behaviour and interactions

At Glyne Gap this species nests in the south-facing sandy cliffs. Here, the females excavate their tunnels into areas of loose friable sand, with the nest entrances often concealed by a covering of sand. As they return to their nests, the females can be seen carrying yellow willow pollen in their pollen baskets. Males can be seen at rest on the cliff face or investigating the surface of the sand, presumably searching for emerging unmated females.

At Rye Harbour Nature Reserve, Peter Greenhalf reports that only a few nests have ever been found and that these are sited on a very small bank. On the continent, however, *A. vaga* can form enormous nesting aggregations with up to 10,000 nests at a density of more than 30 nests per m².

Brood cells are targeted by *Nomada lathburiana*.

Andrena vaga nest site at Rye Harbour Nature Reserve [PG]

On the wing

Sussex records run from late March until the end of April.

Feeding and foraging

The brood cells are provisioned exclusively with willow pollen. Unlike *Colletes cunicularius*, another willow specialist, *A. vaga* does not seem to move on to other pollen sources when willow pollen availability declines.

Andrena varians Blackthorn Mining Bee
FIRST RECORD 1882, Hastings (Edward Saunders) TOTAL RECORDS 70

Geography and history
Most Sussex records for *Andrena varians* are from Blackthorn-dominated areas of scrub on the easternmost South Downs and from along the coast between Pebsham and Camber Sands. Seaford Head, for example, has generated a significant proportion of the county records. Elsewhere, there are sporadic records from scattered locations such as Easebourne, Lyminster and Battle.

Out in the field and under the microscope
A. varians is best searched for by checking flowering Blackthorn scrub in spring. Here, females can be found foraging while males chase each other in and around the bushes. The scrub at South Hill Barn near Seaford Head is a good location to search. Males and females might also be found feeding on other flowering plants such as dandelion and buttercup.

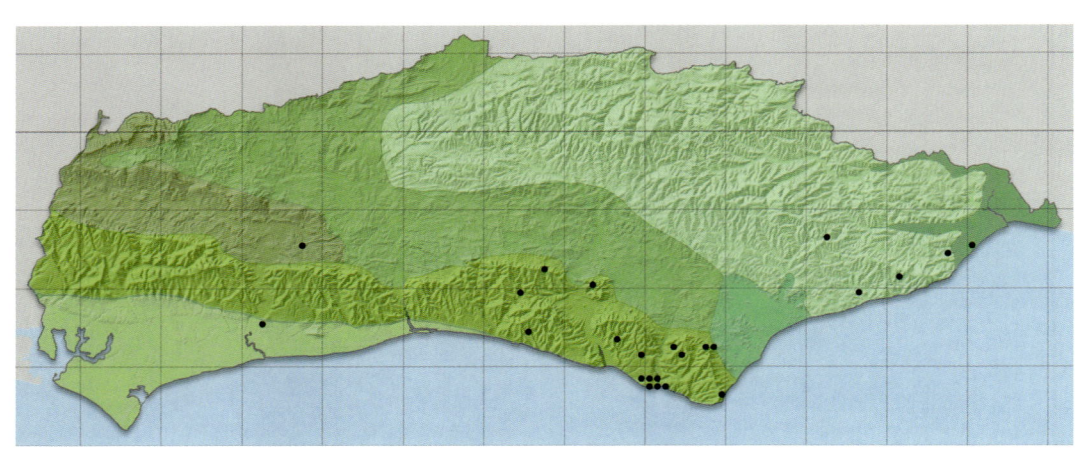

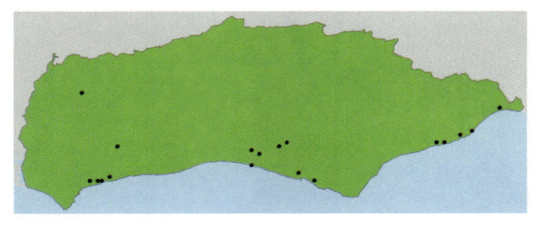

Large map: distribution 2000–2023
Small map: distribution 1844–1999

316 *The Bees of Sussex*

Females are quite variable but typically have dark brown hairs on the face, reddish-brown hairs on top of the thorax and long pale hairs restricted to tergites one and two. Steven Falk has pointed out that on the eastern South Downs a different form of this bee, referred to as var. *mixta*, is often found. Females of this form have pale hairs on the face and long pale brown hairs on the third and fourth tergites, as well as on tergites one and two.

Males are less variable and have extremely long mandibles. They can be separated from the very similar *A. synadelpha* by the absence of any black hairs on the inner margins of the eyes.

Andrena varians var. *mixta* ♀, showing pale hairs on the clypeus and long hairs on tergites one to four [SF]

Behaviour and interactions
A. varians nests either solitarily or in scattered aggregations. Nests tend to be sited in sunny locations, often in areas of short-cropped chalk grassland.

Brood cells are known to be targeted by *Nomada panzeri*, but *N. ferruginata* and *N. fabriciana* may also target this species.

On the wing
Sussex records run from mid March until late May.

Feeding and foraging
While *A. varians* is closely linked to Blackthorn, many other plant species are also visited for pollen. Among these are flowering plants in the Buttercup, Cabbage, Willow, Holly, Maple and Daisy families.

A. varians is an important pollinator of fruit trees.

Blackthorn [PC]

Andrena wilkella Wilke's Mining Bee
FIRST RECORD 1896, Slindon (Henry Guermonprez) TOTAL RECORDS 237

Geography and history
Andrena wilkella is widely distributed across much of Sussex and is found in a range of flower-rich habitats such as species-rich chalk grassland, lowland meadow and pasture, open woodland and heathland.

Locations that have provided recent records include Warnham, Nymans, Chailey Common, Birling Gap and Hastings Country Park.

Andrena wilkella is frequently seen visiting plants in the Pea family such as Kidney Vetch, here growing on the undercliff at Hastings Country Park. [AP]

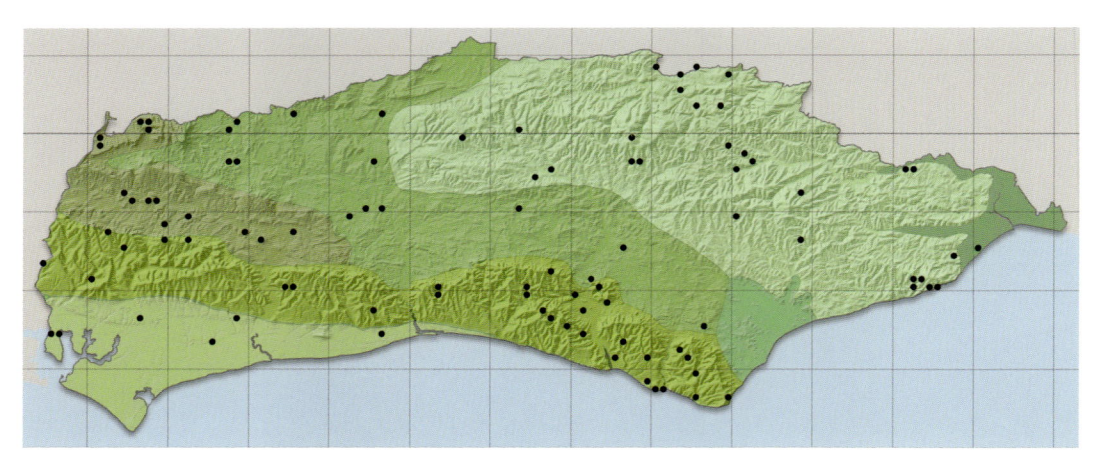

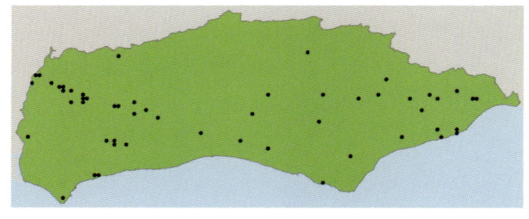

Large map: distribution 2000–2023
Small map: distribution 1844–1999

Out in the field and under the microscope

A. wilkella females most resemble *A. afzeliella*, *A. ovatula* and *A. russula* females, and all have narrow bands of white hairs along the hind margins of the tergites.

A key distinction between *A. wilkella* and the first two of these other species—*A. afzeliella* and *A. ovatula*—is that in *A. wilkella* the hair band on the third tergite is broken in the middle, while in *A. afzeliella* and *A. ovatula* it is continuous.

A. russula also has an interrupted hair band on tergite three, but this is more widely broken than in *A. wilkella*. Fresh specimens of *A. russula* have a covering of reddish-orange hairs on the top of the thorax, while *A. wilkella* has pale brown hairs here.

Males broadly resemble the females—they also have narrow bands of white hairs on the hind margins of the tergites—but are smaller and slimmer. They can be separated from male *A. ovatula* and *A. afzeliella* by the broken or very narrowed hair band on the third tergite. The hairs here also shorten towards the middle of the tergite. *A. russula* males, on the other hand, have very reduced hair bands that are restricted to the sides of tergites two, three and four, although these do wear away with age and may be absent.

In areas of open habitat *A. wilkella* is often found visiting plants such as bird's-foot-trefoil, Kidney Vetch and clover. Males might also be seen flying rapidly around hedgerows and blocks of scrub.

Andrena wilkella ♀ [NO]

Andrena wilkella ♂ [SF]

Behaviour and interactions

This *Andrena* nests either solitarily or in large aggregations.

Brood cells are targeted by *Nomada striata*.

On the wing

Sussex records run from mid April until late September. Most records are from May and June—the few records from July, August and September may indicate a second generation.

Feeding and foraging

While females collect pollen from a wide range of plant families, the Pea family is a particularly important source. Other families visited for pollen include Knotweed, Willow, Cabbage and Maple.

Panurgus – Shaggy Bees

This is a small genus of short-tongued black bees with an unkempt appearance. Pronounced pollen-collecting hairs are restricted to the hind tibiae, and, unlike in any other British species, these are long and threadlike, appearing in the form of a spiral or zigzag. *Panurgus* bees nest in the ground, forming loose aggregations that can be quite extensive. Each brood cell is lined with a wax-like substance. All species collect pollen from yellow-flowered plants in the Daisy family, the females crawling across the flower heads to cover themselves in pollen. There are two species in Britain, both found in Sussex—*Panurgus banksianus* and *P. calcaratus*.

Panurgus banksianus Large Shaggy Bee
FIRST RECORD 1884, Hastings (Edward Saunders) TOTAL RECORDS 32

Geography and history
Panurgus banksianus is a species of dry sandy heathlands and acid grasslands, and is now restricted to two areas of Sussex—the High Weald (especially Ashdown Forest) and heathlands on the Lower Greensand. This bee appears to have become very scarce, and there have been just seven records since 2000. In the late nineteenth century, it was also known from the Hastings area, but it has not been recorded from here since 1884.

The most recent Sussex records are from Weavers Down where it was recorded in 2008, from Old Lodge on Ashdown Forest where it was recorded in 2014, and from a garden in Midhurst where it was recorded in 2018.

Out in the field and under the microscope
A good strategy for finding *P. banksianus* is to look for plants in the Daisy family with yellow flower heads such as Rough Hawkbit and Cat's-ear. Here, this black bee will stand out clearly against the yellow background. It is also possible to come across a nesting aggregation sited along the edge of a footpath.

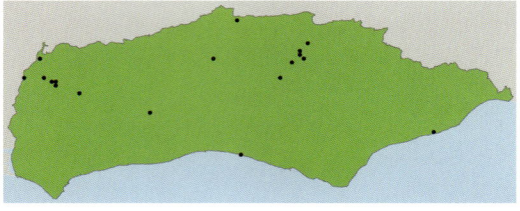

Large map: distribution 2000–2023
Small map: distribution 1844–1999

Panurgus banksianus ♀ [NO]

Panurgus banksianus ♂ [SF]

    This is a very shiny black bee that has a sparse covering of black hairs. Females have yellowish-orange pollen-collecting hairs on the hind legs and many long black hairs on the first four tergites, in contrast to the otherwise similar *P. calcaratus* which has short hairs here.

    Males resemble the females and are only slightly smaller. The short antennae are black in both sexes, in contrast to the antennae of *P. calcaratus* which are partially orange.

Behaviour and interactions

P. banksianus establishes its nests in loose aggregations in areas of firm, sparsely vegetated ground that is often compacted. The aggregations can be extensive. This species is unusual in that each female excavates two U-shaped tunnels that connect below ground. Surplus soil spills out to create a small mound.

    There are no cuckoo bees known to target the brood cells of this species in Sussex.

On the wing

Sussex records run from late June until late July. Elsewhere, this bee is reported to be on the wing from mid June until late August, or occasionally even early September.

Feeding and foraging

Pollen is collected exclusively from yellow-flowered plants in the Daisy family such as hawkweed and Cat's-ear.

Panurgus banksianus **is most frequently found by searching yellow-flowered plants in the Daisy family.** [JP]

The Bees of Sussex

Panurgus calcaratus Small Shaggy Bee
FIRST RECORD 1884, Hastings (Edward Saunders) TOTAL RECORDS 45

Geography and history
Panurgus calcaratus has a similar distribution in Sussex to *P. banksianus*, with the exception that it is still found in the Hastings area. This is where Edward Saunders first reported its presence in Sussex in 1884.

As with *P. banksianus*, *P. calcaratus* is known from Ashdown Forest and from the Lower Greensand. Recent records have come from Gills Lap Quarry, Easebourne and Midhurst. It has also recently been recorded from Great Wood near Battle and from Hastings Country Park.

P. calcaratus is a species of sandy heathlands and acid grasslands.

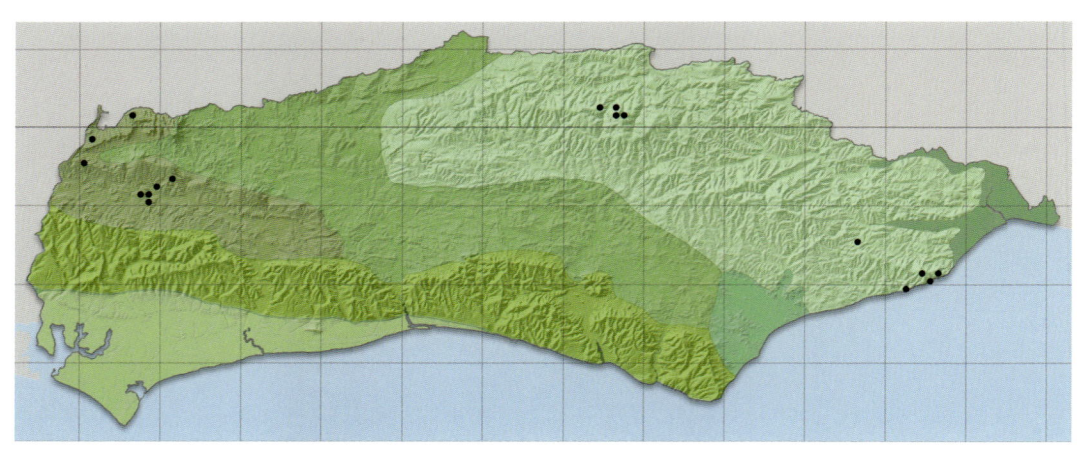

Panurgus calcaratus ♀ [NO]

Panurgus calcaratus ♂ [PB]

Large map: distribution 2000–2023
Small map: distribution 1844–1999

Out in the field and under the microscope

P. calcaratus is the smaller of the two British species in the genus and, like *P. banksianus*, is a black and shiny bee. Females have whitish pollen-collecting hairs, while the males can be separated from *P. banksianus* males by the presence of a small square tooth on the underside of the hind femur—this is absent in *P. banksianus*.

This bee is best searched for by checking yellow-flowered plants in the Daisy family—here females might be seen foraging while males might be found sheltering within the flower heads during bouts of poor weather. These plants often open just in the morning and early afternoon, limiting the time available to search. At nest sites, it is possible to find large numbers of males flying rapidly back and forth just above the ground, looking for an opportunity to mate.

Behaviour and interactions

This species nests communally, with between two and ten females sharing the same nest. They do not, however, appear to share tasks within the nest, each female taking responsibility for excavating and provisioning her own brood cells. These are lined with a wax-like substance, and the access tunnels are blocked with soil once the female has laid her egg on a round ball of pollen and nectar. Nest sites can be used repeatedly over several years.

There are no cuckoo bees known to target the brood cells of this species in Sussex.

Panurgus calcaratus ♀s at communal nest entrance [PB]

Panurgus calcaratus nest site at Hastings Country Park [JP]

On the wing

Sussex records run from the beginning of July until the end of August.

Feeding and foraging

Pollen is collected exclusively from yellow-flowered plants in the Daisy family. These include Cat's-ear, hawkweed, Common Fleabane, hawk's-beard and hawkbit.

Halictus – End-banded Furrow Bees

This is a genus of ground-nesting bees with short pointed tongues and bands of hairs along the hind margins of the tergites. This contrasts with bees in the closely related genus *Lasioglossum* in which all but one species have bands of hairs at the front of the tergites. Females in both genera have a distinctive patch of hairs in the middle of the fifth tergite, known as the 'rima'. Many species are solitary, but others are eusocial. Brood cells of several species are targeted by various species of *Sphecodes*. Six species have been recorded in Britain, of which five have been found in Sussex. Of these, the four still found here are *Halictus confusus*, *H. eurygnathus*, *H. rubicundus* and *H. tumulorum*, with *H. maculatus* long extinct in the county.

Halictus confusus Southern Bronze Furrow Bee
FIRST RECORD 1975, Ambersham Common (Mike Edwards) TOTAL RECORDS 27

Geography and history
Halictus confusus is a very scarce bee in Sussex and is restricted to the light dry sandy soils found on a handful of heathlands on the Lower Greensand. These include Ambersham, Midhurst and Iping Commons. There have been just seven records since 2000, most recently from Iping Common in 2022.

Across Britain, *H. confusus* is restricted to sites with sandy soils within a band stretching from Dorset to Norfolk.

Out in the field and under the microscope
Mike Edwards has reported that *H. confusus* is generally found visiting either Heather or cinquefoil, while Glenn Norris's 2022 record from Iping Common is of a male flying low over a sheltered south-facing area of completely bare sand.

This is one of two species in the genus in which the thorax and abdomen are a metallic bronze colour, the other being *H. tumulorum*.

As with this last species, *H. confusus* females have obvious bands of white hairs on the hind margins of the tergites. These are broad and uninterrupted on the second and third tergites in fresh

Large map: distribution 2000–2023
Small map: distribution 1844–1999

Iping Common [ME]

specimens, while in *H. tumulorum* these bands are narrowed in the centre. Worn specimens of *H. confusus* can particularly resemble this other species.

Males can be separated with confidence from *H. tumulorum* by examining the genital capsule and by checking the colour of the trochanter. In this last species, the trochanters on the middle and hind legs are largely black, while in *H. confusus* they are largely yellowish brown.

Behaviour and interactions
This species is eusocial, the queen rearing a small number of workers which then help her to raise her brood.

In Britain, no species of cuckoo bee is believed to target brood cells of this species.

On the wing
Sussex records run from early April until early September.

Feeding and foraging
Pollen is collected from a wide range of plants, including members of the Bellflower, Cabbage, Pea and Daisy families.

Halictus confusus ♀ [SF]

Halictus confusus ♂ [TF]

The Bees of Sussex 325

Halictus eurygnathus Downland Furrow Bee

FIRST RECORD 1881, Eastbourne and Falmer (Sidney Saunders) **TOTAL RECORDS** 83

Geography and history

The earliest confirmed record for *Halictus eurygnathus* in Britain is from 1881, when Sidney Saunders reported, "On the 3rd of September, I captured four males and about a dozen females of this species on the Brighton Downs near Falmer. Also a few of the latter near Eastbourne." Subsequently, it was also then found in Kent, Dorset and Suffolk.

In Sussex, between that first record in 1881 and the last in 1944, *H. eurygnathus* was widely distributed on the eastern South Downs, with records from places such as Litlington, Seaford and the Brighton Downs. The final Sussex record—from the Brighton Downs—was also the last in Britain, and *H. eurygnathus* was then believed to be extinct.

Southerham Farm, where *Halictus eurygnathus* was rediscovered in Britain after a gap of sixty years. [PG]

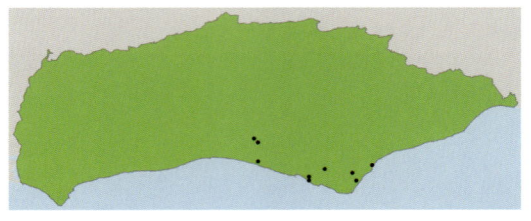

Large map: distribution 2000–2023
Small map: distribution 1844–1999

326 *The Bees of Sussex*

In May 2004, however, Steven Falk rediscovered this species at Southerham Farm and then at Mount Caburn and Beachy Head. By 2008 he had located populations at seven different species-rich chalk grassland sites on the South Downs, all in an area bounded by Brighton, Lewes and Eastbourne. *H. eurygnathus* is now frequently found on a number of different sites within this area, including a vineyard near Alfriston where it was recorded in 2017 and a flower-rich arable field margin near Kingston where it was recorded in 2021.

Until 2019, this area of the South Downs was the national range for this species. However, during a survey near Slindon in 2019, Adam Wright swept a male from a knapweed flower and, in the following year, swept a female from the same site. In 2022, a female was swept from a former reservoir, also near Slindon.

H. eurygnathus is still not known outside Sussex.

Out in the field and under the microscope

Since its rediscovery, this has been an abundant bee in some years. For example, in July 2020 it was very frequent at Castle Hill, with many Greater Knapweed flowers supporting at least one bee. The most reliable technique is to search Greater Knapweed and Field Scabious where males and females are often found head down within the flowers.

H. eurygnathus is slightly smaller than the otherwise similar *H. rubicundus*. Like females of this other species, female *H. eurygnathus* have obvious bands of white hairs on the hind margins of tergites one to four but have dark—as opposed to yellow—hind tibiae.

Males are unusual in that the underside of the head is hollowed out in a very distinctive fashion. One other character that helps to identify the males is the colour of the antennal segments. These are a pale orange on the underside.

Halictus eurygnathus ♀ on Greater Knapweed [PB]

Halictus eurygnathus ♂ [PB]

Behaviour and interactions

George Else has investigated nesting behaviour of *H. eurygnathus* in Sussex. His observations, and the fact that this is a late-emerging bee, support the view that *H. eurygnathus* is a solitary-nesting species in Britain. Elsewhere in Europe, *H. eurygnathus* has been reported as being eusocial.

On the continent, brood cells are targeted by *Sphecodes scabricollis*. This behaviour has not been observed in Britain.

On the wing

Sussex records run from mid May until mid September.

Feeding and foraging

H. eurygnathus is strongly associated with Greater Knapweed but also visits a range of other plants for pollen. These include dandelion, hawk's-beard, Harebell and Clustered Bellflower.

H. eurygnathus is often seen visiting Field Scabious.

Halictus rubicundus Orange-legged Furrow Bee
FIRST RECORD 1858, Lewes (William Unwin) TOTAL RECORDS 397

Geography and history
Halictus rubicundus is found throughout Sussex, occurring in a very wide range of habitats. These include open woodland, heathland, chalk grassland, soft-rock cliffs, lowland meadow and pasture, and saltmarsh. It has also been recorded from gardens.

Out in the field and under the microscope
This dark bee is most often found feeding or foraging.

In common with other species in the genus such as *H. eurygnathus*, females have fine bands of white hairs on the hind margins of the tergites. They can be separated from *H. eurygnathus*, and with care identified in the field, by the colouration of their hind legs. Here, the tibiae are obviously

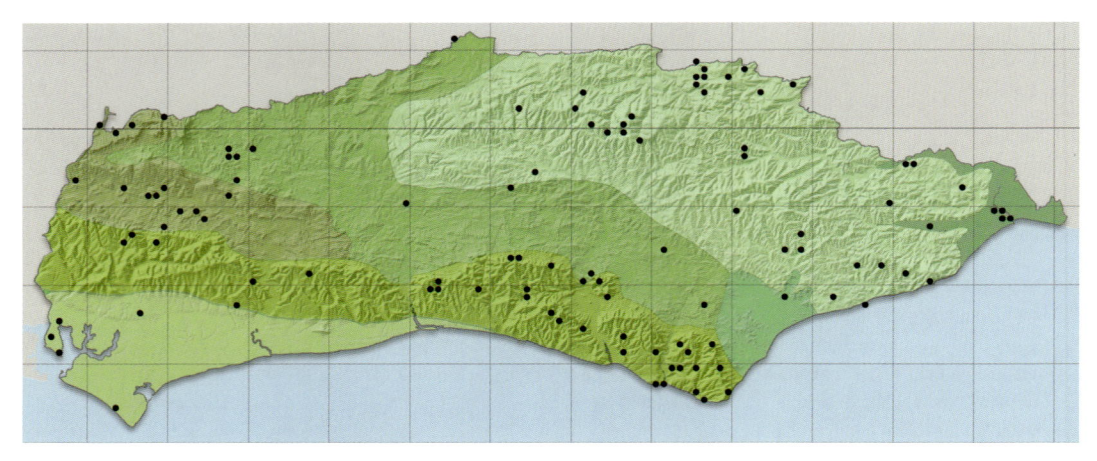

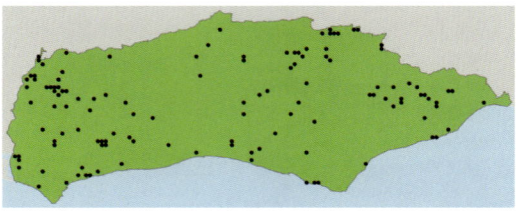

Halictus rubicundus ♀ [PC]

Halictus rubicundus ♂ [PC]

Large map: distribution 2000–2023
Small map: distribution 1844–1999

yellowish orange and contrast strongly with the femora which are very dark. In the otherwise very similar *H. eurygnathus*, both the hind tibiae and femora are dark.

Males have elongated dark abdomens with thin bands of white hairs on the hind margins of tergites one to four. They also have relatively long antennae and resemble male *H. eurygnathus*, but without the hollowed-out underside to the head. Like the females, males have yellowish-orange hind tibiae but with the addition of a black oval spot.

In common with many other species of *Halictus* and *Lasioglossum*, *H. rubicundus* males emerge in late summer and are frequently active until early autumn.

Behaviour and interactions

While in northern Britain *H. rubicundus* is generally a solitary bee, producing a single generation a year, in southern areas this behaviour changes and this species is eusocial.

In eusocial colonies, the queens emerge during April, followed by the first workers in May. There is a division of labour, with, for example, the nest entrance guarded by a worker. Solitary-nesting females tend to be smaller and establish their nests later in the year than those with social colonies.

The new generation of males and females emerge from July onwards. Males patrol around plants near the nest site, looking for an opportunity to mate. Once mated, the females overwinter either in the maternal nest or elsewhere in a new burrow.

H. rubicundus is one of the most frequently targeted of 39 bee species that are attacked by the wasp *Cerceris rybyensis*. The wasp paralyses a bee by stinging it under the thorax before carrying it back to a brood cell as food for a larva. One study from Cambridgeshire showed that brood cells are stocked with up to 11 prey items. *C. rybyensis* generally nests in large aggregations, predating many thousands of bees in the area around the nest site.

H. rubicundus nests are attacked by two species of *Sphecodes*, *Sphecodes gibbus* and *S. monilicornis*.

On the wing

Sussex records run from the beginning of March until mid October.

Feeding and foraging

H. rubicundus forages from a very wide range of plants, collecting pollen from, amongst others, the Buttercup, Pink, Rock-rose, Rose and Heather families.

Halictus tumulorum Bronze Furrow Bee
FIRST RECORD 1900, Eastbourne (Charles Nurse) TOTAL RECORDS 877

Geography and history
Halictus tumulorum is a very abundant and widespread bee. It is one of the most frequently recorded bees in the county and could be found in just about any habitat. As well as on chalk grassland and on lowland meadow and pasture, it can be found on heathland, in open woodland, on scrubby cliff-top grassland and on vegetated shingle. It also frequently occurs in gardens, parks and cemeteries.

South-facing cliff faces, as here at Newhaven, frequently support breeding populations of *Halictus tumulorum*. [WP]

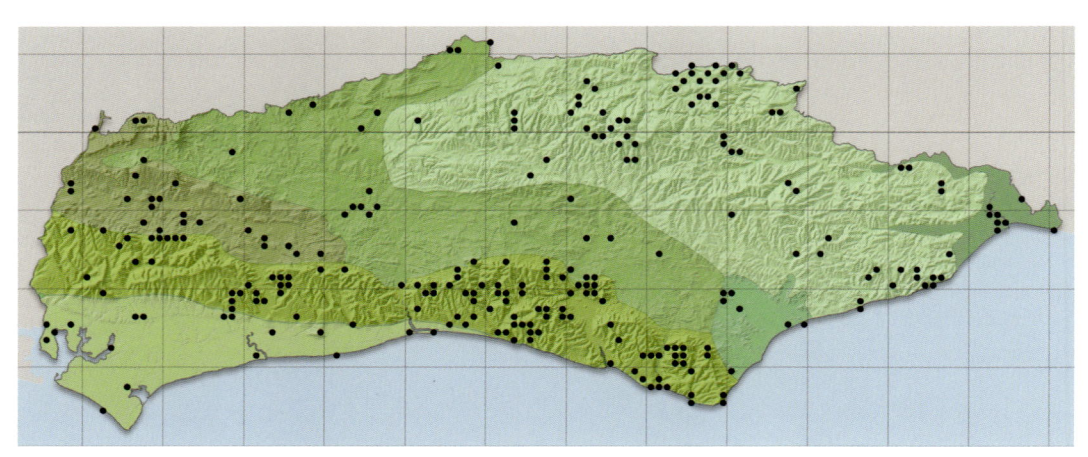

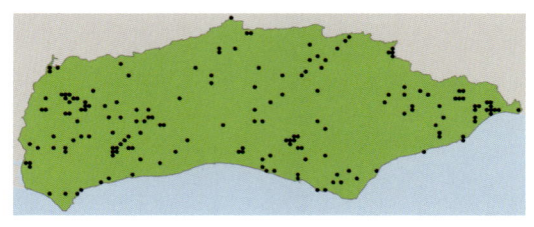

Large map: distribution 2000–2023
Small map: distribution 1844–1999

The Bees of Sussex

Out in the field and under the microscope

This *Halictus* is generally found feeding or foraging but might also be found close to its nests in crumbling wall mortar, on bare south-facing banks or on cliff faces.

H. tumulorum is one of three small metallic bees in this genus, one of which, *H. subauratus*, has not been recorded in Britain since the nineteenth century. *H. tumulorum* is very similar to the other species, *H. confusus*, and worn females of this last species are especially hard to distinguish from *H. tumulorum*. *H. confusus* does, however, have a much more restricted distribution in Sussex—any small metallic bee from this genus away from the Lower Greensand is most likely to be *H. tumulorum*.

In females, one of the key characters to look for is the width of the hair bands on the hind margins of the second and third tergites—in *H. tumulorum* these narrow appreciably in the middle, while in fresh *H. confusus* the band is broad throughout.

Males of the two species can be separated by examining the genital capsule and by checking the colour of the trochanter. In this species, the trochanters on the middle and hind legs are extensively marked with black, while in *H. confusus* they are largely yellowish brown.

Halictus tumulorum ♀ [PC]

Halictus tumulorum ♂ [PC]

Behaviour and interactions

H. tumulorum is eusocial, four or five unmated females remaining with the founding female and helping to rear the brood. Nests generally occur singly, and the main tunnel can descend about 16 cm.

Brood cells are thought to be targeted by *Sphecodes ephippius* and *S. geoffrellus*.

On the wing

Sussex records run from late March until the end of September.

Feeding and foraging

Pollen is collected from various plants, including members of the Daisy, Cabbage, Bellflower, Pea, Buttercup and Speedwell families.

Lasioglossum – Base-banded Furrow Bees

This is a large genus of ground-nesting bees with only very subtle differences between them. They also closely resemble those in the genus *Halictus*. The most consistent difference between these two genera is the presence of weak, rather than strong, outer cross veins on the forewings in *Lasioglossum*. In addition, most species in this genus have bands or patches of hairs at the front of the tergites, although in several species these are absent, and in one, *Lasioglossum sexstrigatum*, they are at the rear of the tergites. Nesting behaviour ranges from solitary to various forms of eusocial behaviour. There are 32 species in Britain, of which 27 are known from Sussex. Many are targeted by species of *Sphecodes*.

Lasioglossum albipes Bloomed Furrow Bee
FIRST RECORD 1877, Worthing (Edward Saunders) TOTAL RECORDS 541

Geography and history
Lasioglossum albipes is a very widely distributed bee that has been recorded in all parts of Sussex. It is found in many different habitats, including woodland, grasslands, heathland, and coastal habitats such as vegetated shingle. It has also been recorded from gardens.

Out in the field and under the microscope
This bee often flies with *L. calceatum* and is most frequently found on yellow-flowered plants from the Daisy family such as ragwort and dandelion.

Females vary in size, with the queens significantly larger than the workers. They most closely resemble *L. calceatum* females but on average are slightly smaller. The two species can

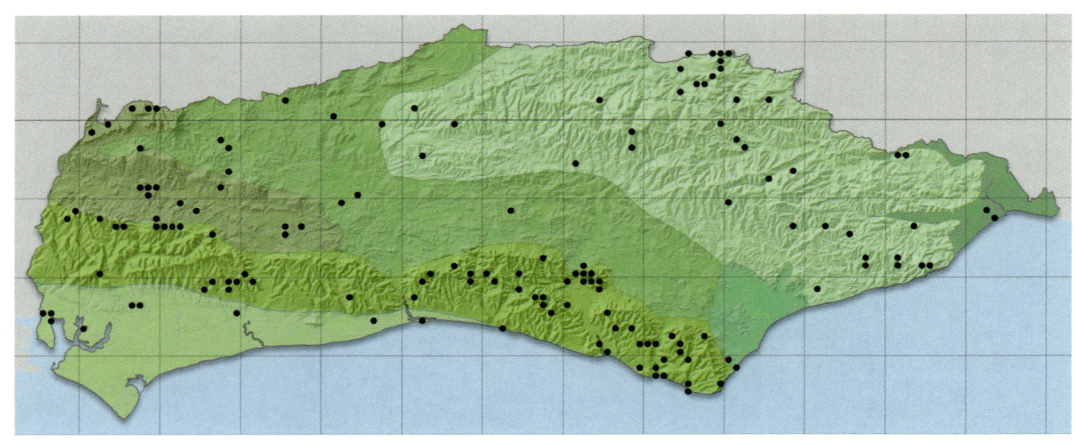

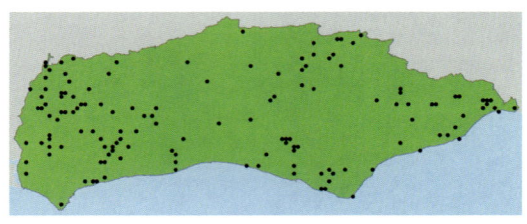

Large map: distribution 2000–2023
Small map: distribution 1844–1999

be tricky to separate, not least as both have a degree of natural variability. One clue can be the presence of a greyish bloom on the tergites of *L. albipes*.

Males are slender and can have a completely black abdomen but this is usually black and red. Like in *L. calceatum*, the antennae are completely dark. However, males of *L. albipes* frequently lack the patches of white hairs always found on the tergites in *L. calceatum*. In contrast to the black labrum and mandibles in *L. calceatum*, males of this species have a yellow labrum and yellow-marked mandibles.

Behaviour and interactions

L. albipes nests solitarily in some parts of its range and is eusocial in others, and it has been suggested that this may indicate the presence of hidden, cryptic species. After mating, females pass the winter in chambers that they excavate within the maternal nest, emerging the following spring to establish their own nests.

Sphecodes monilicornis is known to attack this species on the continent, and it is likely that the same behaviour occurs in Britain.

On the wing

Sussex records run from mid March until late October.

Feeding and foraging

A wide range of plant species are visited for pollen, including those in the Daisy, Carrot, Buttercup and Bellflower families.

Other species visited include Bluebell, Lesser Stitchwort, scabious and cinquefoil.

Lasioglossum albipes ♂ [SF]

Lesser Stitchwort [PC]

Lasioglossum brevicorne Short-horned Furrow Bee
FIRST RECORD 1905, Brighton (William Unwin) **TOTAL RECORDS** 12

Geography and history
Lasioglossum brevicorne is a very scarce bee that has been recorded in Sussex just twelve times. Nationally, its main populations are restricted to heathland, acid grassland, coastal sand dunes and chalk heath. Its populations are concentrated in a band that runs, broadly, from Dorset and on through Hampshire, Surrey, Essex, Suffolk and, finally, Norfolk. It is also known from more northerly counties such as Cheshire, Lancashire and Durham. It is strongly associated with light sandy soils.

Although it has been suggested that it may have expanded its range in recent years, there is no evidence for this in Sussex. It has been known from seven sites in the county and was last recorded

Large map: distribution 2000–2023
Small map: distribution 1844–1999

here in 2006. This last record is from an area of restored heathland near West Chiltington. The other records are from Brighton, Eastbourne, Sullington Warren, Rewell Wood and Bignor Park.

Out in the field and under the microscope
L. brevicorne is best searched for by checking plants in the Daisy family, for example Cat's-ear and ragwort, in areas of hot, almost bare sand or in areas that are sparsely vegetated.

It is a small shiny black bee that most resembles *L. villosulum*. However, female *L. brevicorne* can be distinguished by the patches of fine white hairs to the sides of the front face of the first tergite, and by the yellow stigma with a brown border on each of the forewings—these are a solid brown in *L. villosulum*.

Males are smaller and, like the females, have a yellow stigma with a brown border at the front of each forewing. They also have short antennae and yellowish or white markings on the mandibles and clypeus.

Behaviour and interactions
Little is known of the nesting behaviour of *L. brevicorne* in Britain, although it has been reported on the continent that it is a solitary-nesting species.

In mainland Europe, brood cells are thought to be targeted by *Sphecodes puncticeps* and *S. marginatus* (this last species is restricted to the Channel Islands), but this behaviour has not been confirmed in Britain

On the wing
L. brevicorne is on the wing from the beginning of May until the end of August.

Feeding and foraging
L. brevicorne collects pollen just from plants in the Daisy family such as Common Ragwort, Mouse-ear-hawkweed and Cat's-ear. It has also been observed visiting Wild Mignonette

Common Ragwort [PC]

Cat's-ear [NP]

Lasioglossum calceatum Common Furrow Bee

FIRST RECORD 1877, Worthing (Edward Saunders) **TOTAL RECORDS** 984

Geography and history
Lasioglossum calceatum is found in all corners of Sussex, occurring in various open habitats. It is often found in towns and villages, and is frequently recorded in gardens and parks.

Out in the field and under the microscope
This species is challenging to identify because it is not easy to separate from others in the genus, and especially from *L. albipes*.

One character *L. calceatum* and *L. albipes* females share is the structure of the short lobes on the inner tibial spurs on the hind legs. Taken together, these lobes make the spurs look serrated. Unlike *L. albipes* females, however, typical female *L. calceatum* have a fine ridge that extends over each of the shoulders of the propodeum. Larger specimens also have a rounder face.

Males are a little easier but still tricky. These can be all black or black and red, and can be separated from *L. albipes* males by the pronounced ridge that runs vertically up each side of the propodeum. In *L. albipes* this ridge is weak and generally does not extend far from the base of the propodeum. *L. calceatum* males generally have more obvious patches of white hairs on the second and third tergites, and a dark labrum and dark mandibles.

Checking the flower heads of plants such as dandelion, buttercup or thistle is a good strategy for finding this bee. In late summer, males can also be found in good numbers on garden plants such as Lesser Calamint.

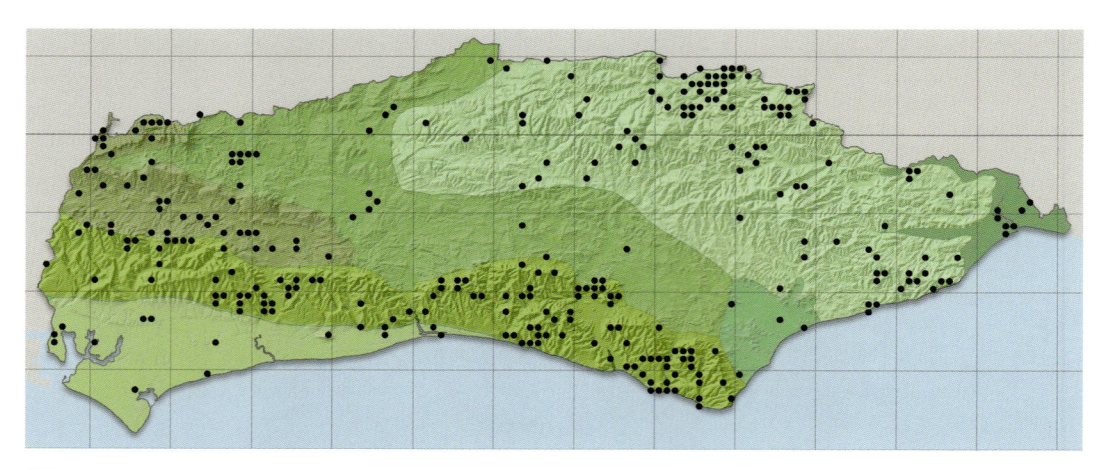

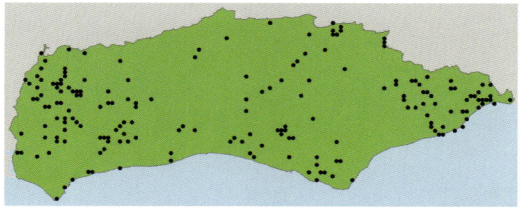

Large map: distribution 2000–2023
Small map: distribution 1844–1999

Lasioglossum calceatum ♂ showing black form [PC]

Lasioglossum calceatum ♂ showing black and red form [PC]

Lasioglossum calceatum ♂'s roosting [PB]

Lasioglossum calceatum nest [JP]

Behaviour and interactions

L. calceatum nests in small aggregations sited in areas of short south-facing grassland, including lawns. The smooth-sided entrance is marked by a small tumulus of loose soil up to about 4 cm across and about 2.5 cm high. Nests have also been observed between paving bricks in urban streets.

In southern areas of the country this is a eusocial species, switching to solitary-nesting behaviour in cooler climates.

In eusocial colonies, mated females emerge from hibernation in the spring and establish a nest. They provision the brood cells and lay a small number of eggs which develop into males and females. The two to three females which emerge support the development of a second brood, with one acting as a guard bee positioned just below the entrance to the nest.

These worker females may also lay eggs or establish their own nests. The second brood (also containing males and females) emerges from the nest from midsummer onwards. After mating, the females frequently pass the winter together, often with the queen living for a second year.

Sphecodes monilicornis—and possibly *S. ephippius*—lays its eggs within *L. calceatum* brood cells. As well as destroying the host's egg, *S. monilicornis* might also kill adult bees within the nest. If an attack is successfully repelled, the nest is quickly repaired by the *L. calceatum* females.

On the wing

Sussex records run from mid March until the beginning of November.

Feeding and foraging

L. calceatum collects pollen from an enormous range of plant families—fifteen have been reported from the continent—and is often seen visiting plants such as Lesser Celandine, Germander Speedwell, dandelion and Creeping Thistle.

Lasioglossum cupromicans Turquoise Furrow Bee
FIRST RECORD 1998 (Newhaven, Colin Plant) **TOTAL RECORDS** 2

Geography and history
This bee was only recognised as being a distinct species in 1970 as it had previously been thought of as a form of *Lasioglossum smeathmanellum*.

The British population of *L. cupromicans* is centred on central and northern England, extending up into Scotland and across Wales. In these areas, it replaces *L. smeathmanellum*, especially at higher altitudes. *L. smeathmanellum* has a generally more southerly distribution.

Large map: distribution 2000–2023
Small map: distribution 1844–1999

There are two records for this species in Sussex. The first is of a male that was netted by Colin Plant on an area of vegetated shingle close to Newhaven in July 1998, while the second is from nearby at Tide Mills where a female was swept by Steven Falk in 2011. These records are unusually far south for this species.

Out in the field and under the microscope

L. cupromicans is one of four species in the genus that are a metallic turquoise or greenish colour, the others being *L. leucopus*, *L. morio* and *L. smeathmanellum*. It is most like this last species, and it is tricky to differentiate between the two, particularly given that there is a degree of variability within populations of each.

In males and females, a good character to look for is the series of fine ridges on the top of the propodeum—these are very weak and peter out well before the hind margin of the propodeal triangle in *L. cupromicans* (they are especially weak in males), whereas they are relatively pronounced and extend almost to the hind margin in *L. smeathmanellum*. There are also differences between the male genital capsules of each species.

It has been suggested that *L. cupromicans* is particularly associated with Rosebay Willowherb.

Behaviour and interactions

This *Lasioglossum* nests in crevices in rocks and stone walls in Britain, while in Germany it is suggested that it nests in the ground. There, it is also thought likely that this species is eusocial, but this has not been demonstrated in Britain.

Sphecodes geoffrellus is thought to target brood cells of this species.

On the wing

Typically, *L. cupromicans* is on the wing from late April until the end of October.

Feeding and foraging

There is no information available on the pollen sources for this species. However, it is known to visit plants such as Turnip, Heather, sedum, cinquefoil, Rosebay Willowherb, Sheep's-bit, thistle, Cat's-ear and hawk's-beard.

Rosebay Willowherb [PC]

Lasioglossum fratellum Smooth-faced Furrow Bee
FIRST RECORD 1919, Hastings (William Butterfield) TOTAL RECORDS 67

Geography and history
In northern and western areas of Britain *Lasioglossum fratellum* is typically a species of moorlands, while in the south and the east it is particularly associated with higher altitude heathlands and open woodlands.

Aside from a solitary record from Flatropers Wood in 2014 and another from Graffham in 2020, all modern Sussex records are from Ashdown Forest, for example Old Lodge, and the countryside south of Tunbridge Wells. The record from Graffham in 2020 was the first from the Lower Greensand since 1993. In the mid 1990s, there were three records from the eastern South Downs.

Old Lodge [WP]

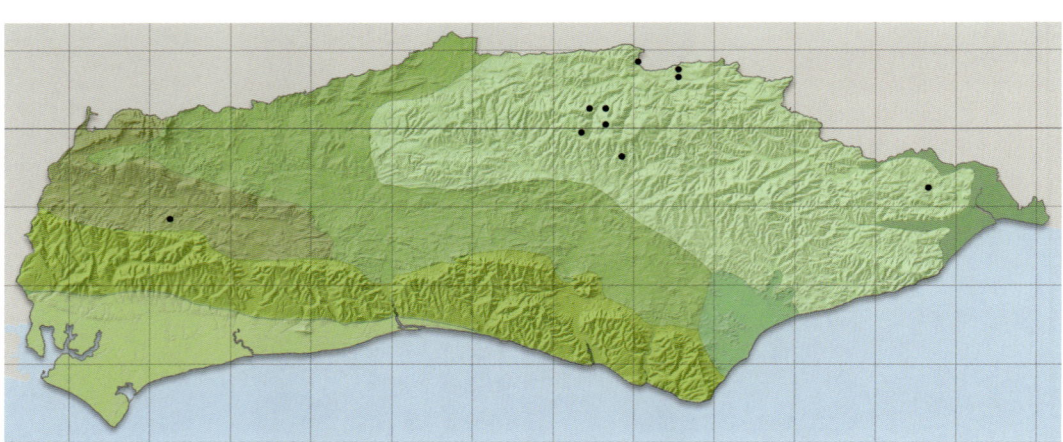

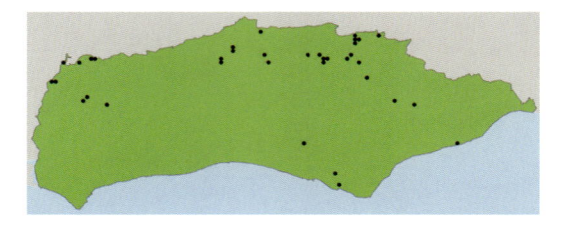

Large map: distribution 2000–2023
Small map: distribution 1844–1999

Out in the field and under the microscope
This species is generally found feeding on the flowers of plants such as dandelion, willowherb or Cow Parsley.

It most closely resembles *L. fulvicorne*, a much more abundant species in Sussex that is particularly associated with chalk grasslands. The males and females of the two species can be differentiated by the clarity of the punctures on the face just above the point where the antennae attach to the head—these are very indistinct in *L. fratellum* and pronounced in *L. fulvicorne*. Males and females of this last species also have a much more rounded face when viewed head-on.

L. fratellum and *L. fulvicorne* are the only two species in which the males have extremely long antennae as well as very short sparse hairs on sternite two.

Lasioglossum fratellum ♀ [SF]

Lasioglossum fratellum ♂ [PB]

Behaviour and interactions
Most *L. fratellum* females establish solitary nests in small aggregations. However, a small number of spring nests contain two females, one to provision the brood cells and the other to guard the nest. The second of these is the larger of the two and is the only one to lay eggs.

In harsher climates, some females can produce eggs in two different years, a strategy that reduces the risk that adverse weather will prevent reproduction.

Brood cells are targeted by *Sphecodes hyalinatus*.

On the wing
L. fratellum is on the wing from mid April until mid October.

Feeding and foraging
While there is no information available on which species are visited specifically for pollen, *L. fratellum* is known to visit a very wide range of plants. These include willow, Heather, Bilberry, bramble, Sheep's-bit, Devil's-bit Scabious, Creeping Thistle, dandelion, Common Ragwort and Goldenrod.

Lasioglossum fulvicorne Chalk Furrow Bee

FIRST RECORD 1900, Eastbourne (Charles Nurse) **TOTAL RECORDS** 416

Geography and history

Lasioglossum fulvicorne is mostly found on areas of species-rich chalk grassland, although it has also been known from sites on the Lower Greensand and scattered locations in the Low and High Wealds, as well as from Pagham Harbour and Chichester.

It is especially abundant on the downs and could be found on just about any area of chalk grassland. Levin Down, Devil's Dyke and Malling Down all regularly generate records. Away from the chalk, there are a handful of modern records from locations such as Sullington Warren, Knepp and Ashburnham Place.

Levin Down [GN/SWT]

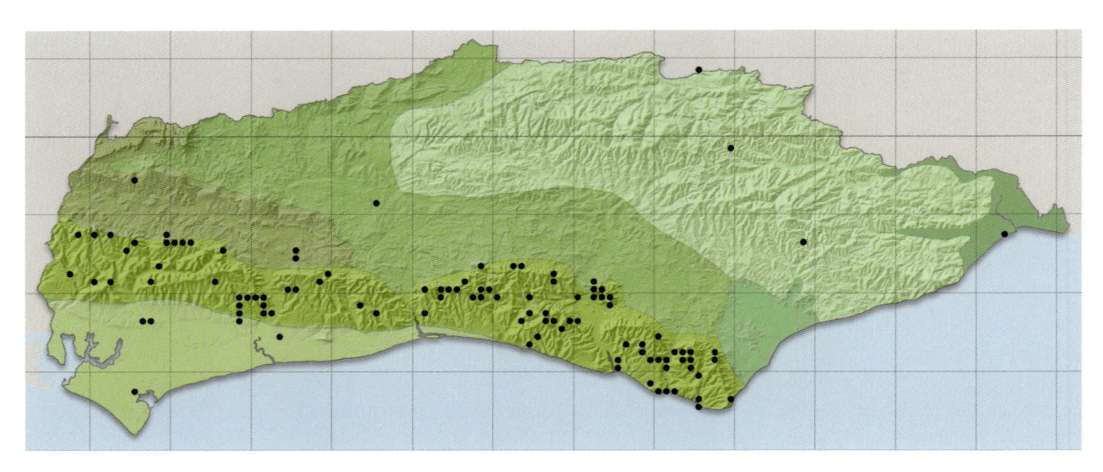

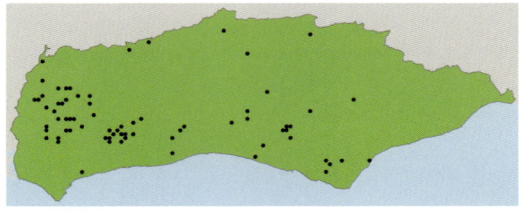

Large map: distribution 2000–2023
Small map: distribution 1844–1999

342 *The Bees of Sussex*

Out in the field and under the microscope

In the spring, females are most likely to be found visiting flowering plants such as dandelion and Ground-ivy, while later in the year large numbers of males might be seen feeding on plants in the Carrot family.

Females closely resemble *L. fratellum*, a species with a much more restricted distribution in Sussex. In contrast to *L. fulvicorne*, this other species has indistinct punctures above the antennal sockets and much reduced patches of white hairs on tergites two and three.

Males are long and slender and have extremely long antennae. They also closely resemble *L. fratellum*, and males of these two species are the only ones in Britain which have a combination of long antennae and a sparse covering of very short hairs on the surface of sternite two. *L. fulvicorne* has a much more rounded face, which is more obviously punctate than in *L. fratellum*. It also has distinct patches of white hairs to the sides of tergites two and three.

Lasioglossum fulvicorne ♀ [ME]

Lasioglossum fulvicorne ♂ [SF]

Behaviour and interactions

L. fulvicorne is believed be a solitary-nesting species and, as far as is known, it does not establish large aggregations, a view that is supported by the lack of observations of nest sites on the downs despite its abundance here. On the continent—in Belgium for example—it is reported to be particularly associated with heathland.

Brood cells are targeted by *Sphecodes hyalinatus* in Britain, with *S. ferruginatus* also given on mainland Europe.

On the wing

Sussex records run from the end of March until late September. Elsewhere, mid October is reported.

Feeding and foraging

Many different plant species are visited, including bramble, strawberry, Ground-ivy, dandelion, ragwort, Wild Marjoram and Wild Parsnip. Pollen sources are species in the Daisy, Cabbage, Heather, Rose and Willow families.

Lasioglossum laevigatum Red-backed Furrow Bee
FIRST RECORD 1896, Shipley (Gorham) TOTAL RECORDS 208

Geography and history
Lasioglossum laevigatum is largely confined to southern counties in England and Wales, and in Sussex it is widely distributed across the county. It is a very localised species, but it can be abundant in some locations.

L. laevigatum is particularly associated with species-rich chalk grassland but also occurs in other habitats. Good populations have been found on a complex of hay meadows and pasture near Warbleton, in Friston Forest and at Ebernoe Common, as well as on the chalk at places such as Blackcap and Mount Caburn.

Hay meadow, Warbleton [NS]

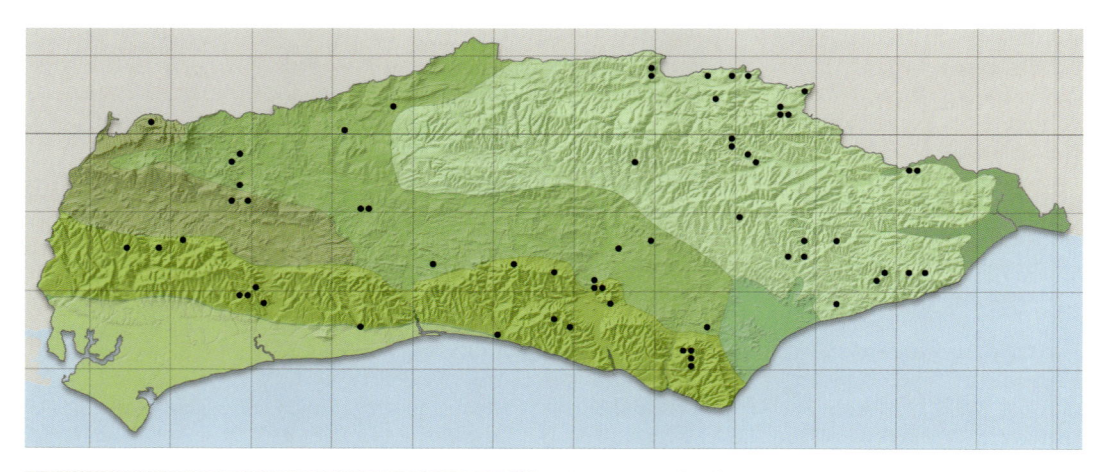

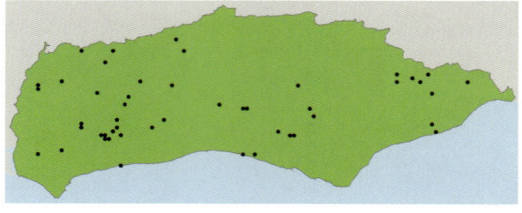

Large map: distribution 2000–2023
Small map: distribution 1844–1999

Out in the field and under the microscope
Females are frequently found on plants such as buttercup, speedwell and yellow-flowered plants in the Cabbage family. There is also a 1996 record by David Porter of this bee visiting Alexanders.

This is one of the larger species in the genus, and it has reddish hairs on the scutum. Females most closely resemble two other species, *L. leucozonium* and *L. zonulum*. In common with a number of others in the genus, all three of these species have an obvious ridge that runs all the way up each side of the propodeum before turning in to run across the top. In *L. laevigatum*, however, the ridge forms a very prominent tooth on the corners at each side of the propodeum.

L. laevigatum females can also be separated from these other species by the large patches of white hairs on the second, third and fourth tergites. Here, long individual hairs stretch back beyond these patches, almost reaching the start of the next tergite.

Males have long antennae and an entirely black clypeus—in the otherwise similar *L. leucozonium* and *L. zonulum* this is usually marked with yellow.

Lasioglossum laevigatum ♀ [PC]

Lasioglossum laevigatum ♂ [TF]

Behaviour and interactions
Little is known about the nesting habits of this species, although it is believed to be a solitary species which establishes its nests in areas of sparsely vegetated ground. Males and females are active until the autumn, and, in common with other species in the genus, mated females overwinter below ground.

It is not known if brood cells of this species are targeted by any species of cuckoo bee.

On the wing
Sussex records run from late March until the beginning of September.

Feeding and foraging
Many different plant families are visited, including the Buttercup, Pink, Cabbage, Carrot, Willow, Rose, Pea, Speedwell and Daisy families. Many of these are also sources for pollen.

Lasioglossum lativentre Furry-claspered Furrow Bee
FIRST RECORD 1899, Bognor Regis (Henry Guermonprez) TOTAL RECORDS 379

Geography and history
Lasioglossum lativentre is very widely distributed across the county, occurring in all sorts of habitats. It is found in open woodlands, sandpits, vegetated shingle, and in various types of grassland. These include scrubby, neglected grasslands as well as flower-rich meadows and chalk grasslands such as Devil's Jumps and Malling Down. It is also known from gardens, churchyards and cemeteries.

Devil's Jumps [ME]

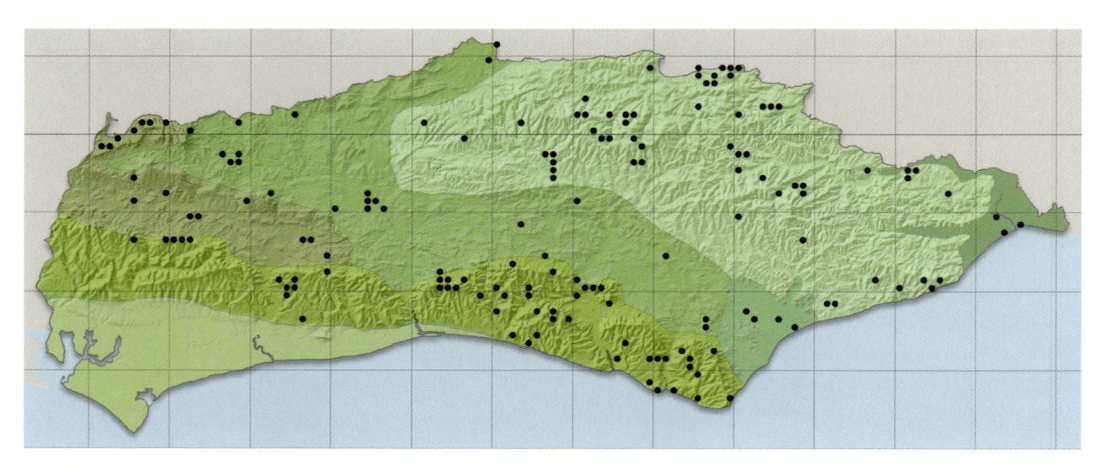

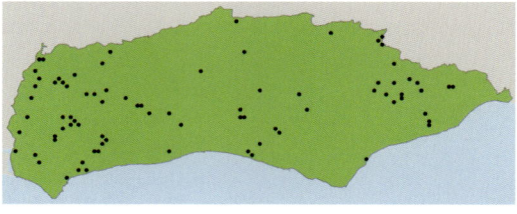

Large map: distribution 2000–2023
Small map: distribution 1844–1999

Out in the field and under the microscope

This species is generally seen as singletons, or sometimes in small numbers, and is often found on plants such as dandelion, speedwell and buttercup.

It is one of two very similar species of *Lasioglossum*, the other being *L. quadrinotatum*, a much scarcer species that is almost entirely restricted to west Sussex. Female *L. lativentre* are part of a large group in the genus that lacks a fine ridge running up each side of the propodeum.

One of the ways to separate females and males of this species from *L. quadrinotatum* is to look at the colour of the shading on the stigma at the front of each forewing—in *L. lativentre*, this is brown with a darkened edge, while in *L. quadrinotatum* it is a yellowish colour.

Males can be separated from *L. quadrinotatum* by examining the genital capsule. In *L. lativentre*, the tips of the capsule are covered in abundant long golden hairs, hence the common name 'Furry-claspered Furrow Bee'. These long hairs are absent in *L. quadrinotatum*.

Behaviour and interactions

In common with other species in the genus, *L. lativentre* has an extremely long flight period, the females emerging in spring and producing the next generation of females and males during the summer. These then persist into the autumn.

This is known to be a solitary-nesting species, but there is little information about where its nests are established apart from a record from Surrey of a nest sited on bare ground.

Brood cells are targeted by two species of *Sphecodes*, *Sphecodes ephippius* and *S. puncticeps*.

On the wing

Sussex records run from late March until the beginning of November.

Feeding and foraging

Visits to species such as Lesser Celandine, willow, Heather, bramble, foxglove, Devil's-bit Scabious, dandelion and Feverfew have been reported. *L. lativentre* collects pollen from several different plant families, including Daisy, Pea, Plantain and Buttercup.

Lasioglossum leucopus **White-footed Furrow Bee**
FIRST RECORD 1882, Littlehampton (Edward Saunders) TOTAL RECORDS 288

Geography and history
Lasioglossum leucopus is widely distributed across Britain but is most abundant in south-east England.

In Sussex, it is widely distributed but with most records from the South Downs, the Wealden Greensand and the High Weald. It does not appear to have a particular habitat preference but can be very abundant on chalk grasslands and has recently been recorded from sites such as Deep Dean, Mount Caburn and Fairmile Bottom. Many of the more recent records are from the eastern portion of the South Downs.

Although it is reasonably common across much of the county, it is not as abundant as the very similar *L. morio*.

Deep Dean [ME]

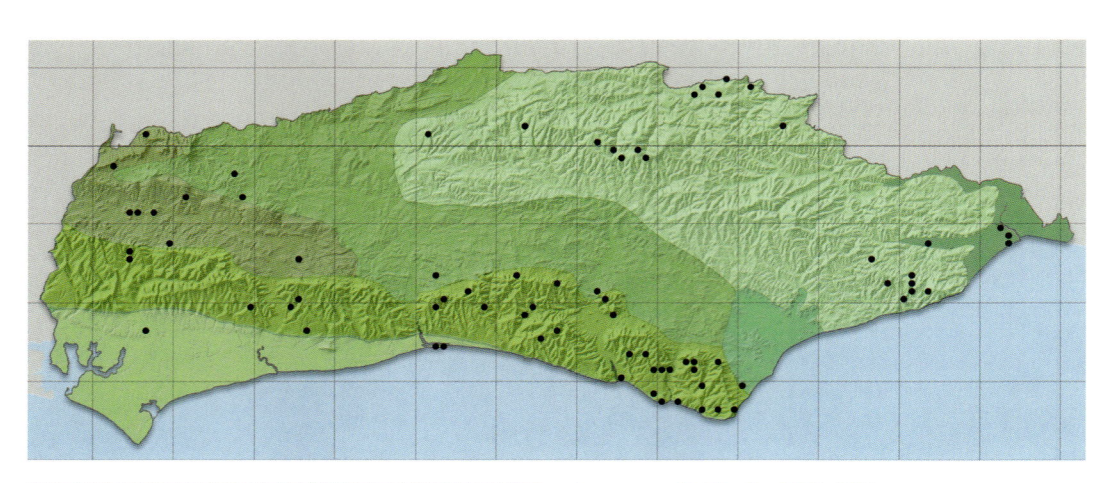

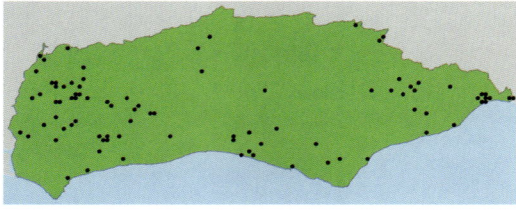

Large map: distribution 2000–2023
Small map: distribution 1844–1999

Lasioglossum leucopus ♀ [SF]

Lasioglossum leucopus ♂ [NO]

Recent work has suggested that *L. leucopus* may in fact prove to be two distinct species, *L. leucopus* and *L. aeratum*. This last species was first described some 200 years ago, but in Britain has been treated as a form of *L. leucopus*. It has been collected by Steven Falk from Beachy Head, Southerham Farm, Birling Gap and Cuckmere Haven.

Out in the field and under the microscope
This is one of four *Lasioglossum* that are a bright metallic blue-green or bronze-green. The others are *L. cupromicans*, *L. smeathmanellum* and *L. morio*.

L. leucopus females can be separated from these other species by the very polished, shiny and smooth marginal areas of tergites one and two—the others all have very fine ridges here. A further thing to note is that, unlike in *L. cupromicans* and *L. smeathmanellum*, the metallic colouring in *L. leucopus* and *L. morio* is restricted to the thorax.

Males are smaller than the females and also resemble these other species. However, *L. leucopus* males can be readily distinguished by the yellow labrum and tarsi, and by the relatively short antennae. In the other three, the labrum and tarsi are dark, while the antennae are comparatively long and reach beyond the end of the thorax.

L. leucopus visits a wide range of plants, and checking flower heads is a good way of finding this species.

Behaviour and interactions
Although *L. leucopus* is believed to be a eusocial species, in some parts of its range it is reported to be a solitary-nesting species. In eusocial colonies, a small number of unfertilised females from the first generation support the development of the founding female's subsequent offspring, including by guarding the nest entrance.

Brood cells are reported to be targeted by *Sphecodes geoffrellus* on the continent, but this relationship has not been confirmed in Britain.

On the wing
Sussex records run from early April until early September. Elsewhere, *L. leucopus* is reported to be active until mid October.

Feeding and foraging
L. leucopus has been reported as visiting a wide range of plants, including Yellow Pimpernel, bramble, cinquefoil, Wood Spurge, Wild Carrot, bellflower, Sheep's-bit and Alder Buckthorn.

On the continent, it is reported that pollen is collected from Autumn Hawkbit, dandelion and cinquefoil.

Lasioglossum leucozonium White-zoned Furrow Bee
FIRST RECORD 1877, Worthing (Edward Saunders) TOTAL RECORDS 627

Geography and history
In Britain *Lasioglossum leucozonium* is a widespread bee found as far north as Yorkshire and Cumbria. It is, however, much more abundant in southerly counties, especially in the south and east.

It is very widely distributed throughout Sussex, although less frequently recorded in the Low Weald and the Pevensey Levels. It uses a broad range of habitats and could be found just about anywhere. There have been, for example, recent records from heathland near Heyshott, hay meadows and pasture in the High Weald, and from saltmarsh at East Head. It has also been recorded from farmland near Stoughton and from gardens such as those at Great Dixter.

Mixed border at Great Dixter
[AP]

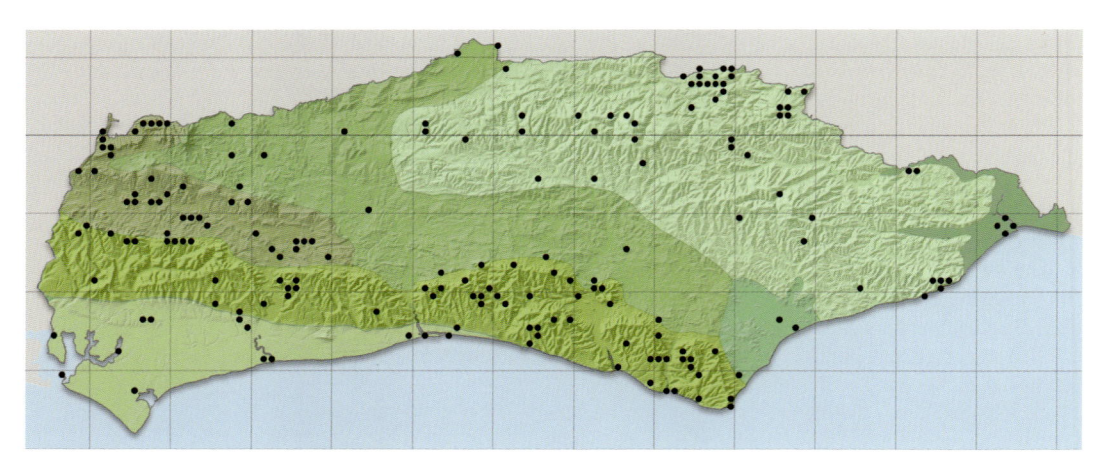

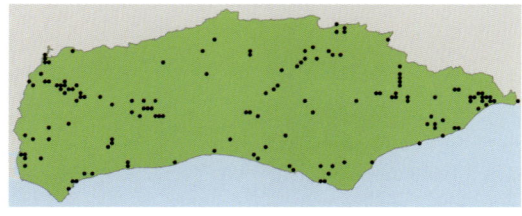

Large map: distribution 2000–2023
Small map: distribution 1844–1999

350 *The Bees of Sussex*

Out in the field and under the microscope
This is one of the larger species in the genus. Fresh females have obvious white hair bands that run all the way across the hind margins of tergites two, three and four, as well as brownish-black hairs on the upper edge of the hind tibiae. These two characters are shared with just one other species, *L. zonulum*, and the two can be readily separated by looking at the surface of the first tergite and the scutum—these are completely smooth and shiny in *L. zonulum* and slightly roughened in *L. leucozonium*.

Males also resemble male *L. zonulum* but have a yellow marking at the top of each of the hind tibiae and a yellow basitarsus—these are both entirely dark in *L. zonulum*.

L. leucozonium can be found by checking known foodplants, especially knapweed, Cat's-ear and other yellow-flowered plants in the Daisy family.

Lasioglossum leucozonium ♀ [PC]

Lasioglossum leucozonium ♂ [SF]

Behaviour and interactions
L. leucozonium is a solitary-nesting bee that establishes its nests in sparsely vegetated or unvegetated areas on south-facing banks or areas of flat ground. The main passageway is up to 15 cm long and has short tunnels branching off this, each of which terminates in a brood cell. The nests are established in the spring by females which mated the previous year, the new generation of males and females appearing from July onwards.

Brood cells are targeted by *Sphecodes ephippius* and possibly *S. pellucidus*.

On the wing
Sussex records run from early April until late September.

In northern counties the flight period is much reduced, and in Lancashire, for example, records run from June until August.

Feeding and foraging
A very wide range of plants are visited for pollen, including members of the Daisy, Bellflower, Rock-rose, Teasel, Bindweed, Crane's-bill, Buttercup and Rose families.

Lasioglossum malachurum Sharp-collared Furrow Bee
FIRST RECORD 1882, Hastings (Edward Saunders) TOTAL RECORDS 509

Geography and history
Lasioglossum malachurum used to be a scarce bee, and there were no records between the first in 1882 and the second, 60 years later, in 1942.

Throughout the 1940s and early 1950s, this bee was frequently recorded from Albourne, Ditchling and Ovingdean, but it was otherwise scarce during the twentieth century until the mid 1970s.

Since then, the number of records has increased rapidly, a pattern that has been mirrored in other southern counties. It is now found throughout the south and east of England. In some locations it can now be very abundant.

Nesting site along worn path, Ditchling Beacon [WP]

BELOW Exposed nest with larvae feeding in their individual brood cells [PW]

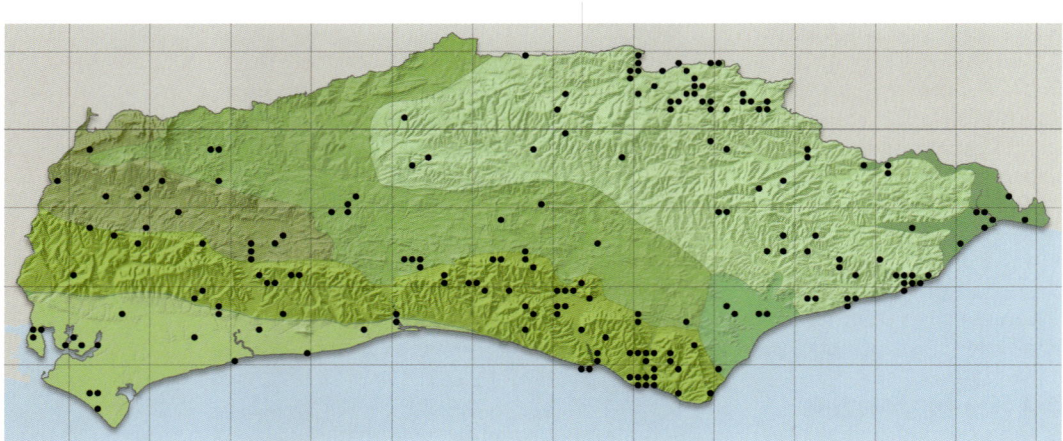

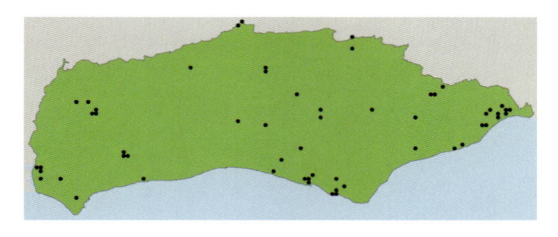

Large map: distribution 2000–2023
Small map: distribution 1844–1999

352 *The Bees of Sussex*

In Sussex, it is found throughout the county, in all sorts of habitats. These include chalk grassland, soft-rock cliffs, sandpits, lowland meadows and pasture, as well as open broadleaved woodland. It is also found in cemeteries, churchyards and gardens.

Out in the field and under the microscope
This is a distinctive bee, with males and females readily distinguishable from the other bees in the genus by the right-angled 'shoulders' in front of the thorax.

The shape of these is less well pronounced in the males, which have yellow tibiae and tarsi, and a distinct black marking on the outer face of each of the tibiae.

While *L. malachurum* is usually found feeding, it is also quite common to come across one of its nesting aggregations. These are often sited on paths or tracks where the ground has been compacted. Where bare ground has been exposed, each tunnel entrance is delineated by a small mound of loose soil. The aggregations can be quite extensive.

Lasioglossum malachurum ♀ [SF]

Lasioglossum malachurum mating pair with ♀ on the left and ♂ on the right [PC]

Behaviour and interactions
L. malachurum is a eusocial species. Nests are established in the spring by a mated queen and can be 20 to 25 cm below ground. The first brood consists of four or five workers. These are smaller than the queen and go on to build and provision cells for a second brood. It is not uncommon for other queens to fight the resident queen and attempt to take over the nest, and, if successful, to either 'adopt' the existing brood or produce their own workers.

In Britain, the second generation is comprised of males and new queens. Elsewhere in Europe, three worker generations are possible, the number of brood cells rising to as many as 200. Once the males and new queens have emerged from the nest, there is a period of mating, after which the females overwinter below ground, ready to emerge the following spring and start the cycle again.

Brood cells are targeted by *Sphecodes monilicornis*, which is known to attack host bees in order to gain entry to a nest. In Hastings, Mike Edwards has observed this species attempting to pull *L. malachurum* guard bees from a nest.

On the wing
Sussex records run from early March until the end of September.

Feeding and foraging
L. malachurum visits a very wide range of plant species, and sources of pollen include members of the Buttercup, Cabbage, Stonecrop, Rose, Carrot, Speedwell and Daisy families.

Lasioglossum minutissimum Least Furrow Bee
FIRST RECORD 1877, Worthing (Edward Saunders) **TOTAL RECORDS** 209

Geography and history
Nationally, most records of *Lasioglossum minutissimum* are from south, central and eastern England, with just a handful from locations in coastal areas of Wales.

In Sussex, it has been found in most areas of the county, although it has been recorded only once from the Pevensey Levels —in 1900—and has been found on just a few occasions in the Low Weald. It is reported to favour sandy or clay soils but does also occur on chalk—Steven Falk recorded it from seven different sites during a survey of the eastern South Downs in the early 2000s, including Blackcap, Birling Gap and Lullington Heath.

L. minutissimum has been found in open woodlands, chalk grasslands, sandpits, heathlands, sand dunes and coastal cliffs.

Blackcap [ME]

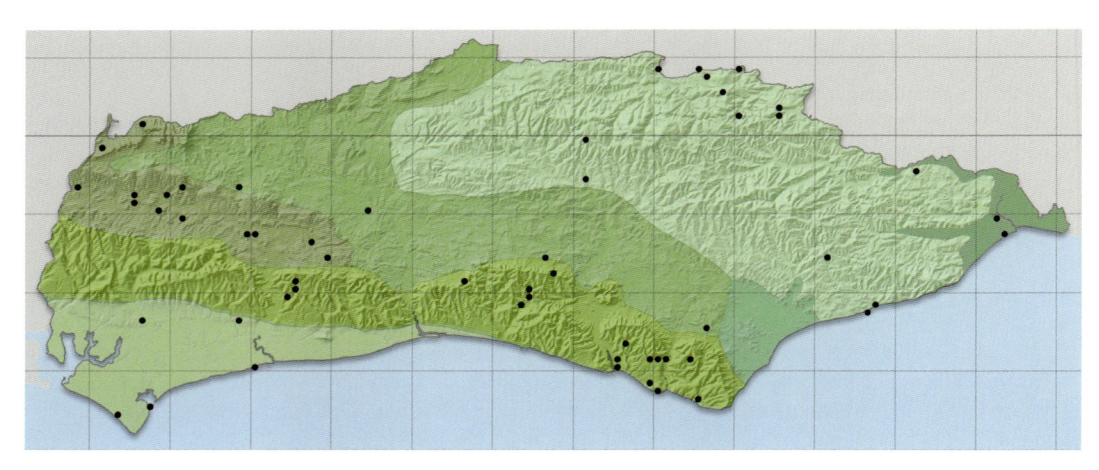

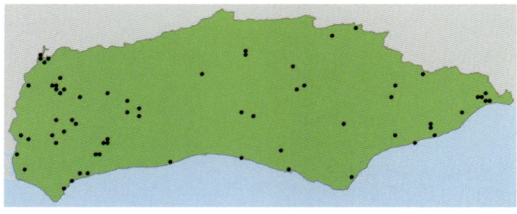

Large map: distribution 2000–2023
Small map: distribution 1844–1999

Out in the field and under the microscope

This is a very small and slender bee—it is one of the smallest in Britain and the smallest in the genus.

In females, one key character to look at is the length of the fine ridges on top of the propodeum—in this species, these stop short of the back edge of the propodeal triangle. In the otherwise similar *L. pauperatum*, the ridges on the propodeal triangle are more obvious and extend all the way back to its hind edge.

Males are smaller than the females and have a shallow dip between the surfaces of tergites one and two where they join and, to a lesser extent, between tergites two and three.

L. minutissimum is very easily overlooked due to its size but can be found by checking foodplants such as Sheep's-bit and forget-me-not. In Pembury in Kent, Ian Beavis reported that "a small group of females was flying low, persistently and evasively over a patch of bare ground at the edge of a sunny clearing".

Lasioglossum minutissimum ♀ [SF]

Lasioglossum minutissimum ♂ [NO]

Behaviour and interactions

L. minutissimum females emerge in the spring and can establish very large aggregations. These are often sited on areas of bare or sparsely vegetated level ground or on banks. It is a solitary-nesting species, with one report suggesting that the main tunnel extends some 9 cm below ground. Off this tunnel, several side tunnels branch off, each ending in a single brood cell.

Males and the new generation of females emerge during July, males flying until late September and females a little later. The mated females pass the winter below ground.

Brood cells are attacked by *Sphecodes longulus*.

On the wing

Sussex records run from late March until early September. Elsewhere, early October is reported.

Feeding and foraging

L. minutissimum has been recorded visiting a wide range of plants, including members of the Buttercup, Daisy, Cabbage, Borage, Spurge, Primrose and Speedwell families.

Pollen is collected from the Dead-nettle and Daisy families.

Lasioglossum morio Green Furrow Bee
FIRST RECORD 1877, Worthing (Edward Saunders) TOTAL RECORDS 979

Geography and history
Lasioglossum morio is found throughout the county and occurs in just about any habitat where there are patches of soil exposed to the sun. It is often found in gardens and parks.

Out in the field and under the microscope
Females of this metallic-looking *Lasioglossum* closely resemble three other species in the genus, *L. cupromicans*, *L. smeathmanellum* and *L. leucopus*. As in this last species, the thorax is a shiny bluish green, while the abdomen is black—in *L. cupromicans* and *L. smeathmanellum* the head, thorax and abdomen are all bluish green.

Nest site, Seven Sisters [JP]

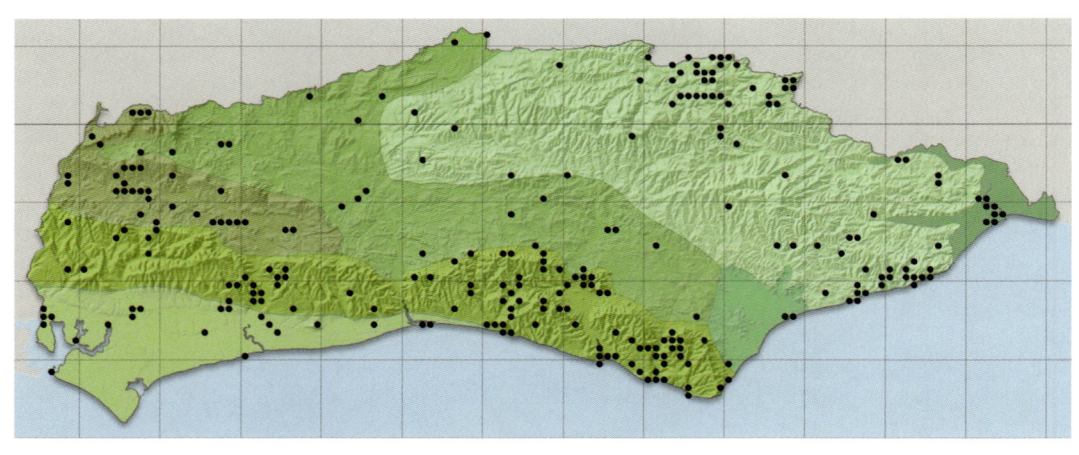

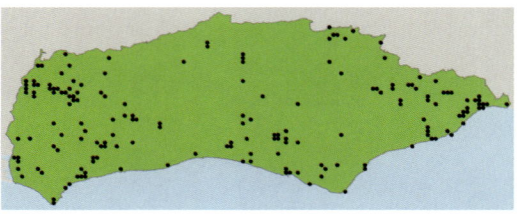

Large map: distribution 2000–2023
Small map: distribution 1844–1999

Lasioglossum morio ♀ [PB]

Lasioglossum morio ♂ [PB]

Female *L. morio* can be separated from *L. leucopus* by the presence of very fine ridges running across the marginal areas of tergites one and two, and by the texture of an area on each side of the thorax, immediately below the wings. These have roughened surfaces and fine, indistinct punctures, whereas in *L. leucopus* these areas are smooth and clearly punctate.

Males are tiny and very slender. They can be separated from *L. leucopus* males by the dark tarsi and labrum—there is no yellow here. They also have a slightly roughened surface to the scutum which helps to separate them from *L. cupromicans* and *L. smeathmanellum*, both of which have a completely smooth and shiny scutum.

As well as finding bees feeding on plants such as bramble, forget-me-not and speedwell, it is not uncommon to come across very large nesting aggregations, each nest entrance marked by a small mound of soil. Writing in 1945, Alfred Brazenor reported finding a male of this species on hawkweed.

Behaviour and interactions

L. morio females are first active in spring when they emerge to establish their nests. These are generally established on warm banks and slopes, large numbers of females frequently nesting close together. This is a eusocial species, and, although little information is available on the social organisation of the nest, it is thought that just one worker generation is raised. Each nest might contain between ten and eighteen brood cells, each of these either directly adjoining the main tunnel or sited at the end of a short side tunnel. Nests can run to a depth of 21 cm.

A new generation of females and males emerges from the nest during June and July, the males flying rapidly over nest sites and over sunlit rocks or stone walls as they seek a mate. The males die off during the autumn, while after mating females pass the winter below ground within a burrow. Two or more females often share the same space.

Brood cells are attacked by *Sphecodes niger*. Other species suggested are *S. geoffrellus* and *Nomada sheppardana*, but this has not been confirmed.

On the wing

Sussex records run from early February until the beginning of October. Elsewhere, *L. morio* is reported to be active until the end of October.

Feeding and foraging

A very wide range of plant families are visited by this bee, including Buttercup, Pink, Willow, Cabbage, Mignonette, Rose, Daisy and Borage.

Pollen is collected from many of these, including Daisy, Cabbage, Pink and Buttercup.

Lasioglossum nitidiusculum Tufted Furrow Bee
FIRST RECORD 1899, Bognor Regis (Henry Guermonprez) TOTAL RECORDS 45

Geography and history
From the late 1800s until the 1950s *Lasioglossum nitidiusculum* was a widespread species, and in 1896 Edward Saunders described it as "a very common species everywhere". From the 1950s onwards, however, it declined, and it has now become very scarce. It is generally confined to the coasts of England, Wales and south-east Scotland, and it is most often found in areas with sandy soils.

In Sussex, it was only regularly recorded in the county for a brief period between 1975 and 1994. Even then, the number of records in any one year was low, reaching a maximum of three in 1979 and 1988. It was last recorded in the county on the border with Kent in 2007, the only record this century.

Out in the field and under the microscope
This bee may be extinct in Sussex but should be looked for in areas with sandy soils and with plentiful yellow-flowered plants from the Daisy family.

This is a tricky bee to identify. Females are dark and most closely resemble *L. parvulum*, a relatively widespread and abundant species. To help separate the two, it is necessary to examine the ridges on the top of the propodeum. In *L. nitidiusculum* these are relatively weak and peter out before the back edge of the propodeal

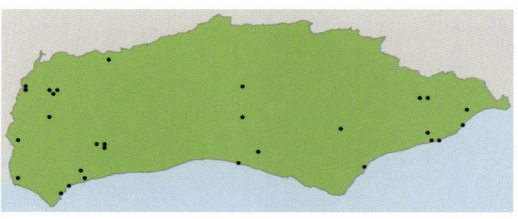

Lasioglossum nitidiusculum ♀ [OB]

Large map: distribution 2000–2023
Small map: distribution 1844–1999

358 *The Bees of Sussex*

triangle, but in *L. parvulum* these ridges are more pronounced and clearly extend all the way to the back edge here.

Males are more straightforward to identify and have tufts of long silvery hairs to the sides of the third, fourth and fifth sternites.

Behaviour and interactions

L. nitidiusculum is a solitary-nesting species which was known for establishing very large nesting aggregations in Britain, many of which would persist for several years. This is still the case in Scotland and elsewhere in Europe. Away from Scotland, large aggregations are now rare in Britain.

Nests are sited on areas of sandy ground and descend at an angle to the surface, with the brood cells established in a line within short spurs off this main tunnel.

The flight season runs from March or April, with the new generation of males and females appearing from mid June or July and remaining active until October.

Three different species lay their eggs within *L. nitidiusculum* brood cells—*Sphecodes geoffrellus*, *S. crassus* and *Nomada sheppardana*. This is also one of the few species in the genus to be attacked by *Stylops*, an internal parasite.

On the wing

L. nitidiusculum is on the wing from mid March until October.

Feeding and foraging

L. nitidiusculum visits a wide range of plant families but seems to have a particularly close association with yellow-flowered plants in the Daisy family.

Pollen is collected from plant families such as Buttercup, Spurge, Carrot, Speedwell and Daisy.

Lasioglossum parvulum Smooth-gastered Furrow Bee
FIRST RECORD 1898, Arundel (Henry Guermonprez) TOTAL RECORDS 443

Geography and history
Lasioglossum parvulum is a locally common bee in southern and eastern England, but it is becoming scarcer in more northerly parts of the country. It is found in many different habitats.

It is widely distributed in Sussex and could be found just about anywhere. As well as in gardens, it has been found in areas of woodland, heathland, hay meadows and pasture, coastal grassland and scrub, sandpits, soft-rock cliffs and chalk grassland. There have been, for example, recent records from West Dean Woods, Truleigh Hill and from the gardens at Great Dixter.

West Dean Woods [ME]

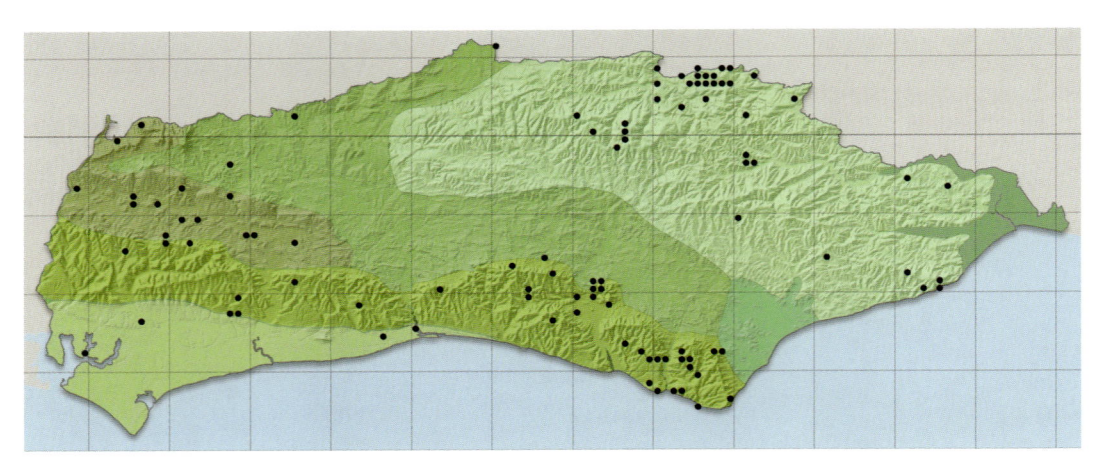

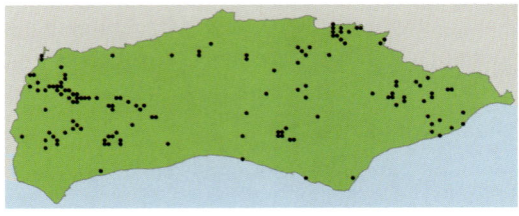

Large map: distribution 2000–2023
Small map: distribution 1844–1999

360 *The Bees of Sussex*

Out in the field and under the microscope

This *Lasioglossum* is easily overlooked but is best searched for by checking the flower heads of its known foodplants, including dandelion, ragwort and Hogweed.

L. parvulum females most closely resemble the much scarcer *L. nitidiusculum*. However, the two species can be separated by examining the ridges on the top of the propodeum. In the case of *L. parvulum*, these are pronounced and extend all the way to the back edge of the propodeal triangle, but they are weak and stop just short of the hind margin in *L. nitidiusculum*.

Males can be more readily separated as they lack the tufts of hair found on the sternites of *L. nitidiusculum*.

Both sexes have a dark brown stigma with a darker hind margin on each of the forewings, as well as a densely punctate, shiny scutum.

Lasioglossum parvulum ♀ [NO]

Lasioglossum parvulum ♂ [TF]

Behaviour and interactions

This is a solitary-nesting species which establishes its nests in small, compact aggregations. The nests are excavated in a variety of situations, including vertical root plates and steep unvegetated or sparsely vegetated banks. The nests can descend 10 to 15 cm below ground with side branches ending in a single brood cell. Within these, the larvae feed on top of a spherical ball of pollen.

On emergence in July, females mate and then excavate a burrow in a bank or other suitable location. This is where they spend the winter.

Brood cells are targeted by *Sphecodes geoffrellus*, *S. crassus* and *Nomada sheppardana*.

On the wing

Sussex records run from late March until mid September.

Feeding and foraging

L. parvulum visits plants from a wide range of plant families, including Pink, Willow, Rose, Daisy, Carrot and Spurge.

Pollen is collected from the Daisy, Bellflower, Buttercup and Willow families.

Lasioglossum pauperatum Squat Furrow Bee
FIRST RECORD 1906, Beeding Down (Henry Guermonprez) TOTAL RECORDS 15

Geography and history
Lasioglossum pauperatum is a very scarce and localised bee which is largely restricted to the south and east of England. Its most northerly location is Cromer on the north Norfolk coast and its most westerly the Dorset coast near Weymouth. The one part of the country where it can be abundant is Essex.

In Sussex, it has been recorded intermittently over the years with records in 1906, 1919 and then, after a gap of 72 years, in 1991. Altogether, *L. pauperatum* has been recorded on just 15 occasions, with 11 of these records since 2000.

Southerham Farm [PG]

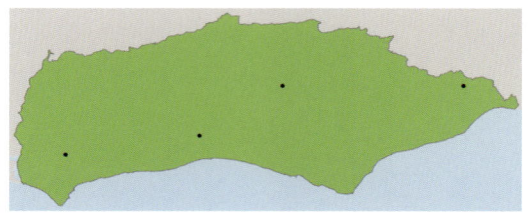

Large map: distribution 2000–2023
Small map: distribution 1844–1999

362 *The Bees of Sussex*

This more recent flurry of records started in 2006 when *L. pauperatum* was recorded in Wadhurst by Ian Beavis. It was most recently recorded by Graeme Lyons in 2021 and 2022 on five separate occasions from two locations, Ovingdean and near Amberley.

L. pauperatum does not appear to have specific habitat preferences. It is associated with sandy soils and occurs on cultivated farmland, in orchards and on species-rich chalk grassland. A record from Alfriston is from a vineyard.

Out in the field and under the microscope

In Kent Ian Beavis has recorded "very large numbers of males [. . .] visiting umbellifers growing along drainage channels through arable fields [. . .]. A few were also seen on the same day at ragwort around gravel pits." At Southerham Farm, Graeme Lyons recorded this species from an area of very species-rich short-cropped chalk grassland.

This is a small dumpy bee. In females, the ridges on the top of the propodeum are very pronounced, clearly separated, regularly spaced, and reach the hind margin of the propodeal triangle. In between the ridges, the surface is very shiny.

The males have yellow tarsi as well as yellow markings at the tip of the clypeus, on the labrum and on the mandibles.

Lasioglossum pauperatum ♀ [SF]

Lasioglossum pauperatum ♂ [SF]

Behaviour and interactions

There is little information about the nesting behaviour of this species, and it is not known if it is attacked by any species of parasitic bee.

Overwintered females emerge in April, followed by a new generation of males and females in the middle of July. These remain active until September.

On the wing

Sussex records run from mid April until early September. Elsewhere, early October is reported.

Feeding and foraging

This bee has been recorded visiting Viper's-bugloss, Common Ragwort, hawk's-beard, Field Fleawort, and white-flowered plants in the Carrot family.

A preference for yellow-flowered plants in the Daisy family is thought likely on the continent.

Lasioglossum pauxillum Lobe-spurred Furrow Bee
FIRST RECORD 1896, Hastings (Edward Saunders) TOTAL RECORDS 613

Geography and history
Lasioglossum pauxillum used to be a scarce bee, and in 1896 Edward Saunders described it as "not common". Since then, however, it has significantly increased in abundance.

In Sussex, *L. pauxillum* can now be found just about anywhere, including in gardens. It is abundant on the South Downs, and, in his survey of the eastern downs between 2003 and 2008, Steven Falk commented that this species is "frequent and often the most abundant bee at a site". Mike Edwards reports that, along with *L. malachurum*, it is frequently the commonest bee in arable fields.

Arable field margin, Birling Gap [PG]

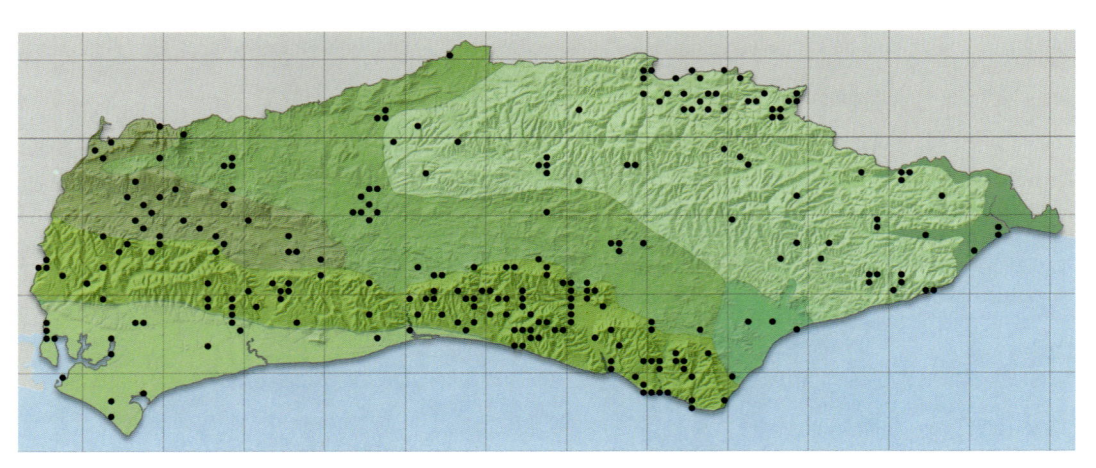

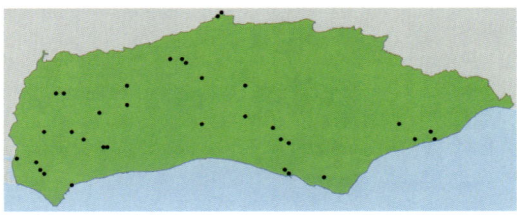

Large map: distribution 2000–2023
Small map: distribution 1844–1999

364 *The Bees of Sussex*

Out in the field and under the microscope

There is quite a bit of variation in the size of females of this species, largely due to caste. However, the shape of the small lobes on the innermost spine on the hind tibiae is a very reliable identification guide. These lobes are broad and clearly rounded and are unlike the spurs on any other species in the genus.

Males are more difficult to identify but have very short unbranched hairs on the second sternite, a yellowish-white marking on the clypeus, and yellowish colouring on the undersides of the antennae.

L. pauxillum is best searched for by checking flowering plants such as dandelion, thistle and members of the Carrot family. Populations appear to be strongest on dry south-facing slopes on the downs, with Mike Edwards reporting that a clay layer is needed to make a nest.

Lasioglossum pauxillum ♀ [NO]

Lasioglossum pauxillum ♂ [SF]

Behaviour and interactions

This *Lasioglossum* can be both eusocial and solitary nesting. The nests themselves can be established in aggregations that range in size from a handful of nests to several hundred.

The nests are excavated in the ground with the main tunnel descending vertically. At the entrance, there is a small dome of soil that is smoothed on the inner surface. At about the mid-point, a cluster of brood cells is created around the main tunnel.

Nests might be established by a single female or by as many as six females working together. Eusocial nests are guarded and produce many more offspring than those established by a single bee.

Two species of *Sphecodes* may attack brood cells—*Sphecodes crassus* and *S. ferruginatus*.

On the wing

Sussex records run from mid March until late September.

Feeding and foraging

Pollen is collected from members of many plant families, including the Buttercup, Daisy, Carrot, Maple and Pea families.

Lasioglossum prasinum Grey-tailed Furrow Bee
FIRST RECORD 1974, Ambersham Common (Mike Edwards) TOTAL RECORDS 71

Geography and history
Lasioglossum prasinum is very closely associated with areas of dry heathland, especially in Dorset, Hampshire, Surrey and west Sussex. It is also known from south Wales, Cornwall, Devon, areas of chalk heath in Norfolk and Suffolk, as well as east Sussex. It can be locally common.

In west Sussex, its populations are centred on heathlands on the Lower Greensand, including Weavers Down, Chapel Common, Iping Common, Ambersham Common and Lavington Common.

In the east of the county, it is known from Isle of Thorns on Ashdown Forest, where it has been recorded twice (most recently in 2019, 15 years after the first record), and from Hastings Country Park, where it was recorded for the first time in 2022. These are the only three records for east Sussex.

Iping Common [RC/SWT]

Large map: distribution 2000–2023
Small map: distribution 1844–1999

Out in the field and under the microscope

L. prasinum often nests in areas of exposed compacted sand and is therefore frequently found along footpaths or banks. Here, it is possible to find females establishing or provisioning their nests, or males flying low over nearby vegetation. Foodplants such as Heather, Bell Heather and Common Fleabane are also worth checking.

Female *L. prasinum* are distinctive. They are larger than most species in the genus and unique in that the front face and sides of the first tergite have a covering of white hairs that are flattened against the surface. Similar flattened hairs are also present on the second, third and fourth tergites, and give this bee a greyish appearance. The scutum is densely punctate and can have a dull bluish-green tinge.

Males are also distinctive. As well as being slender and having very long antennae, they are black overall except on the seventh tergite which is a reddish-orange colour—no other male *Lasioglossum* has this colouring.

Lasioglossum prasinum ♀ [NO]

Lasioglossum prasinum ♀, showing distinctive flattened hairs on the abdomen [NO]

Lasioglossum prasinum ♂ [TF]

Behaviour and interactions

L. prasinum is probably a solitary-nesting species. Little is known of its nesting behaviour aside from the fact that it establishes compact aggregations in areas of exposed sand.

In Britain, brood cells are targeted by *Sphecodes reticulatus* and possibly by *S. pellucidus*.

On the wing

Sussex records run from late May until the end of September. Elsewhere, *L. prasinum* has been recorded from late April until early October.

Feeding and foraging

L. prasinum collects pollen from a small number of plant families, including Heather, Rock-rose and Pine.

Lasioglossum punctatissimum Long-faced Furrow Bee
FIRST RECORD 1878, Hastings (Edward Saunders) TOTAL RECORDS 221

Geography and history
Lasioglossum punctatissimum is widely distributed in Britain, reaching as far north as south-west Scotland. Most records, however, are concentrated in the south and east of England.

In Sussex, *L. punctatissimum* has been recorded in many parts of the county but especially from the heathlands of the Lower Greensand and the High Weald. There is a concentration of records from Ashdown Forest and the High Weald south of Tunbridge Wells.

It is closely associated with light sandy soils and is most frequently found on heathland, acid grassland, sandpits and cliffs.

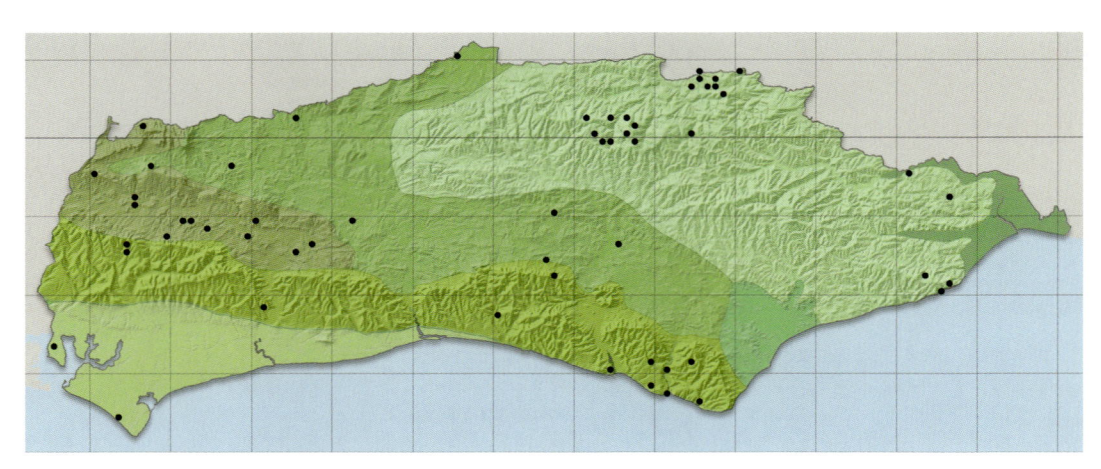

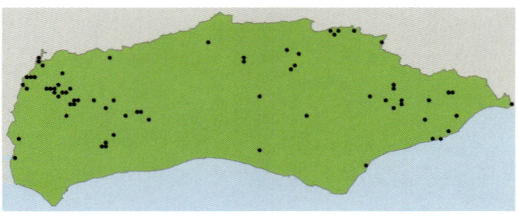

Large map: distribution 2000–2023
Small map: distribution 1844–1999

Out in the field and under the microscope

L. punctatissimum females have long, distinctly oval faces and very densely punctate tergites. They closely resemble *L. angusticeps*, which does not occur in Sussex.

Males have similarly long oval faces, as well as a yellow labrum and a yellow marking at the end of the clypeus. The tarsi on the legs are pale.

This *Lasioglossum* is most likely to be found feeding or foraging on plants such as dandelion, Cat's-ear, forget-me-not and, in particular, Ground-ivy

Behaviour and interactions

Little is known about the behaviour of this bee—it is not known, for example, if this is a solitary-nesting or eusocial species. Given the difficulty of locating nests, it is likely that nests are dispersed rather than established in aggregations.

While no parasitic bees have been reported from Britain, on the continent this *Lasioglossum* is targeted by *Sphecodes crassus*.

On the wing

Sussex records run from mid April until the end of October.

Feeding and foraging

It has been suggested that this species specialises in collecting Dead-nettle pollen, but it has also been observed collecting pollen from other plant families, including Daisy, Heather, Buttercup, Rose and Speedwell.

Other plant families visited include the Violet, Pea, Spurge and Borage families.

Lasioglossum punctatissimum is frequently found feeding or foraging from Ground-ivy. [NS]

Lasioglossum puncticolle Ridge-cheeked Furrow Bee
FIRST RECORD 1878, Hastings (Edward Saunders) TOTAL RECORDS 76

Geography and history
Although *Lasioglossum puncticolle* is found as far north as the Midlands and Norfolk and as far west as south Wales, most records are from the south and east of England, with a concentration of records from south Essex and north Kent.

The earliest Sussex record is from 1878, Edward Saunders writing, "In a lane near Guestling, Hastings, in October. I found the ♀ pretty commonly, but I found only one ♂.

Today, *L. puncticolle* is largely restricted to the east of the county, although there are several records from scattered locations in west Sussex such as Ebernoe Common, Knepp and Gatwick Airport. Most Sussex records are from the eastern High Weald, with Great Dixter generating a high proportion of modern records.

In Sussex *L. puncticolle* has been recorded recently from gardens, soft-rock cliffs, a sandpit, species-rich chalk grasslands, and a scrubby flower-rich grassland.

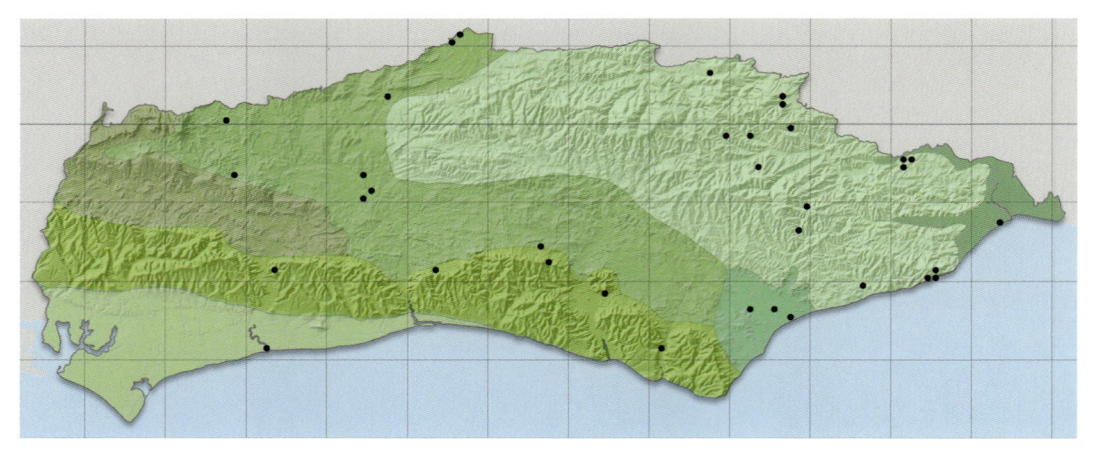

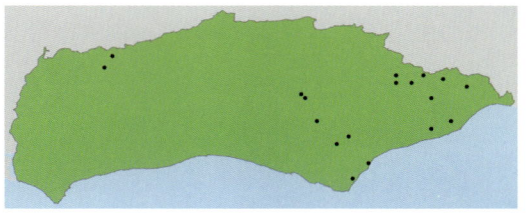

Lasioglossum puncticolle ♀ [IT]

Lasioglossum puncticolle ♀ [IT]

Large map: distribution 2000–2023
Small map: distribution 1844–1999

Out in the field and under the microscope

Females of this species can be identified by checking the underside of the head. This area is very shiny and has a series of very bold, clearly-defined ridges. While other species also have ridges here, these are always much weaker. Females also have a clearly punctate, smooth and shiny scutum.

Males are slightly more challenging to identify. They also have ridges on the underside of the head, but these are much weaker than on the females. They have yellowish-white markings at the tip of the clypeus, on the labrum and on the mandibles, as well as yellow tarsi.

L. puncticolle is often found feeding or foraging, especially from plants in the Daisy family such as dandelion and Common Fleabane.

Lasioglossum puncticolle ♂ [IT]

Lasioglossum puncticolle ♂ [IT]

Behaviour and interactions

This is believed to be a eusocial species, but the details are little understood. It often nests in aggregations in steep unvegetated banks, although single females have also been reported.

There are no known cuckoo bees targeting brood cells of this species.

On the wing

Although most Sussex records are from June to August, *L. puncticolle* is on the wing from the end of March until early October.

Feeding and foraging

Pollen is collected from a range of plants, but especially from members of the Buttercup and Daisy families.

Lasioglossum quadrinotatum Four-spotted Furrow Bee
FIRST RECORD 1908, Sompting (Henry Guermonprez) TOTAL RECORDS 10

Geography and history
Lasioglossum quadrinotatum is a widely distributed species which has been found as far north as Yorkshire and Lancashire and as far west as Dorset. Within this area, however, it is a very scarce species.

There are just ten Sussex records, the first from Sompting in 1908 and the second a male found feeding on fleabane at Bosham Hoe in 1951. It was subsequently found on a handful of occasions between 1979 and 1992 at Wiggonholt Common, at a site between Rogate and Trotton, and at Ebernoe Common. The only other county records are from near Slindon in 1996, Whitehawk in 2001, Horncroft in 2008 and, finally, Burpham in 2021.

This species is largely restricted to sandy sites such as heathland and acid grassland, but it is also known from chalk grasslands and woodlands.

Out in the field and under the microscope
L. quadrinotatum is very similar in appearance to *L. lativentre* (the two species were only recognised as distinct in 1913). The differences between females of these two species are very subtle. They include the colour of the stigma at the front of each forewing, which in *L. quadrinotatum* is a pale yellowish brown throughout but in *L. lativentre* is a darker brown with a dark rear edge.

Males can only be separated confidently by examining the genital capsule. *L. quadrinotatum*

Lasioglossum quadrinotatum ♀ [SF]

Large map: distribution 2000–2023
Small map: distribution 1844–1999

Lasioglossum quadrinotatum ♂ [NO]

males lack the long dense hairs on the tips of the genital capsule found in *L. lativentre*. In males of both species, the stigma on the front of each forewing is similarly marked to that of the females.

The specimen found near Burpham in 2021 was seen "nectaring on tall herbs".

Behaviour and interactions
Having mated the previous year, females emerge in the spring. Other information on their nesting habits is limited, although it is thought most likely that this is a solitary-nesting species.

Sphecodes ephippius and *S. puncticeps* are believed to target brood cells of this species.

On the wing
L. quadrinotatum is on the wing from late March until the end of September.

Feeding and foraging
L. quadrinotatum visits many different species of flowering plant, including plants in the Buttercup, Heather, Spurge, Speedwell and Daisy families.

In Germany, it has been confirmed that pollen is collected from the Cabbage and Daisy families.

Lasioglossum semilucens Small Shiny Furrow Bee
FIRST RECORD 1985, Aldsworth (Mike Edwards) TOTAL RECORDS 7

Geography and history
Lasioglossum semilucens is a very scarce bee restricted to the south and south-east of England. A high proportion of records are from west Kent, with only a small number from other counties supporting this species.

In Sussex, it has been recorded from a handful of scattered locations across the county. These are Aldsworth in 1985, near Slindon in 1999, Chichester in 2008, and Hargate Forest in 2005. The most recent records are from Fore Wood in 2021 and Broadwater Warren in 2022.

Fore Wood [IB]

Large map: distribution 2000–2023
Small map: distribution 1844–1999

Out in the field and under the microscope

This is a tiny bee, no more than 5 mm long, which is very easily overlooked.

Females resemble *L. rufitarse* (which is not known from Sussex), but they lack the very faint ridges that run across the marginal areas of tergites two and three in *L. rufitarse*. In *L. semilucens* the tergites are very shiny.

L. semilucens males have a smooth and very shiny marginal area on the first tergite. The mandibles, labrum and tip to the clypeus are yellow, and the hind legs are marked with yellow.

In his surveys of sites in the High Weald close to the county boundary with Kent, Ian Beavis reports finding this bee foraging for pollen from Tormentil and flying rapidly in large numbers close to the ground near its nest sites.

Lasioglossum semilucens ♀ [TF]

Lasioglossum semilucens ♂ [TF]

Behaviour and interactions

Very little is known of the life history of *L. semilucens*, although it is thought most likely that it is a solitary-nesting species. Observations by Ian Beavis suggest that this species forms small aggregations with "a number [of females] found nesting in the bare vertical side of a path". In other places, he has also found several females nesting beside a woodland track.

No species of cuckoo bee is confirmed from Britain, although *Sphecodes marginatus*—which in the British Isles is restricted to the Channel Islands—is likely to target this species. *S. longulus* may also attack its brood cells.

On the wing

L. semilucens is on the wing from mid May until early September.

Feeding and foraging

It is thought that *L. semilucens* collects pollen from a range of plant families. There are a few reports of this bee visiting cinquefoil for pollen.

Lasioglossum sexstrigatum Fringed Furrow Bee
FIRST RECORD 2016, Stedham Common (George Else)　TOTAL RECORDS 2

Geography and history
Lasioglossum sexstrigatum was added to the British list in 2011, with the earliest British record now known to have been collected five years earlier, on Blackheath Common in Surrey in 2006. It is currently known from Sussex and Kent, as well as from Surrey.

In Sussex, it has been recorded twice, both times in 2016 at Iping and Stedham Commons. These remain the only county records and are the only known British male specimens.

L. sexstrigatum is a species of sandy habitats, especially sandpits, heathland and, on the continent, sand dunes.

Iping and Stedham Commons [NS]

Large map: distribution 2000–2023
Small map: distribution 1844–1999

376　*The Bees of Sussex*

Out in the field and under the microscope

This bee is most likely to be found flying near vertical exposures of sand, or feeding and foraging nearby.

Unlike the other species in the genus, female *L. sexstrigatum* have patches of hairs along the hind margins of the tergites, as in the genus *Halictus*. They also have very pointed lobes on the innermost tibial spurs, and the marginal area of tergite one lacks punctures. There are very faint ridges running across the surfaces of the marginal areas.

Males lack the patches of white hairs on the tergites but can be distinguished from all other species in the genus by the shape of the back of the head—in side view, this has a very sharp-angled lower hind margin.

Lasioglossum sexstrigatum ♀ [SF]

Lasioglossum sexstrigatum ♂ [TF]

Behaviour and interactions

L. sexstrigatum generally nests in small to large aggregations, with up to 34 nests per m² reported from Germany. It may be eusocial.

On the continent brood cells are targeted by *Sphecodes miniatus* and *Nomada sheppardana*, and this may be the case in Britain.

On the wing

L. sexstrigatum is on the wing from mid April until mid September.

Feeding and foraging

On the continent this species visits plants in the Daisy, Willow, Rose and White Bryony families, including Cat's-ear, Goat Willow and cinquefoil.

Lasioglossum smeathmanellum Smeathman's Furrow Bee
FIRST RECORD 1882, Hastings (Edward Saunders) TOTAL RECORDS 198

Geography and history
Lasioglossum smeathmanellum has been recorded from across the county and is found in a wide range of habitats, especially those with exposed soil or cliff faces. These habitats include chalk grassland, heathland and soft-rock cliffs. It is also frequently found in gardens where it often nests in crumbling mortar joints.

Lasioglossum smeathmanellum ♀ [PB]

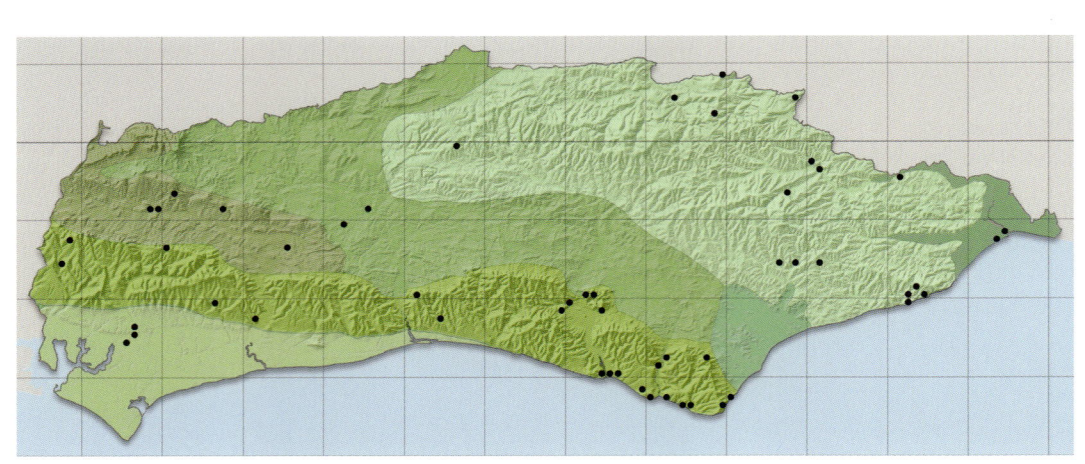

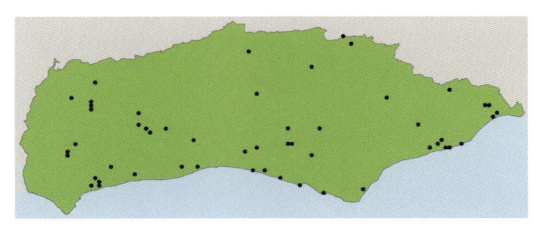

Large map: distribution 2000–2023
Small map: distribution 1844–1999

378 *The Bees of Sussex*

Out in the field and under the microscope
One of the most effective ways of finding this bee is to search along garden walls where the mortar is failing. Here nesting females and patrolling males might be found, or both might be seen feeding or foraging nearby.

L. smeathmanellum is one of four species in this genus with a metallic colouring on the head and thorax. The other three are *L. cupromicans*, *L. leucopus* and *L. morio*. It most closely resembles *L. cupromicans* (which generally has a more northerly distribution in Britain but has been recorded twice in Sussex) and, like this last species, the metallic blue-green colouring extends to the abdomen. The other two species have non-metallic, black abdomens.

Distinguishing between females of these two species on physical characters alone is not straightforward. Like *L. cupromicans* females, *L. smeathmanellum* females have very fine, barely discernible ridges running across the marginal areas of tergites one and two. One subtle character to help separate them is the surface of the propodeal triangle. In *L. smeathmanellum* this has ridges that extend all the way to its back edge, while in *L. cupromicans* these ridges are shorter and peter out before this.

Male *L. smeathmanellum* have a propodeal triangle that is similar to the females, and they have a yellow marking at the tip of the clypeus. The top of the scutum is clearly punctate, very shiny and almost completely smooth. To be sure of an identification, however, it is necessary to examine the genital capsule

Lasioglossum smeathmanellum ♂ [SF]

Behaviour and interactions
This species nests in aggregations in locations with vertical exposures of soil, cracks in rock or crumbling mortar, and it is probably a solitary-nesting species.

No species of cuckoo bee is known to target brood cells of this species although *Nomada sheppardana* has been suggested as a possibility.

On the wing
Sussex records run from mid April until the beginning of October.

Feeding and foraging
The pollen preferences of this species are not well known, although it is believed to forage from several plant families. There is a 2022 record from Batemans where females of this species were observed foraging from Kale, in the process gathering copious quantities of pollen.

Lasioglossum villosulum Shaggy Furrow Bee
FIRST RECORD 1899, Bognor Regis (Henry Guermonprez) TOTAL RECORDS 493

Geography and history
Lasioglossum villosulum is a very widely distributed bee. It is found in many different habitats throughout Sussex, including chalk grassland, heathland, woodland, soft-rock cliffs, lowland meadows and pasture, and coastal dunes. It also occurs in gardens.

Out in the field and under the microscope
Females have a smooth and shiny scutum that is clearly punctate, and they most closely resemble *L. brevicorne*. Females of the two species can be separated by checking the colour of the stigma on

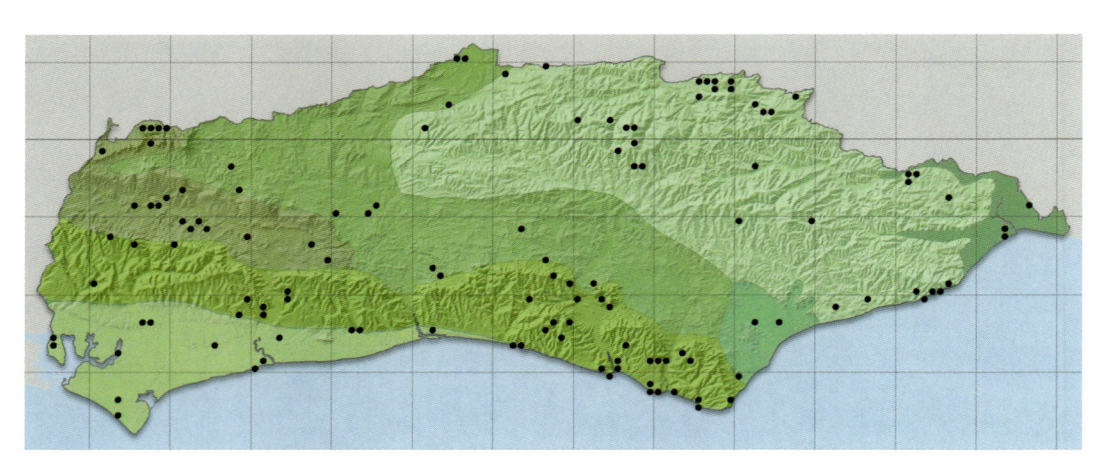

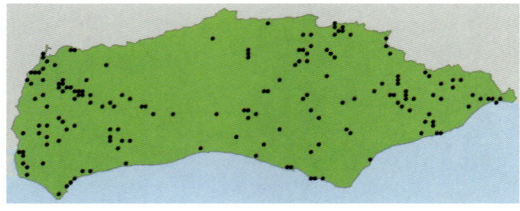

Large map: distribution 2000–2023
Small map: distribution 1844–1999

each forewing and the patches of fine hairs on the front face of tergite one. In *L. villosulum* the stigma on the front of each forewing is dark brown throughout and the hair patches are much reduced or even absent, while in *L. brevicorne* the stigma is yellow with a dark edge and the hair patches are extensive.

Like the females, males have a smooth and shiny scutum with widely separated punctures, but to be sure of an identification it is necessary to examine the genital capsule.

L. villosulum is most often found foraging or feeding, often from yellow-flowered plants in the Daisy family.

Behaviour and interactions
L. villosulum is seemingly able to provision brood cells more rapidly than closely related species such as *L. fratellum* and *L. calceatum*, particularly during periods of poor weather.

This is thought to be a solitary-nesting species that can sometimes form large aggregations.

L. villosulum is one of a small number of bees in this genus that are attacked by the conopid fly *Thecophora atra*. This parasitoid waits for its host either at a nest site or on a flowerhead, launching itself towards its target before quickly inserting an egg into the host's abdomen. The fly larva starts by feeding on non-essential tissues, allowing the host bee to continue to feed. The host eventually dies before the fully formed adult fly emerges.

While no cuckoo bees have been reported from Britain, elsewhere in Europe brood cells are thought to be targeted by *Sphecodes puncticeps* and *S. geoffrellus*. *S. marginatus*, which is not found in Sussex, is another possibility.

Thecophora atra ♀ [PC]

On the wing
Sussex records run from early April until mid October. Elsewhere, *L. villosulum* is reported to occasionally be active into early November.

Feeding and foraging
Pollen is collected from several plant families, including the Daisy, Buttercup and Borage families.

L. villosulum is also frequently found visiting yellow-flowered plants in the Daisy family such as dandelion, Cat's-ear, oxtongue, sow-thistle and hawk's-beard. Other species visited include hawthorn, spurge, Hogweed and Creeping Thistle.

Lasioglossum xanthopus Orange-footed Furrow Bee
FIRST RECORD 1848, Hastings (Frederick Smith) and Littlehampton (S. Stevens) TOTAL RECORDS 65

Geography and history
In Britain, *Lasioglossum xanthopus* is restricted to central and southern England and south Wales. Within this area, it is generally a scarce species and is associated with soft-rock cliffs, landslips and chalk grassland.

In Sussex, it is a scarce and very local bee. It is most closely associated with species-rich chalk grasslands, with most records from the eastern South Downs. Very occasionally it has been found away from the downs, including at Combe Haven in 2015.

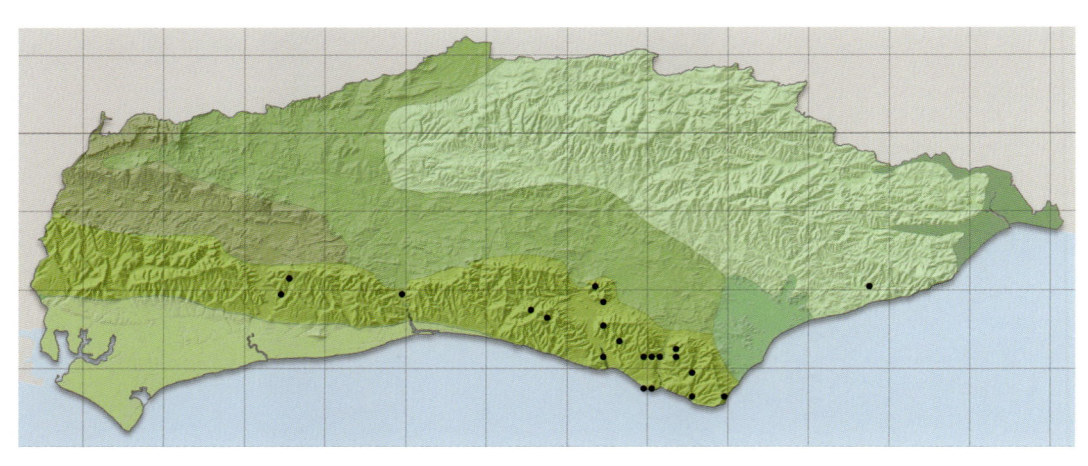

Large map: distribution 2000–2023
Small map: distribution 1844–1999

Out in the field and under the microscope
This is the largest of the *Lasioglossum*. Females can be readily identified by both their size and the orange-yellow colouring of the hind tibiae and tarsi. The hind tibiae and the top of the thorax have golden-yellow hairs, while fresh specimens have obvious bands of white hairs on tergites two, three and four.

Males are also large (though slightly smaller than the females) and have orange-yellow hind tibiae and tarsi, as well as obvious bands of white hairs on tergites two, three and four. The genital capsule is very distinctive.

L. xanthopus is most often found feeding or foraging, particularly on plants such as Greater Knapweed, Germander Speedwell and Viper's-bugloss. In his survey of sites within the eastern South Downs, Steven Falk reports that males "are most typically found on [Perennial] Sow-thistle and late thistles from August onwards", while David Porter reported collecting a female from a scabious flower during a visit to chalk grassland near Denton. There is also a recent record of a female swept from Greater Knapweed at Red Lion Pond on the downs above Southease.

Behaviour and interactions
This is a solitary-nesting species which generally establishes its nests at a low density, although large aggregations are known. The main tunnel descends almost half a metre, and there are between seven and eight brood cells within each nest.

Within each cell, an egg is laid on top of the small spherical ball of pollen. Once the nest has been completed, the female seals the entrance with soil and remains within the nest. In Britain, males die off during the autumn, while in southern Europe they live through the winter below ground.

Brood cells are attacked by *Sphecodes spinulosus* and possibly *S. monilicornis*.

On the wing
Sussex records run from the beginning of April until late September. Elsewhere, males are reported to be active well into October.

Feeding and foraging
In east Sussex pollen is known to be collected from Greater Knapweed, bramble and Germander Speedwell. Other sources of pollen reported from elsewhere include Weld and Viper's-bugloss.

A wide range of plants are visited at other times, including members of the Cabbage, Pink, Pea and Dead-nettle families.

Lasioglossum zonulum Bull-headed Furrow Bee
FIRST RECORD 1897, Hastings (Edward Saunders) TOTAL RECORDS 454

Geography and history
Lasioglossum zonulum is widely distributed in Sussex but has been found most frequently in the High Weald and on the Wealden Greensand. It has been recorded in a variety of different habitats in the county, including heathland, hay meadows and pasture, open woodland, and species-rich chalk grassland. It has also been recorded in gardens, including at Great Dixter.

Wildflower lawn, Great Dixter [PG]

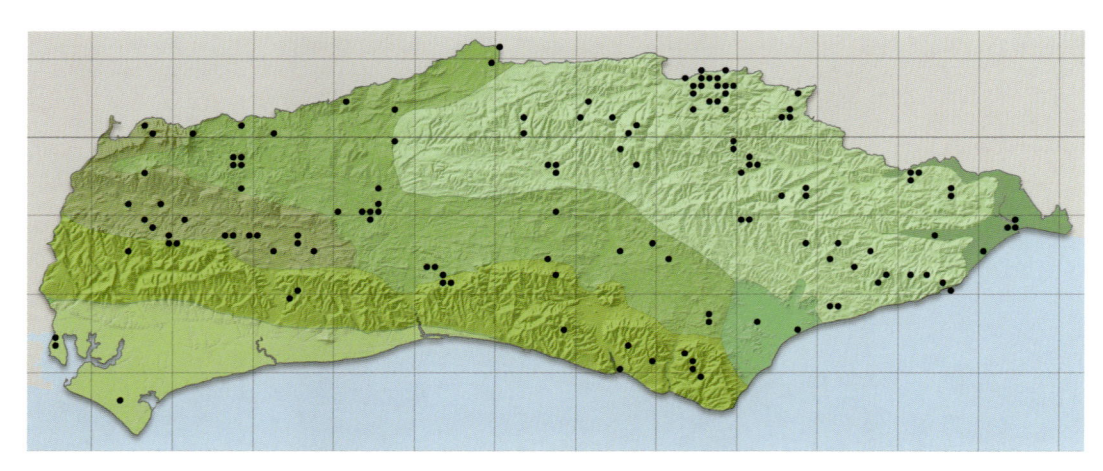

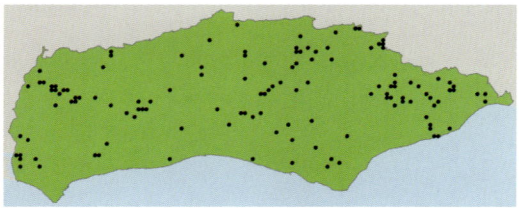

Large map: distribution 2000–2023
Small map: distribution 1844–1999

Out in the field and under the microscope

This bee has a very shiny abdomen and scutum. Females have pronounced vertical ridges on each side of the propodeum and brown hairs on the upper surfaces of the hind tibiae. The combination of these last two characters is only shared with *L. leucozonium*, a species which differs in having a slightly roughened surface to tergite one and to the scutum—*L. zonulum* is entirely smooth and shiny here.

Males are smaller than the females but share the same smooth and shiny surface to the scutum and to tergite one. They can also be separated from the similar *L. leucozonium* by the all-black hind legs—in *L. leucozonium* there are yellowish-white spots at the top of each of the hind tibiae.

This species is usually found foraging or feeding.

Lasioglossum zonulum ♀ [SF]

Lasioglossum zonulum ♂ [SF]

Behaviour and interactions

This is a solitary-nesting species with females capable of provisioning brood cells and laying eggs over two successive years. Nests are excavated to a depth of some 20 cm, with short lateral tunnels excavated off the main tunnel. At the end of each of these is a single brood cell. On completion of egg-laying, the female seals herself inside the nest and awaits the emergence of her brood.

On emergence, the new generation of females is sought out by the males which patrol close to flowers frequented by the females. After mating, females return to the nest. Here, they extend the main tunnel before constructing individual cells within which they pass the winter.

Brood cells are targeted by *Sphecodes scabricollis*. *S. monilicornis* also targets this species on the continent.

On the wing

Sussex records run from the beginning of April until early October.

Feeding and foraging

L. zonulum visits many different plants for pollen, including members of the Daisy, Cabbage, Bellflower, Teasel, Buttercup and Plantain families.

Other flowering plants visited are bramble, Ragged-Robin and Germander Speedwell.

Sphecodes – Blood Bees

The *Sphecodes* are among the most challenging of bees to identify. The females are recognisable from the black thorax and shiny red and black abdomen. Most males are also largely red on the abdomen, although one, *Sphecodes niger*, is entirely black. This is a genus of cuckoo bees, females laying their eggs in brood cells of bees in the genera *Andrena*, *Colletes*, *Halictus* and *Lasioglossum*. Many species target several different hosts. As the larvae do not have mandibles with which to destroy the host's egg or larva, it is believed that, unusually, adult females destroy the host's egg prior to laying their own egg in a cell. In some species, the *Sphecodes* will also attack, and may kill, the female of the host species. Mated females (and, in two species, males) overwinter as adults. In Britain 17 species have been recorded, with 16 known from Sussex.

Sphecodes crassus Swollen-thighed Blood Bee
FIRST RECORD 1896, Hastings (Edward Saunders) and Seaford (Ramsden) TOTAL RECORDS 181

Geography and history
Sphecodes crassus is a widely distributed bee in Sussex and is found in all sorts of different habitats. A high proportion of Sussex records are from the South Downs and from the Wealden Greensand.

Out in the field and under the microscope
This species is most often found close to the nest of one of its host species, where it will fly just above the surface before landing to investigate a nest entrance. In the case of one of its suspected hosts, *Lasioglossum parvulum*, nests are often sited on steep banks of bare or sparsely vegetated ground, and it is here that *S. crassus* might be found. Males might also be found feeding on plants such as Creeping Thistle and Yarrow.

Although some females can be distinguished by the shape of the hind femora—the upper surfaces of these can be particularly swollen—this is not a consistent character. Much more reliable is the length of the labrum, which in this species is relatively long, a character shared only with *S. ferruginatus*. This last bee is larger and has a very densely punctate face, with the individual

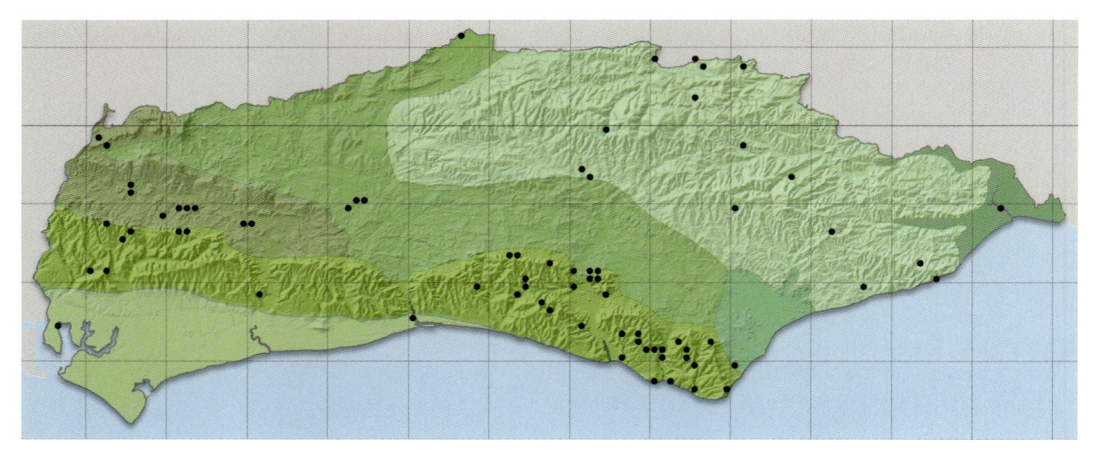

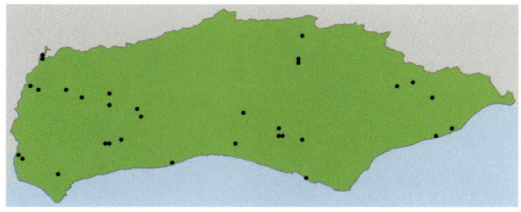

Large map: distribution 2000–2023
Small map: distribution 1844–1999

Sphecodes crassus ♀ [TF]

Sphecodes crassus ♂ [TF]

Lasioglossum nitidiusculum ♀, a suspected host [TF]

Lasioglossum parvulum ♀, a suspected host [TF]

Lasioglossum pauxillum ♀, a suspected host [TF]

punctures almost touching. In contrast, *S. crassus* has smooth and shiny areas between the punctures.

Males can only be identified with confidence by examining the genital capsule.

Behaviour and interactions
S. crassus targets brood cells of several different species of *Lasioglossum*, although the precise relationships are not always confirmed. Suspected hosts in Britain are *L. nitidiusculum*, *L. parvulum* and *L. pauxillum*. This last species is believed to be the major host, with male and female *S. crassus* frequently found around nesting aggregations of this species. As with *L. pauxillum*, *S. crassus* has become much more common than was previously the case.

On the continent, other species have also been suggested, including *L. prasinum* and *L. punctatissimum*. Host-cuckoo relationships may vary across different parts of the range of this species.

On the wing
Sussex records run from early April until late September.

Feeding and foraging
Various plant species are visited, including Creeping Thistle, mayweed, Yarrow and Heather.

Sphecodes ephippius Bare-saddled Blood Bee
FIRST RECORD 1882, Worthing (Edward Saunders) TOTAL RECORDS 401

Geography and history
Sphecodes ephippius is the most frequently recorded *Sphecodes* in Sussex. It is found in most parts of the county, in a wide variety of habitats.

Out in the field and under the microscope
This species is best searched for by checking nest sites being used by one of its host species. Here it will often be found flying close to the surface, or on the ground, often in the company of other species of *Sphecodes*. These nest sites are frequently sited on sunny and sheltered banks and on slopes with bare or sparsely vegetated ground—any such areas are always worth checking.

Females belong to a group of *Sphecodes* lacking punctures behind the ocelli on the top of the head. A key character to look for is the length of the hairs on the front face of the first tergite, which should be very short and sparsely distributed, a feature which is only shared with one other species, *S. rubicundus*.

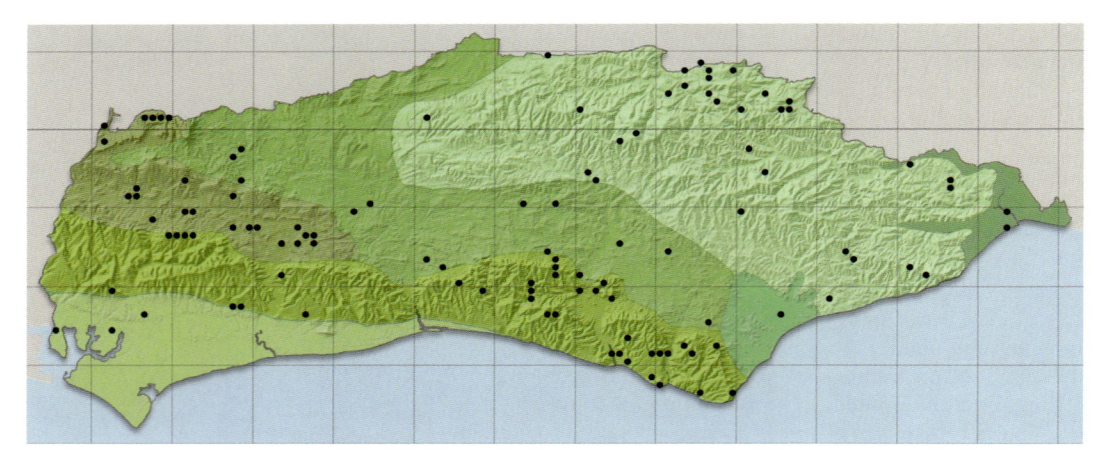

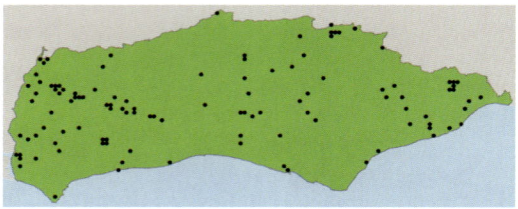

Sphecodes ephippius ♀ [PB]

Sphecodes ephippius ♂ [TF]

Large map: distribution 2000–2023
Small map: distribution 1844–1999

The females of these two species can be separated by looking at the colour of the fourth and fifth tergites, which are almost entirely black in *S. ephippius* (there may be some small patches of red at the corners of tergite four) and largely red in *S. rubicundus*.

Males can only be identified with confidence by examining the genital capsule.

Behaviour and interactions

Throughout its range, *S. ephippius* has an enormous number of host species, with 18 potential and confirmed hosts cited from across Europe. These include species of *Andrena* as well as various species of *Lasioglossum*. The size of this *Sphecodes* varies considerably because of differences in the size of the various host species.

In Britain, probable hosts are *Lasioglossum calceatum*, *L. lativentre*, *L. leucozonium*, *L. quadrinotatum* and *L. laticeps* (not known from Sussex).

Lasioglossum calceatum ♀, a probable host [PB]

Lasioglossum lativentre ♀, a probable host [SF]

Lasioglossum leucozonium ♀, a probable host [PC]

Lasioglossum quadrinotatum ♀, a probable host [SF]

On the wing

Sussex records run from mid March until late August. Elsewhere, *S. ephippius* is reported to be active until October.

Feeding and foraging

A very wide range of plants is visited, including members of the Buttercup, Rose, Pea, Spurge, Daisy and Carrot families.

Sphecodes ferruginatus Dull-headed Blood Bee
FIRST RECORD 1908, Rye (Leopold Vidler) TOTAL RECORDS 39

Geography and history
Despite being widely distributed across England and parts of Wales, and despite the abundance of its presumed hosts, *Sphecodes ferruginatus* is a scarce bee. Most records are concentrated in southern England.

In Sussex, records are thinly scattered across the county but with a concentration of these on the eastern South Downs, especially since 2000. Here, locations such as Seaford Head, Birling Gap and Beachy Head have all generated recent records.

S. ferruginatus is most strongly associated with species-rich chalk grassland but has also been recorded in other habitats such as sandpits and open woodland.

Out in the field and under the microscope
Females most closely resemble *S. crassus* but are larger. Like this last species, they have relatively long hairs on the front face of tergite one and a relatively long labrum. However, *S. ferruginatus* has a very

Sphecodes ferruginatus ♀ [TF]

Sphecodes ferruginatus ♂ [SF]

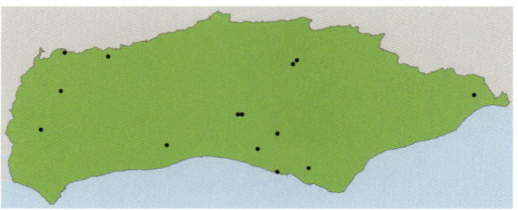

Large map: distribution 2000–2023
Small map: distribution 1844–1999

densely punctate face—the individual punctures above the antennae are almost touching—while *S. crassus* has smooth shiny areas between the punctures. *S. ferruginatus* females also lack the swollen femora found on the hind legs of many *S. crassus* females.

Males are more distinctive than most other species of *Sphecodes*—they are red on tergites one, two and three, and black on the remaining segments—but can still only be identified with confidence by examining the genital capsule.

This species is likely to be found close to the nest of one of its hosts.

Behaviour and interactions

S. ferruginatus lays its eggs within the brood cells of species of *Lasioglossum*. These are believed to include *Lasioglossum fratellum*, *L. fulvicorne*, *L. pauxillum*, *L. laticeps* and *L. rufitarse* (these last two species are not known from Sussex). Of these, *L. fulvicorne* is likely to be the main host.

Lasioglossum fulvicorne ♀, the probable main host [SF]

Lasioglossum fratellum ♀, a probable host [SF]

Lasioglossum pauxillum ♀, a probable host [SF]

On the wing
Sussex records run from the beginning of May until the end of August.

Feeding and foraging
There are few records of this bee visiting flowers, with just Fennel, Wild Carrot and cinquefoil listed in Britain, and Hogweed and Heather cited in Germany.

Sphecodes geoffrellus Geoffroy's Blood Bee

FIRST RECORD 1900, Hastings (Rosse Butterfield) TOTAL RECORDS 299

Geography and history
Sphecodes geoffrellus is a widely distributed bee which is probably under-recorded due to its very small size. It is found in a wide range of different habitats and could occur anywhere in Sussex.

Out in the field and under the microscope
This bee is best searched for by checking warm, south-facing banks of bare or sparsely vegetated ground that are being used as a nest site by one of its hosts. It often flies with other species of *Sphecodes*, including *S. ephippius* and *S. monilicornis*.

S. *geoffrellus* falls into a group which lacks punctures on the top of the head behind the ocelli and has relatively long hairs on the front face of the first tergite and a relatively short labrum. In Britain the only other species in this group is *S. miniatus*.

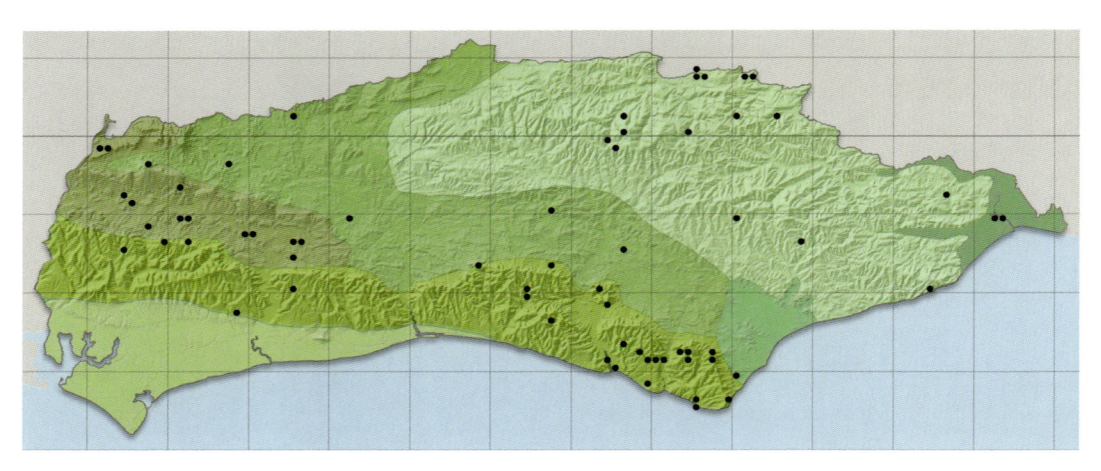

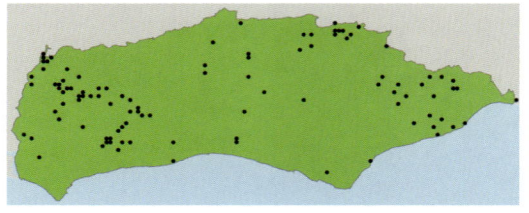

Sphecodes geoffrellus ♀ [PB]

Sphecodes geoffrellus ♂ [NO]

Large map: distribution 2000–2023
Small map: distribution 1844–1999

392 The Bees of Sussex

Although there is some variability, one good character to use to separate *S. geoffrellus* from *S. miniatus* is the structure of the ridges on the top of the propodeum. In this species, they are mostly longitudinal, have very few cross ridges, and often do not reach the back edge of the propodeal triangle. In *S. miniatus*, by contrast, there are several cross ridges and the longitudinal ridges generally reach the back edge of the propodeal triangle.

In males, the front face of each of the antennal segments has extensive patches of pale hairs, a character shared only with *S. miniatus*. Identification should be confirmed by examining the genital capsule.

Behaviour and interactions
S. geoffrellus targets the brood cells of several different species of *Halictus* and *Lasioglossum*. Despite this, individual females may be faithful to a specific species, travelling to nests of its preferred host even if alternative hosts are close by. It is thought to parasitise *Halictus tumulorum, Lasioglossum morio, L. nitidiusculum, L. parvulum, L. villosulum* and *L. rufitarse* (not known from Sussex).

Halictus tumulorum ♀, a probable host [SF]

Lasioglossum villosulum ♀, a probable host [SF]

Lasioglossum morio ♀, a probable host [SF]

On the wing
Sussex records run from late March until late August. Elsewhere, *S. geoffrellus* is reported to be active into early October.

Feeding and foraging
This species has been reported visiting plants from a variety of different families, including Rose, Pea, Carrot and Daisy. Specific species visited include bramble, Creeping Thistle, Wild Carrot and Wild Parsnip.

Sphecodes gibbus Dark-winged Blood Bee
FIRST RECORD 1891, Littlehampton (Edward Saunders) TOTAL RECORDS 170

Geography and history
Sphecodes gibbus has been widely distributed across Sussex but may recently have declined. Its main host *Halictus rubicundus* occurs in a wide range of different habitats, including chalk grassland, open woodland, lowland meadow and pasture, heathland, soft-rock cliffs and saltmarsh.

Out in the field and under the microscope
This *Sphecodes* was first recorded in Sussex when Edward Saunders found a nest site for its host on a bank in Littlehampton in 1891. Here it was flying with *S. pellucidus* and *Lasioglossum leucozonium*. Finding nest sites of its host is still one of the most effective ways of locating this bee.

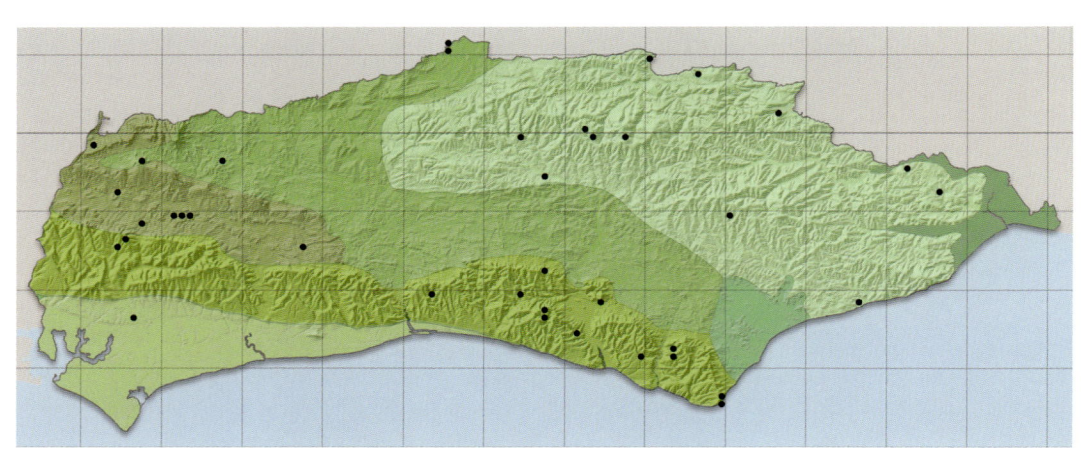

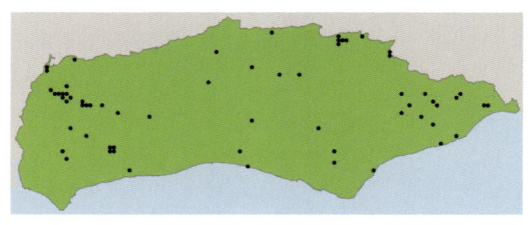

Large map: distribution 2000–2023
Small map: distribution 1844–1999

Typical females are among the larger *Sphecodes*, although there is considerable size variation. The tips of the wings are darkened, while the top of the head behind the ocelli has a deep area of five to six rows of strong punctures. There are only two other species that have punctures here, *S. monilicornis* and *S. reticulatus*. However, *S. monilicornis* has a distinctly box-shaped head, while *S. reticulatus* has just two to three rows of dense, weak punctures behind the ocelli.

Males are smaller and, like the females, have a wide margin with strong punctures on the top of the head behind the ocelli—it is the only species of *Sphecodes* where males have such a deep area of punctures. The punctures in the males are less distinct than in the females.

Behaviour and interactions
S. gibbus attacks the brood cells of *H. rubicundus*, a eusocial species in the southern part of its range. On the continent, other species in the same genus such as *H. maculatus* and *H. quadricinctus* are also suggested as hosts, along with *L. malachurum*. The variation in size of the host species may explain the size variation of individuals in *S. gibbus* populations.

It is likely that individual females specialise in parasitising just one specific species.

On the wing
Sussex records run from late April until the end of August. Elsewhere, it is reported to be on the wing until mid September.

Feeding and foraging
S. gibbus has been recorded visiting a variety of different plants, including Creeping Thistle, Sneezewort, ragwort, hogweed and other members of the Spurge, Carrot and Daisy families.

Halictus rubicundus ♀, the host [SF]

Sphecodes hyalinatus Furry-bellied Blood Bee
FIRST RECORD 1919, Hastings (William Butterfield) TOTAL RECORDS 93

Geography and history
Sphecodes hyalinatus is mostly confined to the South Downs, mirroring the distribution of its main host *Lasioglossum fulvicorne*. There are smaller populations in the High Weald, an area where its other host *L. fratellum* may also be found.

Within these areas, *S. hyalinatus* is generally found on species-rich chalk grasslands and in open woodlands. It can be abundant in some locations. Places such as Mount Caburn, Castle Hill, Seaford Head and Blackcap have all generated recent records.

Castle Hill [ME]

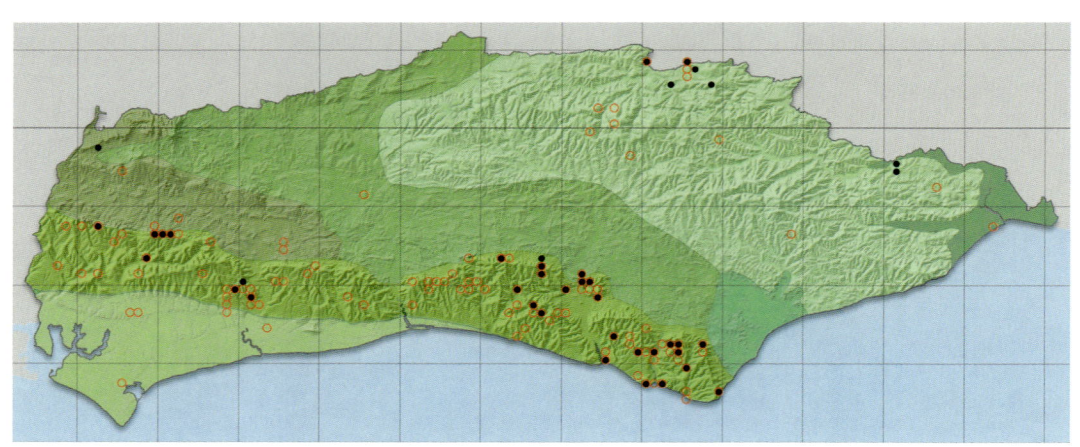

Large map: distribution 2000–2023, with orange circles showing distribution of *Lasioglossum fratellum* and *L. fulvicorne*
Small map: distribution 1844–1999

Sphecodes hyalinatus ♀ [ME]

Sphecodes hyalinatus ♂ [TF]

Out in the field and under the microscope
S. hyalinatus is usually found close to the nest sites of its hosts, often flying with other species of *Sphecodes*. Females are among the smaller bees in the genus and are within a large group that has weak ridges on the top of the head behind the ocelli, long hairs on the front face of tergite one, and two teeth on the mandibles.

They are the only females in this group, however, which have an extensive covering of short dense white hairs on the underside of the thorax, although these do wear away with age. Another character to look for is the relatively smooth and shiny upper sides to the propodeum. This species also has very few cross struts within the propodeal triangle—the ridges here are largely longitudinal.

Males lack the dense covering of hairs found on the underside of the thorax in females, and identification is only possible by checking the genital capsule.

Behaviour and interactions
S. hyalinatus targets the nests of two species, *L. fratellum* and *L. fulvicorne*. Mated females overwinter and emerge in the spring, with males and the new generation of females emerging in July.

Lasioglossum fulvicorne ♀, a host [TF]

Lasioglossum fratellum ♀, a host [SF]

On the wing
Sussex records run from mid April until early September. Elsewhere, *S. hyalinatus* is reported to be active until the end of September.

Feeding and foraging
S. hyalinatus visits plants in the Carrot, Bellflower and Daisy families.

Sphecodes longulus Little Sickle-jawed Blood Bee
FIRST RECORD 1899, Stedham and Stopham (Henry Guermonprez) TOTAL RECORDS 38

Geography and history
Sphecodes longulus is found in dry sandy places, including heathland, acid grassland and sandpits. It is sometimes also found in open broadleaved woodlands. In Britain it is known from the south and east of England, plus from a handful of locations on the south Wales coast. Within these areas, it occurs very sparingly.

Most Sussex records are from heathlands on the Lower Greensand, with outlying records from Rewell Wood and Broadwater Warren. Between 1993 and 2014, there was a gap of 21 years when this bee was not recorded in Sussex. It was most recently recorded in 2022 at Lord's Piece.

Out in the field and under the microscope
This is one of the smallest British bees—only 4 to 6 mm long—and therefore it is very easily overlooked. It is most likely to be found close to its hosts' nest sites.

Large map: distribution 2000–2023, with orange circles showing distribution of *Lasioglossum minutissimum*
Small map: distribution 1844–1999

As well as by their small size, females can be readily distinguished from most other species of *Sphecodes* by their sickle-shaped mandibles which lack a second tooth. The only other species with similar mandibles is *S. puncticeps*, but this is a larger bee with a shallow, shiny, longitudinal furrow at the front of the thorax, a character that is absent in *S. longulus*.

The males have similar mandibles but are more challenging to separate from *S. puncticeps*, particularly as the largest *S. longulus* male can approach the same size as the smallest *S. puncticeps* male. For this reason, it is necessary to examine the genital capsule to confirm an identification.

Behaviour and interactions
S. longulus is believed to lay its eggs within the brood cells of several species of small *Lasioglossum*. In Britain *Lasioglossum minutissimum* has been confirmed as a host, while on the continent species found in Sussex such as *L. leucopus*, *L. morio*, *L. punctatissimum*, *L. semilucens* and *L. sexstrigatum*, have been reported.

Lasioglossum minutissimum ♀, a host [SF]

On the wing
Sussex records run from mid April until early August. Elsewhere, *S. longulus* is reported to be on the wing until mid September.

Feeding and foraging
S. longulus visits plants in the Carrot and Daisy families, including Wild Carrot, Creeping Thistle and mayweed.

Wild Carrot [PC]

Sphecodes miniatus False-margined Blood Bee
FIRST RECORD 1905, Worthing (Edward Saunders) **TOTAL RECORDS** 18

Geography and history
Given the challenge of identifying *Sphecodes miniatus*, it is not surprising that it is under-recorded. This means that its range in Britain is not fully understood. However, there are many records from the counties in the east and south-east of the country, including Norfolk, Suffolk, Essex, Surrey and Kent, as well as Sussex.

Here in Sussex, records are very thinly scattered from a handful of locations in the High Weald and from a small number of heathlands on the Lower Greensand. In the High Weald, it has been recorded close to the Kent/Sussex border near Tunbridge Wells and in Battle, Combe Haven and Hastings. The last of these records, in 2021, is the most recent for *S. miniatus* in the county.

In the west of the county, it was recorded at Ambersham Common and Weavers Down in 1993, and was then unrecorded until 2016 when it was found at Stedham Common. It has previously been known from Bepton, Midhurst Common and from near Petersfield.

Large map: distribution 2000–2023
Small map: distribution 1844–1999

Out in the field and under the microscope
S. miniatus is a very difficult bee to identify. Females closely resemble *S. geoffrellus*, but checking the structure of the ridges on the top of the propodeum will help to distinguish them. In *S. miniatus*, the propodeal triangle has pronounced longitudinal ridges and cross-struts, while in *S. geoffrellus* the ridges are weak, have few cross struts, and generally peter out before the back edge of the propodeal triangle.

Small males are even more challenging to identify but can be tackled by examining the genital capsule.

S. miniatus is best searched for close to the nest sites of its host species.

Behaviour and interactions
This is a species of warm and dry sandy sites that lays its eggs within the brood cells of small *Lasioglossum*, although the precise relationships are not fully resolved. *Lasioglossum nitidiusculum* has been suggested in British literature, with species such as *L. morio*, *L. pauxillum* and *L. sexstrigatum* reported from the continent.

On the wing
Typically, *S. miniatus* is on the wing from mid May until mid September.

Feeding and foraging
S. miniatus has been recorded visiting plants in the Daisy family, with Creeping Thistle and hawk's-beard reported in Britain. On the continent, plants such as dandelion, hawkweed, Yarrow and Cow Parsley are given.

Lasioglossum nitidiusculum ♀, a possible host [OB]

Sphecodes monilicornis Box-headed Blood Bee
FIRST RECORD 1900, Eastbourne (Charles Nurse) TOTAL RECORDS 336

Geography and history
Sphecodes monilicornis is widely distributed in Sussex, occurring in a very wide range of habitats—it could be found just about anywhere in the county.

Out in the field and under the microscope
S. monilicornis is nearly always found close to the nest site of one of its host species, often flying with other species of *Sphecodes*. Females are usually found flying just above a nest, or landing and then crawling inside. They might also be seen being driven away by a guard bee if they attempt to enter a nest.

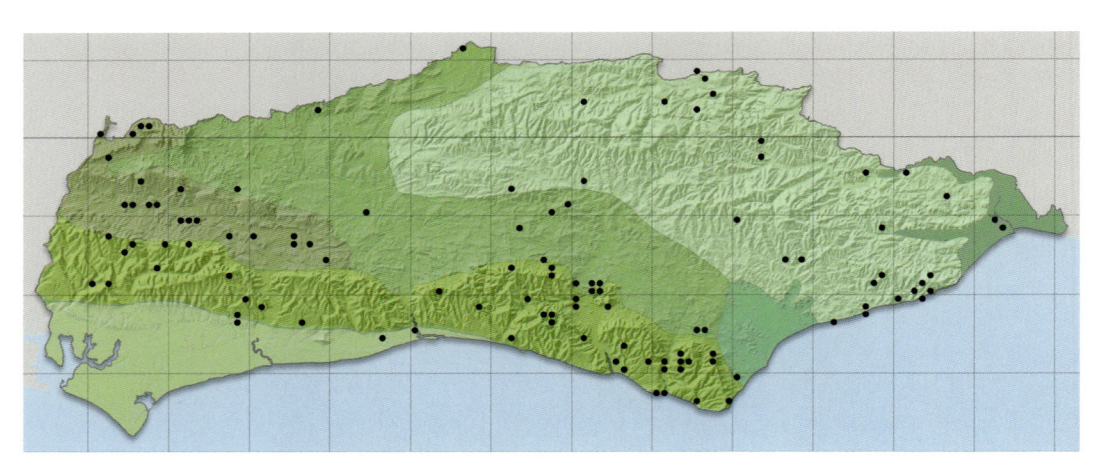

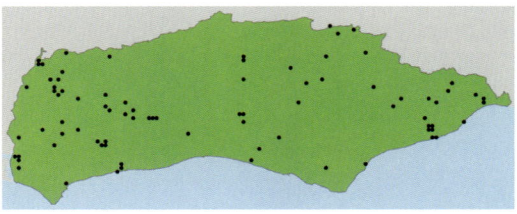

Large map: distribution 2000–2023
Small map: distribution 1844–1999

Females have a characteristically box-shaped head. It is one of just three species that have punctures on the top of the head behind the ocelli, and it can be separated from the other two, *S. gibbus* and *S. reticulatus*, by the shape of the head.

Males are best identified by examining the genital capsule.

Behaviour and interactions

S. monilicornis is a generalist parasite, laying its eggs within the brood cells of quite a few different species, including *Halictus rubicundus*, *Lasioglossum albipes*, *L. calceatum*, *L. malachurum*, *L. xanthopus* and, potentially, *L. laticeps* (though this species is not known in Sussex). Other species are also listed on the continent, including *H. tumulorum* and *L. leucozonium*.

Halictus rubicundus ♀, a host [NO]

Many of the host species are eusocial, and this *Sphecodes* will attack nests both during the founding phase when just a single female is present and later when the workers are present. Despite having a range of potential hosts, individual females are thought to specialise by targeting a specific species.

It is an aggressive species, and a female will potentially kill all the bees present in a nest before laying eggs in each of the brood cells, including those that have been sealed and those that are only partially provisioned. It is known to sometimes decapitate the bees present in a nest, or to eject them.

Mike Edwards has reported seeing a *S. monilicornis* female attacking a nest in Hastings. Here, a female *L. malachurum* guarding the nest entrance was clasped around the head by the *Sphecodes* using its mandibles and pulled from the tunnel.

Lasioglossum cakceatum ♀, a host [NO]

Lasioglossum malachurum ♀, a host [TF]

On the wing

Sussex records run from the beginning of April until the end of September.

Feeding and foraging

S. monilicornis has been reported visiting plants in several different plant families, including Spurge, Carrot, Bellflower and Daisy. Examples include Creeping Thistle, Wild Carrot, water-dropwort, Sneezewort and Hemp-agrimony.

Sphecodes niger Dark Blood Bee
FIRST RECORD 1993, Northiam (Alfred Jones) TOTAL RECORDS 67

Geography and history
Across Britain, *Sphecodes niger* used to be a very scarce species. In Surrey, for example, it was first recorded in 1930, again in 1950, and then not until 1996. It was first recorded in Sussex in 1993 when it was found near Northiam.

Since the 1990s *S. niger* has become a more widespread species, and in Surrey David Baldock has described it as "widespread" and "locally common". In Sussex, records are thinly distributed across much of the county. It can be locally common but the number of records in any one year is generally low, with between none and four per year typical.

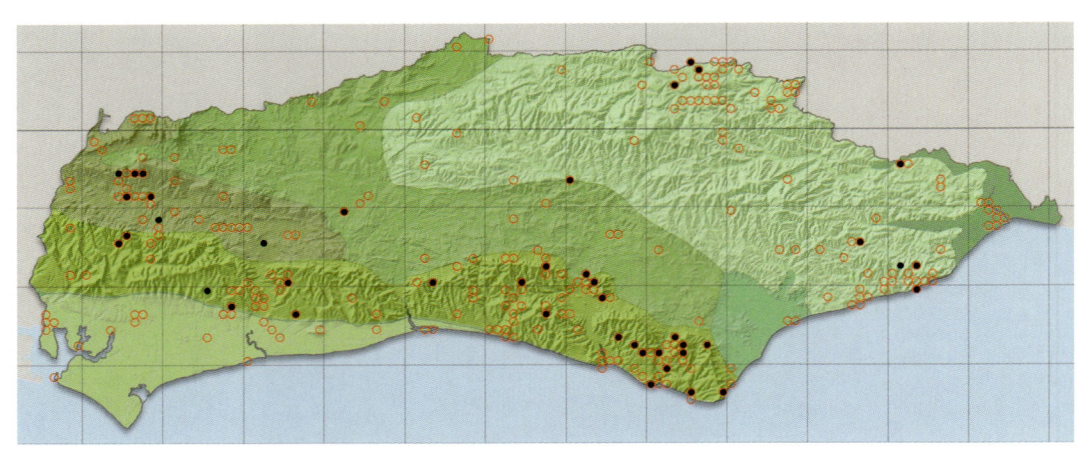

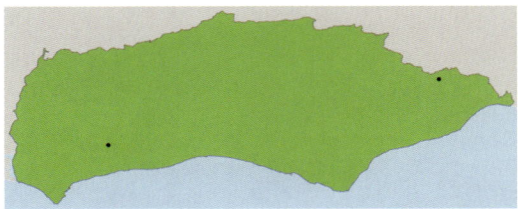

Sphecodes niger ♀ [PB]

Sphecodes niger ♂ [PB]

Large map: distribution 2000–2023, with orange circles showing distribution of *Lasioglossum morio*
Small map: distribution 1844–1999

404 *The Bees of Sussex*

S. niger is a species of warm, sunny locations in a variety of different habitats. These include chalk grassland and chalk heath, soft-rock cliffs, heathland, sandpits and open woodland.

Out in the field and under the microscope

S. niger is usually found at nesting aggregations of its only host, *Lasioglossum morio*. Here, females might be seen flying just above the nest or crawling along the surface before entering a nest burrow.

There are records of this bee from 2021 and 2023 at Seven Sisters. Here, it was found on an extensive south-west-facing vertical bank of exposed chalky soil with *L. morio*. On the first occasion, it was flying among huge numbers of the bee *Colletes hederae* and good numbers of the Eumenid wasp *Ancistrocerus parietinus*, both of which were nesting in the cliff face.

Females have a smooth shiny area on the sides of the thorax just below the wing bases, and it is the only *Sphecodes* with this character. They also have extensive areas of black on tergite three—all other species are mostly red here.

Males are most likely to be confused with male *Lasioglossum* but have a denser and more uniform covering of flattened hairs on the clypeus. They also have shorter antennae and, like the females, have smooth shiny areas beneath the wing bases. Unlike females, however, males have a completely black abdomen.

Behaviour and interactions

This bee attacks the nests of *L. morio*, its only known host.

On the wing

Sussex records run from the beginning of April until late September.

Feeding and foraging

S. niger has been recorded visiting plants in the Carrot family such as Wild Carrot, angelica and hogweed, as well as plants in the Daisy family such as oxtongue and mayweed.

Sphecodes pellucidus Sandpit Blood Bee
FIRST RECORD 1891, Littlehampton (Edward Saunders) TOTAL RECORDS 136

Geography and history
Sphecodes pellucidus is largely restricted to the dry sandy heaths of the Lower Greensand where its host *Andrena barbilabris* is found. In this area, places such as Wiggonholt Common, Weavers Down and Stedham Common have all generated recent records.

Elsewhere in Sussex, there have been recent isolated records from scattered locations such as the sand dunes at Littlehampton, a wide sunny ride in Rewell Wood, and the cliff-top grassland at Seaford Head. During the 1940s, it was regularly recorded from a sandpit near Ditchling.

Large map: distribution 2000–2023, with orange circles showing distribution of *Andrena barbilabris*
Small map: distribution 1844–1999

406 *The Bees of Sussex*

Out in the field and under the microscope
This *Sphecodes* is usually found close to the nesting aggregations of its host. These are sited in areas of bare sand and can be very extensive. Here, a female might be seen either flying slowly close to the ground or crawling along the surface, investigating tunnel entrances.

Females fall into a group of *Sphecodes* without punctures on the top of the head behind the ocelli and with relatively long hairs on the front face of tergite one. They can be separated from the others, however, by the very broad pygidium and by a right-angled structure on the side of the pronotum.

Males also have this right-angled structure, but separation from other species requires examination of the genital capsule.

Behaviour and interactions
This bee lays its eggs within the brood cells of *A. barbilabris*. This is the only confirmed host in Britain, although others are possible. Females that mated the previous summer emerge from hibernation and seek out the hosts' nests in the spring.

On locating a nest, a female enters the tunnel whether or not the host is present. If the host is within the nest, she may be either expelled from the nest or killed by the *Sphecodes*. After egg-laying, the female may pass the night within the nest.

The new generation of males and females emerge from July onwards, and, after mating, the males die off by winter while the females overwinter below ground.

Sphecodes pellucidus mating pair, ♂ on the left and ♀ on the right [PC]

Andrena barbilabris ♀, the host [PC]

On the wing
Sussex records run from late April until late August. Elsewhere, *S. pellucidus* is reported to be active until mid October.

Feeding and foraging
S. pellucidus feeds on plants in the Buttercup, Carrot and Daisy families.

Sphecodes puncticeps Sickle-jawed Blood Bee
FIRST RECORD 1885, Hastings (unknown) TOTAL RECORDS 122

Geography and history
Sphecodes puncticeps occurs throughout Sussex, mirroring the distribution of its various hosts. These are found in a variety of open habitats, including chalk grassland, heathland, woodland, soft-rock cliffs, lowland meadows and coastal dunes. It has been most frequently recorded on the heathlands of the Lower Greensand.

Out in the field and under the microscope
Females are most often found close to a nest site in the spring, while males will be seen from late summer onwards and are generally found feeding. Either side of the Kent/Sussex border, Ian Beavis has found males feeding on plants in the Carrot family, on mayweed and on yellow-flowered members of the Daisy family.

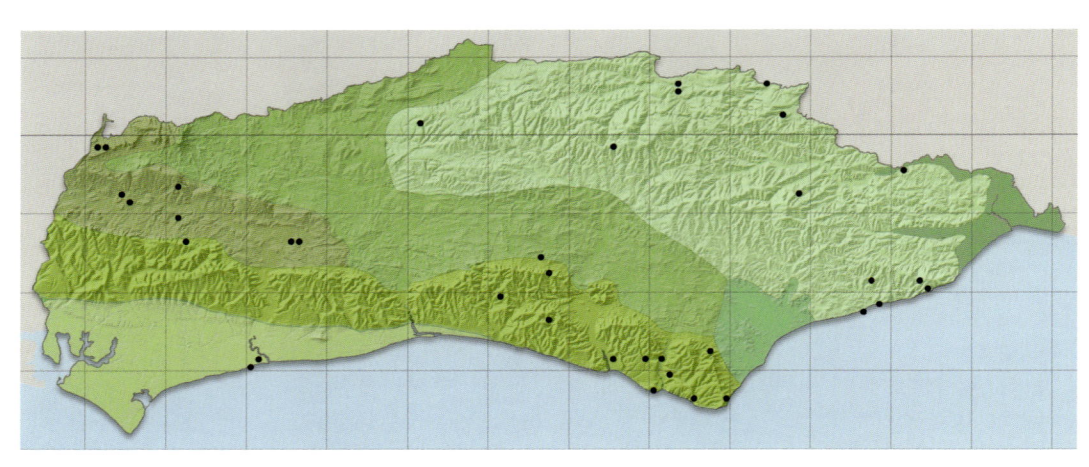

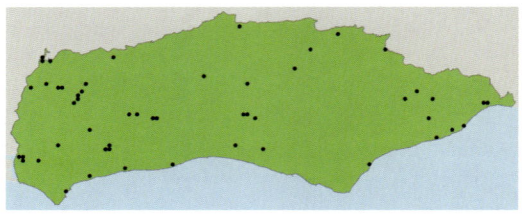

Sphecodes puncticeps ♀ [NO]

Sphecodes puncticeps ♂ [TF]

Large map: distribution 2000–2023
Small map: distribution 1844–1999

408 *The Bees of Sussex*

Female *S. puncticeps* have sickle-shaped mandibles which lack a second tooth. This character is shared only with *S. longulus*, a much smaller species. The two can be separated by the presence of a shallow, shiny, longitudinal furrow at the front of the thorax in *S. puncticeps*.

Males of the two species can only be distinguished from each other by checking the genital capsule.

Behaviour and interactions

S. puncticeps lays its eggs within the brood cells of several species of *Lasioglossum*, although the precise relationships have not been confirmed in all cases. Likely hosts include *Lasioglossum lativentre*, *L. quadrinotatum* and *L. villosulum*. On the continent, *L. brevicorne* is also given.

Overwintered females are on the wing from the spring until July, with the new generation of males and females emerging from July.

Lasioglossum villosulum ♀, a likely host [IT]

Lasioglossum lativentre ♀, a likely host [SF]

On the wing

Sussex records run from late April until late September. Elsewhere, it is reported to be active until early October.

Feeding and foraging

S. puncticeps visits plants from the Carrot and Daisy families, including Fennel, Wild Parsnip, Creeping Thistle and mayweed.

Wild Parsnip [PC]

Sphecodes reticulatus Reticulate Blood Bee
FIRST RECORD 1905, Harting (Beaumont) TOTAL RECORDS 42

Geography and history
Most populations of *Sphecodes reticulatus* are concentrated in the east, south and south-east of the country, but its range extends as far as north-west Wales, Yorkshire and Devon. It is usually found on heathlands, on acid grasslands, in sandpits and in other sandy habitats, but it is also known from chalk grasslands.

In Sussex, modern records are mostly from the heathlands of the Lower Greensand. There are also a handful of records from the heathlands in the High Weald, from a sandpit in the Low Weald near Lewes, and from the chalk on the South Downs. The last of these was from a wide verge on the approach to Devil's Dyke.

Wide chalk grassland road verge on the approach to Devil's Dyke [WP]

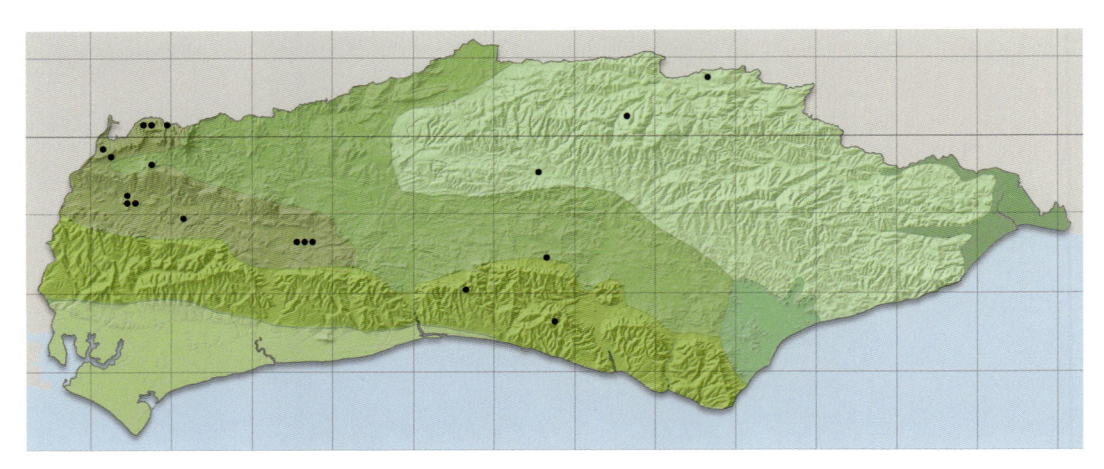

Large map: distribution 2000–2023
Small map: distribution 1844–1999

Sphecodes reticulatus ♀ [SF]

Sphecodes reticulatus ♂ [NO]

Out in the field and under the microscope
This is one of the larger *Sphecodes*. Females are one of three species—*S. gibbus* and *S. monilicornis* are the other two—with punctures on the top of the head behind the ocelli. *S. reticulatus* is closer to *S. gibbus* in its head shape, but it has just two or three rows of weak, indistinct punctures while *S. gibbus* has several rows of distinct punctures. *S. monilicornis*, meanwhile, has a very obviously box-shaped head.

Males also have very indistinct punctures on the top of the head but can only be separated from other species by examining the genital capsule.

Behaviour and interactions
It is likely that *S. reticulatus* lays its eggs within the brood cells of more than one species of bee, although the relationships between species have not been confirmed. Those believed to be hosts are *Andrena argentata*, *A. barbilabris* and *A. dorsata*, plus *Lasioglossum prasinum*.

Andrena barbilabris ♀, a probable host [PC]

Andrena dorsata ♀, a probable host [PB]

On the wing
Sussex records run from mid June until the end of August. Elsewhere, *S. reticulatus* is reported to be active from late May into October.

Feeding and foraging
S. reticulatus visits plants from the Carrot, Borage, Dead-nettle and Daisy families, including Wild Parsnip, forget-me-not, Creeping Thistle and mayweed.

Sphecodes rubicundus Red-tailed Blood Bee

FIRST RECORD 1890, Rustington (Edward Saunders)　TOTAL RECORDS 48

Geography and history

Sphecodes rubicundus is a scarce bee restricted to central and southern England and to south Wales. In Sussex, records are thinly scattered across many parts of the county, but most are from the High Weald.

Within these areas, it has been recorded from soft-rock cliffs, coastal grasslands, open woodlands, a sandpit, and a steep south-facing grassland bank. Recent records are from Rye, Sheffield Park, Seaford Head and Bodiam Castle.

Out in the field and under the microscope

S. rubicundus females are most often found prospecting for suitable nests, and they might be seen flying just above the surface or crawling on the ground near nest entrances, either entering or exiting

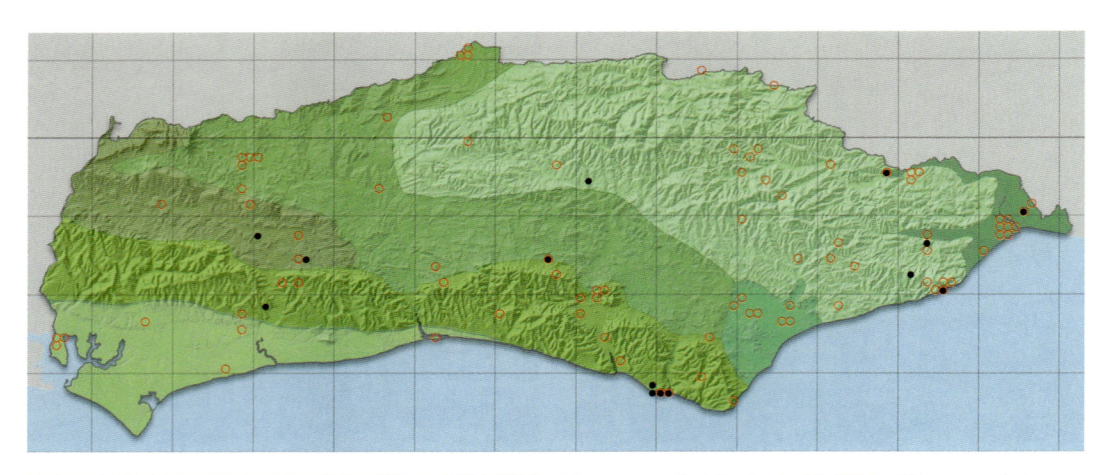

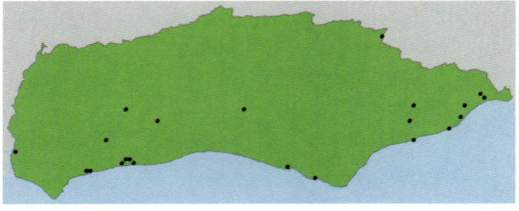

Sphecodes rubicundus ♀ [ME]

Sphecodes rubicundus ♂ [SF]

Large map: distribution 2000–2023, with orange circles showing distribution of *Andrena labialis*
Small map: distribution 1844–1999

from openings. A 2018 record from Sheffield Park is of a female investigating potential nests along a newly re-engineered bank of the River Ouse. Males are most often encountered feeding.

This is one of the larger *Sphecodes*. Females can be identified by the lack of punctures on the top of the head behind the ocelli, the short hairs on the front face of tergite one, and the extensive areas of red on tergite four. It is most like *S. ephippius*, but this last species has largely black markings on tergite four.

Males should be identified by examining the genital capsule.

Behaviour and interactions

This *Sphecodes* lays its eggs within the brood cells of *Andrena labialis*, with one of the first reports in the country of this behaviour coming from Littlehampton. On an otherwise disappointing stay in the town in 1891 ("six species only ought to make any respectable locality blush"), Edward Saunders did make one notable record: "The other rarity I found was *Sphecodes rubicundus* (♂ and ♀) flying about, and on one occasion entering, the burrows of *Andrena labialis* [. . .]. I think it may be taken for granted that *labialis* is its host."

It is also thought likely that *S. rubicundus* lays its eggs within the brood cells of another species of *Andrena*, *A. flavipes*. Mike Edwards has observed males and females emerging from *A. flavipes* nests close to the cliffs to the west of Newhaven.

Unusually, males as well as females overwinter as adults before emerging to mate the following spring, an aspect of their behaviour shared with just one other species in the genus, *S. spinulosus*.

With both males and females spending the winter in their brood cells, this species has a shortened flight season, and most activity is before mid July.

Andrena labialis ♀, a confirmed host [SF]

Andrena flavipes ♀, a probable host [SF]

On the wing

Sussex records run from mid May until late August.

Feeding and foraging

S. rubicundus visits plants from the Spurge, Carrot and Daisy families.

Sphecodes scabricollis **Rough-backed Blood Bee**
FIRST RECORD 1949, Chailey (M. Bacchus) TOTAL RECORDS 52

Geography and history
Sphecodes scabricollis has a very southerly distribution in Britain, with south Essex and south Wales the most northerly extent of its range. Within this area, most records are restricted to Surrey, Kent, Hampshire and Sussex, but there are additional records from Dorset, Somerset and Devon. It was recognised as a British species in 1917.

It was not known from Sussex until 1949 when it was found at Chailey by M. Bacchus. Since then, it has been found most frequently in the High and Low Weald, occurring in scattered locations such as Knepp, Ebernoe Common, Great Dixter and woodland near Hastings. There have also been several records from flower-rich hay meadows and pasture (for example, near Fletching and Burwash Weald), as well as from heathlands and areas of open woodland such as Hargate Forest.

On the continent, this species is reported to be rare and declining in some parts of its range.

Hargate Forest [IB]

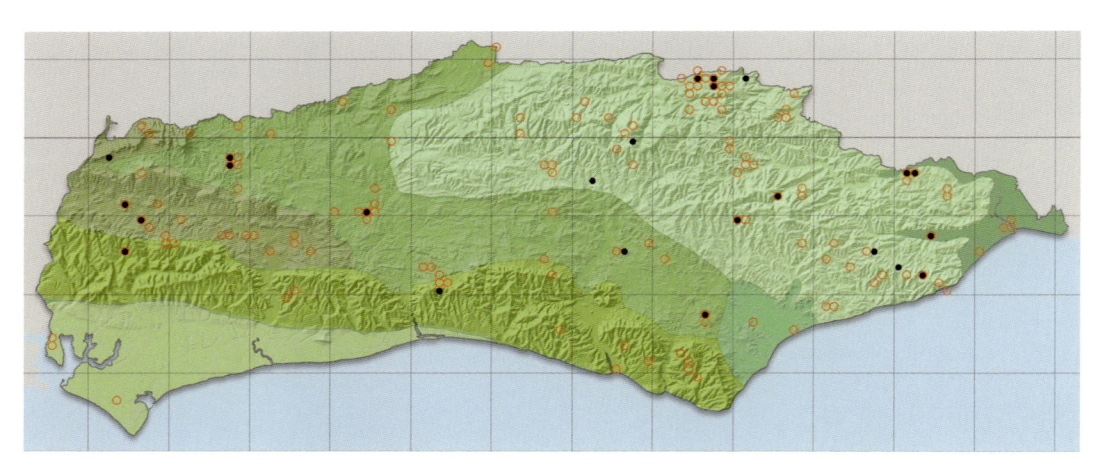

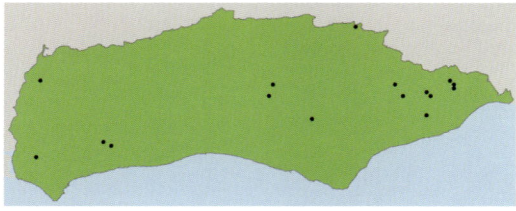

Large map: distribution 2000–2023, with orange circles showing distribution of *Lasioglossum zonulum*
Small map: distribution 1844–1999

414 *The Bees of Sussex*

Sphecodes scabricollis ♀ [TF]

Sphecodes scabricollis ♂ [SF]

Out in the field and under the microscope
S. scabricollis females are most likely to be found close to the nest sites of its host *Lasioglossum zonulum*, although there are also many reported observations of this bee visiting flowers. Howard Hallett found a male on fleabane during a visit to Bosham Hoe in 1951, while more recently Ian Beavis has made records of males on fleabane, thistle and Goldenrod either side of the Kent/Sussex border.

This is one of the larger *Sphecodes*, and both males and females have a sharp ridge that runs along the back edge of the head. This character is best viewed from the side. In females, tergite three is largely black.

Behaviour and interactions
This *Sphecodes* lays its eggs within the brood cells of *L. zonulum*. *Halictus eurygnathus* is also reported as a host on the continent.

On the wing
Sussex records run from late May until the end of August. Elsewhere, *S. scabricollis* is reported to be active until late September.

Feeding and foraging
S. scabricollis visits plants from the Daisy family such as Creeping Thistle, Common Fleabane, golden-rod and Yarrow.

Lasioglossum zonulum ♀, the host [SF]

Sphecodes spinulosus Spined Blood Bee

FIRST RECORD 1919 (Chichester, Philip Harwood) TOTAL RECORDS 17

Geography and history

Sphecodes spinulosus is a scarce and very local bee that is restricted to southern England and south Wales. Its range extends as far north as Northamptonshire and Suffolk and as far west as Cornwall and Pembrokeshire.

Within Sussex, it is a very scarce bee that has been recorded just 17 times, most recently in 2022 from near Burpham. In 2006 it was recorded at Birling Gap by Steven Falk, the first county record in 58 years. Prior to this record, it had been recorded just five times in Sussex.

There are nine recent records from chalk grasslands in the area between Denton Downs and Birling Gap, a significant concentration for such a scarce bee. This area is also where the most records for its host *Lasioglossum xanthopus* have been made.

In 2021 it was also recorded on two occasions in the Littlehampton area.

Cliff top grassland and scrub, Birling Gap [SF]

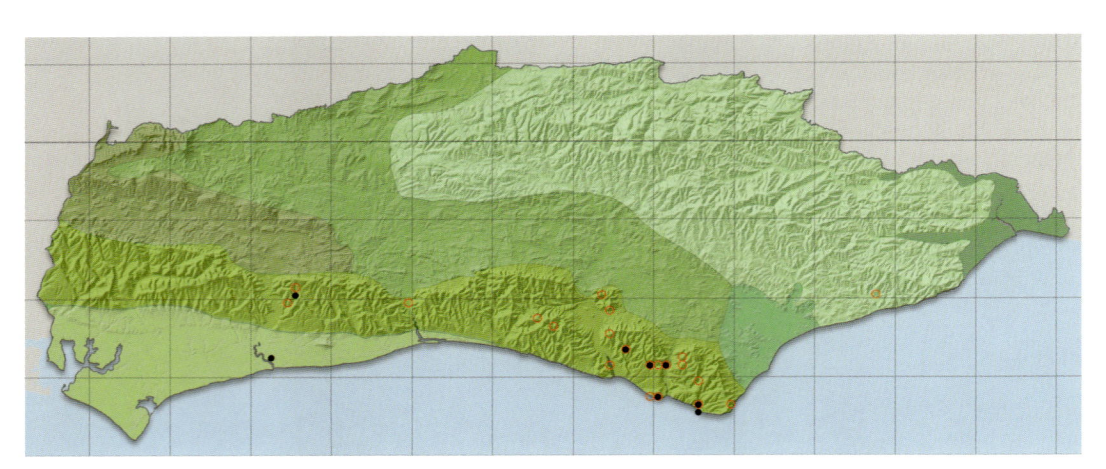

Large map: distribution 2000–2023, with orange circles showing distribution of *Lasioglossum xanthopus*
Small map: distribution 1844–1999

416 *The Bees of Sussex*

Sphecodes spinulosus ♀ [TF]

Sphecodes spinulosus ♂ [ME]

Out in the field and under the microscope
This is an elusive bee. Given that its host nests at a low density, females are most likely to be found feeding. Males can also be found feeding or patrolling around isolated bushes. At Birling Gap Steven Falk has reported that males and females were found on the bare cliff tops.

Typical females are among the largest and most robustly built of the *Sphecodes* and, like the males, have two distinctive characters that help with their identification. These are a shallow trough that runs across the base of the back of the head (this is best viewed from above) and a furrow that runs across sternite two.

Behaviour and interactions
S. spinulosus lays its eggs within the brood cells of *L. xanthopus*. This *Sphecodes* is one of two species in the genus—the other being *S. rubicundus*—where the males as well as the females overwinter as adults in their brood cells before emerging to mate in the spring. As a result, *S. spinulosus* has a short flight season for a species in this genus.

Lasioglossum xanthopus ♀, the host [PB]

On the wing
Typically, *S. spinulosus* is on the wing from mid May until late June.

Feeding and foraging
S. spinulosus has been reported visiting Bulbous Buttercup, Hawthorn, plants in the Spurge family, Field Maple and Wild Carrot.

Colletes – Plasterer Bees

These are solitary ground-nesting bees which are distinguished from superficially similar genera such as *Andrena* by an S-shaped outer wing vein. Females lack a pygidium and have short bilobed tongues which are used to line (or 'plaster') the brood cells with a cellophane-like substance. Brood cells are stocked with a liquid mix of pollen and nectar, and the egg is attached to the upper wall of the cell. In Britain, all but one species fly during the summer or autumn. Nine species have been recorded in Britain, with eight known from Sussex.

Colletes cunicularius Early Colletes
FIRST RECORD Rye Harbour Nature Reserve (Chris Bentley) TOTAL RECORDS 109

Geography and history
For many years, this species was regarded as a specialist of coastal dune systems in north Wales, south Wales, Lancashire and Cumbria. From 1997 onwards, however, there were increasing records from inland locations. Then, in 2012, it was discovered in Hampshire, and it is now also known from other counties in the south-east, including Sussex, Surrey and Kent. It is likely that records for the south-east were initially due to colonisations taking place from across the English Channel.

Since reaching Sussex, *Colletes cunicularius* has spread widely across the county. The earliest record is from Rye Harbour Nature Reserve in 2015, with additional recent records from Galley Hill in 2019, Rewell Wood in 2020, East Head in

Colletes cunicularius nesting site, Galley Hill [AP]

Large map: distribution 2000–2023
Small map: distribution 1844–1999

418 *The Bees of Sussex*

2019, Graffham in 2020, and a sandpit near Lewes in 2021. Most recently, a colony was discovered near Wadhurst in 2022, and a specimen was also swept from the beach near Kemp Town in Brighton in early 2023.

Out in the field and under the microscope
C. cunicularius is best looked for in early spring in areas with exposed sandy soils and abundant Creeping or Goat Willow. Sandpits and soft-rock cliffs are known to be readily colonised.

This is the largest of the *Colletes*, and it is also the only *Colletes* to fly in the spring. The S-shaped outer wing vein and spring flight period mean that this species is unmistakable.

Both males and females are covered in predominantly brown hairs, and the males are slightly smaller than the females.

Colletes cunicularius ♀ [PC]

Colletes cunicularius mating pair, ♂ on the left and ♀ on the right [PC]

Behaviour and interactions
This is a pioneer species, able to colonise new sites and build up significant populations rapidly when conditions are favourable. It will even fly on cold days, and Peter Greenhalf reports that males were on the wing the day after a snow fall at Rye Harbour Nature Reserve in April 2022.

Males emerge first. After a few days feeding, they are generally seen in enormous numbers flying rapidly just above the ground, pouncing on females as they emerge from their brood cells. If a male does not immediately appear, a female will pause at the surface until one does.

Nests are excavated in areas of loose sand, often in large aggregations. Graeme Lyons reports that the colony near Wadhurst was "sited on a sandy, south-facing, sheltered bank in an area of heathy vegetation surrounded by trees".

This bee's emergence coincides with the production of pollen and nectar by Goat Willow and Grey Willow, with which the females initially provision their brood cells. Later in the season females forage from other plant species.

Brood cells are targeted by *Sphecodes albilabris* (not currently known from Sussex).

On the wing
Sussex records run from late February until late June.

Feeding and foraging
Pollen is primarily collected from willow. Species visited at other times include dandelion and Alexanders.

Colletes daviesanus Davies' Colletes
FIRST RECORD 1873, Littlehampton (Edward Saunders) **TOTAL RECORDS** 124

Geography and history
Records for *Colletes daviesanus* are thinly scattered across much of the county but with a concentration on the Wealden Greensand and some coastal areas, including Seaford Head and Hastings Country Park. It is found in a wide range of different habitats, and is regularly recorded from gardens.

Seaford Head [SF]

Out in the field and under the microscope
This bee is best searched for by checking the flower heads of plants in the Daisy family, especially Tansy, Corn Marigold and Yarrow.

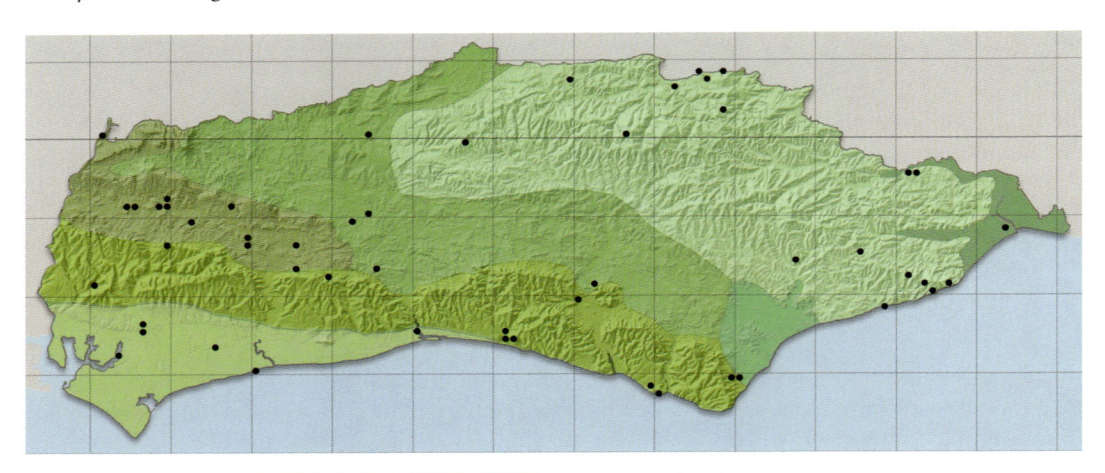

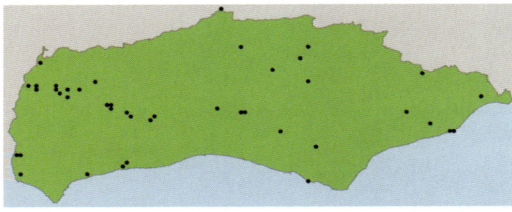

Large map: distribution 2000–2023
Small map: distribution 1844–1999

Colletes daviesanus ♀ [PC]

Colletes daviesanus ♂ [PC]

This is not a straightforward bee to identify. Females have pale hairs on the clypeus and bands of white hairs which fill the marginal areas of tergites two, three and four. Tergite one is shiny and relatively sparsely punctate. Females are close in appearance to *C. floralis* females, but this last species is not known from Sussex.

Males resemble females but are slightly smaller. Key characters to look for are the presence of tufts of long hairs on the sides of sternite six and a shiny, sparsely punctate tergite one.

Behaviour and interactions

Nest sites can be established within the walls of buildings where the mortar is soft and crumbly, but *C. daviesanus* more typically nests in vertical rock faces and the steep sides to sand and gravel pits.

Females are often faithful to the same nest site from one year to the next, using their mandibles and saliva to create a tunnel within the nesting substrate. At the end of the tunnel, there might be a single brood cell or a cluster of some two to six cells (and occasionally as many as ten).

C. daviesanus is one of at least five species in the genus that are attacked by the shadow fly *Miltogramma punctata*. This fly follows a female bee as she returns to her nest, before laying her eggs within the bee's brood cells.

Brood cells are also attacked by the cuckoo bee *Epeolus variegatus*. On the continent, female *C. daviesanus* have been observed fighting off this cuckoo.

Miltogramma punctata ♀ [PB]

On the wing

Sussex records run from the end of May until the middle of August. Elsewhere, *C. daviesanus* is reported to be active until mid September.

Feeding and foraging

Pollen is collected entirely from plants in the Daisy family, with other plant families visited for nectar. Commonly used sources of pollen include Yarrow, Tansy, Corn Marigold, ragwort and Common Fleabane.

Colletes fodiens Hairy-saddled Colletes
FIRST RECORD 1845, Littlehampton (Sydney Stevens) TOTAL RECORDS 88

Geography and history
Colletes fodiens is closely associated with dry sandy habitats, and in Sussex it is most frequently found on the heathlands and acid grasslands of the Lower Greensand, plus in sandy habitats along the coast. It is seemingly absent from the cooler heathlands of Ashdown Forest and the High Weald.

Along the coast, it has been found in areas of vegetated shingle, as at Norman's Bay and Rye Harbour Nature Reserve, and on sand dunes at Littlehampton and Camber Sands. On the Lower Greensand, Iping and Stedham Commons have both generated recent records. In these areas, it can be locally abundant.

Vegetated shingle, Norman's Bay [WP]

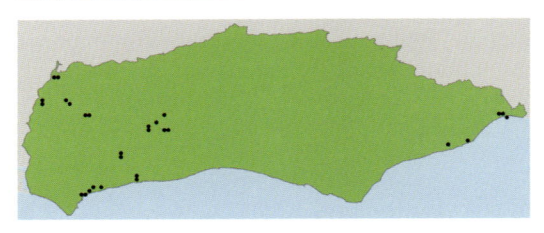

Large map: distribution 2000–2023
Small map: distribution 1844–1999

Out in the field and under the microscope
The most successful way to find this bee is by checking ragwort flowers where females can be seen collecting pollen. Males are also usually found close to foodplants, often in good numbers.

C. fodiens superficially resembles a few other species of *Colletes*, especially worn examples of *C. similis*. Like this species, *C. fodiens* has bands of pale hairs that fill the marginal areas of tergites two, three, four and five. Differences between these two species are slight, but one character to look for is the surface at the front of tergite one. This is hairy across its entire width in *C. fodiens*, whereas in *C. similis* the central area lacks hairs. These hairs can wear away with time.

Males represent a bigger challenge but can be separated from *C. similis* by checking the structure of the genital capsule.

Colletes fodiens ♀ [SF]

Colletes fodiens ♂ [PB]

Behaviour and interactions
This *Colletes* is understood to nest at low densities, either singly or in small aggregations. Nests are established in the ground in areas with light sandy soils that are bare or sparsely vegetated. The passages to the brood cells are filled with soil once the cell has been sealed.

Brood cells are attacked by the cuckoo bee *Epeolus variegatus* and possibly by *E. cruciger*.

On the wing
Sussex records run from late June until late September.

Feeding and foraging
While *C. fodiens* visits a range of plant families, it is known to collect pollen just from plants in the Daisy family, particularly ragwort but also fleabane, Tansy and mayweed.

Other plant species visited include bramble, Bog Pimpernel, Creeping Thistle, Yarrow and Sheep's-bit.

Colletes halophilus Sea Aster Bee
FIRST RECORD 1989, Rye Harbour Nature Reserve (Roger Morris) TOTAL RECORDS 41

Geography and history
Colletes halophilus has an unusual worldwide distribution. It is endemic to the North Sea as far north as the Durham coast, to the English Channel, coastal areas of Somerset, Cheshire and Lancashire and the Bay of Biscay. There is strong evidence that this species is currently expanding its range.

C. halophilus was first recorded in Sussex in 1989 at Rye Harbour Nature Reserve, but it was only in 2003 that it was then found at Cuckmere Haven. Subsequently, it has spread along the Sussex coast and reached West Wittering in 2018. It is currently known from just six locations in the county, namely Rye Harbour Nature Reserve, Hastings, Norman's Bay, Cuckmere Haven, Medmerry and West Wittering.

Sea Aster, Rye Harbour Nature Reserve [BY]

Large map: distribution 2000–2023, with white circles showing distribution of Sea Aster
Small map: distribution 1844–1999

424 *The Bees of Sussex*

Out in the field and under the microscope
In Sussex, as elsewhere, *C. halophilus* is most likely to be found close to where Sea Aster grows. This is a plant of saltmarshes and brackish, muddy tidal sections of rivers and natural harbours, and therefore the River Rother near Rye and the Cuckmere, particularly in their lower reaches, are good places to look for this bee.

Female *C. halophilus* are usually found foraging for pollen from Sea Aster or provisioning their nests, while males might be seen feeding or in flight close to the nest sites.

C. halophilus can occur in the same localities as other species of *Colletes*, including the very similar *C. hederae*—the two species were observed nesting in the same low bank at Cuckmere Haven in 2023. While the choice of pollen source helps to distinguish between these two species, this is not an absolute guide.

Among the characters which help to separate both male and female *C. halophilus* from *C. hederae* is the colour of the hair bands that fill the marginal areas on the tergites. These are white in *C. halophilus* and yellow in fresh specimens of *C. hederae* (this colour can fade with time).

Behaviour and interactions
C. halophilus is one of the last bees to emerge each year, the females timing their emergence to coincide with Sea Aster coming into flower in early September. Males are on the wing a little earlier.

Females tend to nest close to each other, forming small aggregations with the nest chambers excavated in areas of exposed soil. Nests can withstand some flooding by the sea.

The aggregations can be in a variety of locations, including on sand dunes, on the edges of saltmarshes, and in one case at Rye Harbour Nature Reserve on a 12-metre-long low bank of lightly vegetated sandy soil and shingle.

In contrast, at Cuckmere Haven a colony is located along a ten-metre section of a vertical muddy bank about 0.5 m high. Immediately adjacent to this nest site is a large brackish pool, and in 2022 it was possible to see large numbers of males and females trapped on the surface of the water, presumably having fallen in after mating.

Males can often be seen feeding on flower heads or flying just in front of the surface of the nest site, diving onto females to mate.

Brood cells are attacked by the cuckoo bee *Epeolus variegatus*.

On the wing
Sussex records run from the end of July until mid October. Elsewhere, *C. halophilus* is reported to occasionally be active into November.

Feeding and foraging
Sea Aster is the principal source of pollen, although other plants in the Daisy family are also visited.

Adult bees will also visit a range of other plant species, including Bristly Oxtongue, Common Mallow, ragwort, sea-lavender and Sea Rocket.

Colletes hederae Ivy Bee
FIRST RECORD 2004, Hastings (Andrew Grace) **TOTAL RECORDS** 563

Geography and history
Colletes hederae was only recognised as a distinct species in 1993. At that time it was only confirmed from the Channel Islands, France, Germany and Croatia, but since that time it has spread rapidly across Europe.

It was first recorded in Britain in 2001, and it is now a very widely distributed species across the country. As of 2021, it had been found as far north as Dunbar on the south-east Scottish coast.

It was first recorded in Sussex in 2004 in Hastings and has since colonised most parts of the county. It can be found anywhere with flowering Ivy.

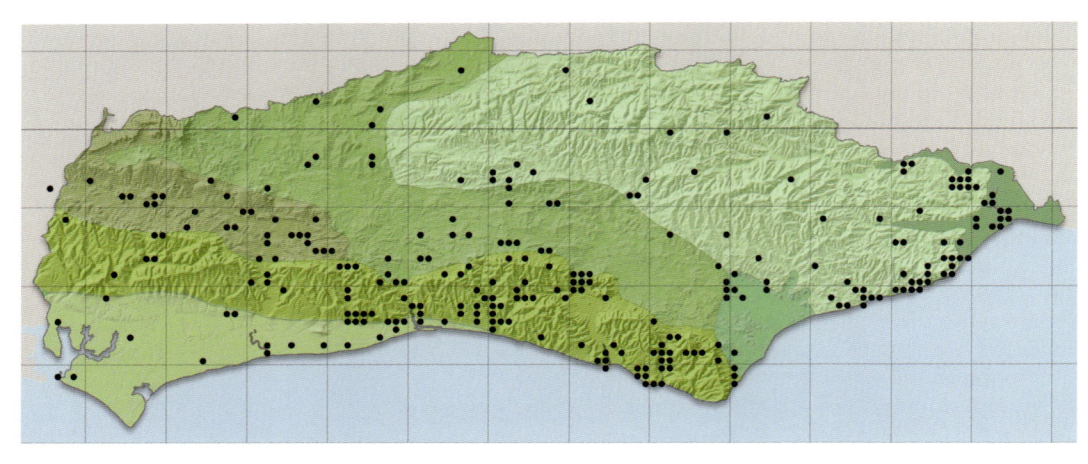

Colletes hederae ♀ [PC]

Colletes hederae ♂ [PG]

Large map: distribution 2000–2023
Small map: distribution 1844–1999

426 *The Bees of Sussex*

Out in the field and under the microscope

This species of *Colletes* is usually seen foraging on Ivy, often among wasps and hoverflies feeding on the flowers. It is very dependent on flowering Ivy for pollen, and any patch is worth checking.

C. hederae forms large nesting aggregations in a variety of locations, including lawns, so these can also be good places to find it.

Females are among the larger of the *Colletes*, and fresh specimens have yellow hair bands on the tergites and are therefore very distinctive. Once these colours fade, they could be confused with *C. halophilus* (in areas with Sea Aster) and *C. succinctus* (in areas with Heather).

Males resemble the females but are smaller and slimmer.

Behaviour and interactions

In Sussex, nests have been observed at various locations, including in banks of bare vertical soil and in areas of short vegetation such as lawns or mown roadside verges. Nests are established in very large aggregations, with thousands of bees potentially active in any one location. Favourable sites will be used repeatedly over several years, females either re-using old tunnels or excavating new ones. These can be to a depth of 45 cm.

Males emerge from the nests several days ahead of the females and, after a period of feeding, start patrolling rapidly just above the ground and testing the soil to locate emergent, unmated females.

On the continent—for example, in Germany in 2020—there have been recent observations of *Epeolus variegatus* investigating *C. hederae* nests, which are evidence of a new host-parasite relationship. This behaviour was first reported from Britain in 2022 in Suffolk and was observed at two sites in Sussex in 2023—Seven Sisters and a site near Plumpton.

In 2022, *E. cruciger* was observed targeting *C. hederae* nests in a garden in Midhurst, a relationship that is known on the continent, but that had not been reported from Britain before.

In addition to these two cuckoo bees, *E. fallax* (not known from Britain) is reported as a parasite of *C. hederae* on the continent.

Colletes hederae mating ball, Seven Sisters [PG]

On the wing

Sussex records run from early August until early November.

Feeding and foraging

While Ivy is the principal source of pollen, contributing 98.5% of the pollen collected in one Sussex-based study, pollen is also collected from plants such as White Clover, bramble, Traveller's-joy, and plants in the Daisy family. Once Ivy pollen is available, however, female *C. hederae* turn to this plant for much of the pollen that they collect.

Ivy is also the principal source of nectar.

Colletes marginatus Margined Colletes
FIRST RECORD 1845, Littlehampton (Sydney Stevens) TOTAL RECORDS 27

Geography and history
The very first record for *Colletes marginatus* in Sussex was in 1845 from Littlehampton, where it was also found in 1875. It was then recorded at Pagham in 1907, but after that date it went unrecorded in the county until it was found at Rye in 1959, 52 years later.

Subsequently, it has been recorded intermittently in Chichester Harbour, on the sand dunes at Littlehampton, and, most frequently, at Camber Sands and Rye Harbour Nature Reserve. This last location is where it was most recently recorded in 2022.

Across Britain, this is largely a coastal species, and it is found as far north as the Sefton coast in Lancashire. There is also a strong population in The Brecks in East Anglia, its only inland location.

Vegetated shingle, Rye Harbour Nature Reserve [PG]

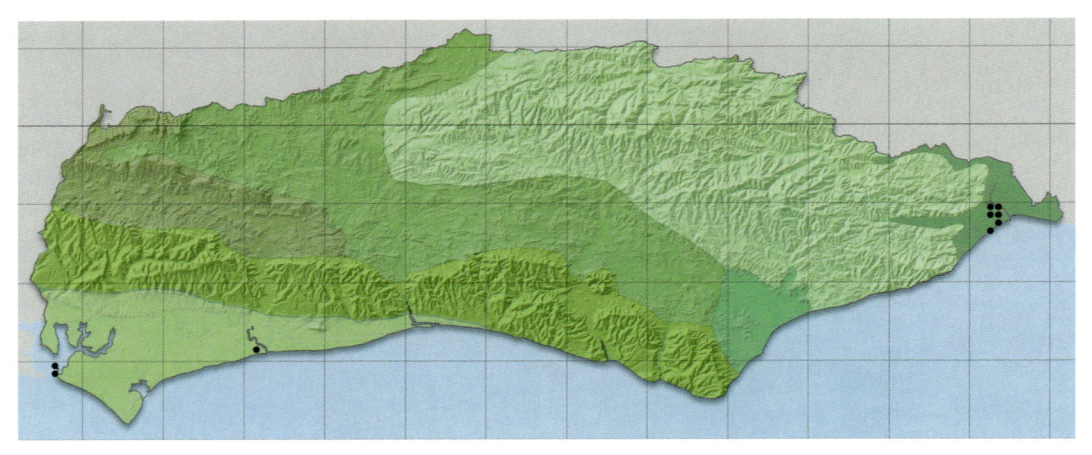

Large map: distribution 2000–2023
Small map: distribution 1844–1999

Colletes marginatus ♀ [SF]

Colletes marginatus ♂ [SF]

Out in the field and under the microscope
C. marginatus is the smallest of the *Colletes*. Females have white hairs along the upper margins of the hind tibiae and have a narrow straight hind margin to the head at the upper corners. Otherwise similar species such as *C. similis* and *C. fodiens* have a thicker, obviously rounded hind margin here.

Sternite seven in males is characteristically fan-shaped, while both males and females have dense bands of white hairs on sternites two, three and four.

As nests are rarely located, it is most often found feeding or foraging.

Behaviour and interactions
This bee is understood to form small aggregations, which are generally sited in areas of sparsely vegetated, sandy soils. Unusually for a *Colletes*, this species appears to forage for pollen from a range of plants species from several different families.

Brood cells are believed to be attacked by two species of cuckoo bee, *Epeolus cruciger* and *E. variegatus*.

On the wing
Sussex records run from late June until early August. Elsewhere, *C. marginatus* is reported to be on the wing until mid August.

Feeding and foraging
C. marginatus has been observed visiting plants such as Thrift, Wild Parsnip, Wild Carrot and ragwort, with pollen collected from the Pea family, as well as from bramble and Weld.

Colletes similis Bare-saddled Colletes

FIRST RECORD 1879, Hastings (Edward Saunders)   TOTAL RECORDS 60

Geography and history

Colletes similis is a local bee that can occasionally be abundant. It is found in warm, dry habitats, and although it is most frequent in sandy areas it does occur more widely. Across Sussex, it has been found on heathlands, arable field margins, chalk grasslands and vegetated shingle, and in open woodlands.

Most modern records are from the coast near Rye Harbour, Chichester Harbour, and a couple of locations on the eastern South Downs. It has been recorded with decreasing frequency in the county in recent years.

Corn Chamomile growing in a field margin near Denton [SF]

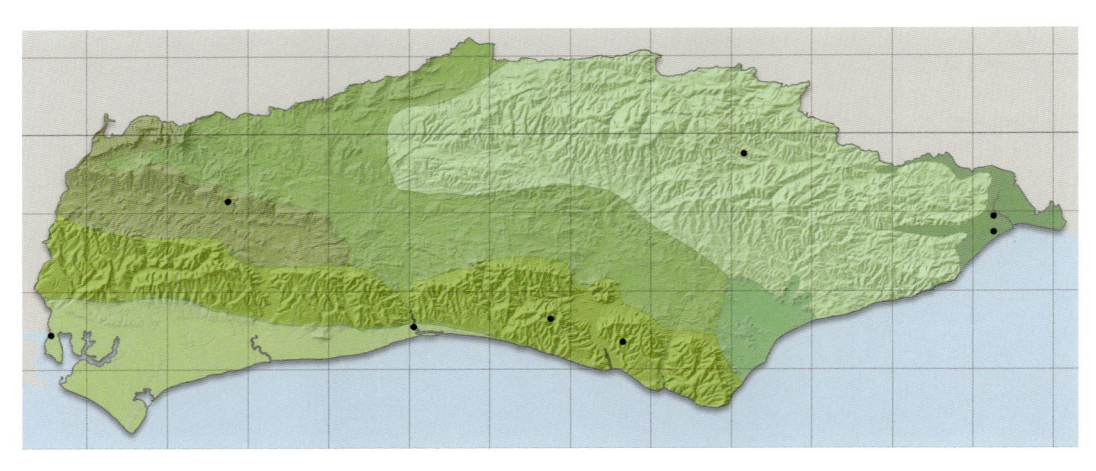

Large map: distribution 2000–2023
Small map: distribution 1844–1999

Colletes similis ♀ [PB]

Colletes similis ♂ [SF]

Out in the field and under the microscope
C. similis is best searched for by checking its known foodplants, and Steven Falk has reported finding it on Corn Chamomile on field margins near Denton.

Females most closely resemble *C. fodiens*. The common name refers to the absence of hairs on the front, upper surface of tergite one, which helps to separate this bee from fresh specimens of the more extensively haired *C. fodiens*. Fresh specimens of *C. similis* have fox-red hairs on the top of the thorax and bands of white hairs which fill the marginal areas of the tergites.

Males are more challenging to identify and are most reliably separated from *C. fodiens* by examining the structure of the genital capsule.

Behaviour and interactions
This *Colletes* is one of the earliest in the genus to emerge (though not as early as *C. cunicularius*) and flies from June onwards. It nests in small aggregations, with a handful of scattered burrows located in areas of bare or sparsely vegetated ground.

Brood cells are targeted by the cuckoo bee *Epeolus variegatus*. There is an observation of a female *C. similis* successfully forcing the parasite from a nest.

On the wing
Sussex records run from late June until mid August. Elsewhere, *C. similis* is reported to be on the wing until mid September.

Feeding and foraging
Pollen is collected from plants in the Daisy family such as ragwort, mayweed, Tansy and chamomile. It has also been reported that *C. similis* collects pollen from Wild Carrot.

There are other records of visits to plant species such as Creeping Thistle, Cat's-ear, Autumn Hawkbit, Hogweed, water-dropwort and White Bryony.

Colletes succinctus Heather Colletes
FIRST RECORD 1905, Hastings (Edward Saunders) TOTAL RECORDS 202

Geography and history
Colletes succinctus is very strongly associated with plants in the Heather family, and most Sussex records are from the heathlands and acid grasslands of the Lower Greensand and High Weald.

It has also been recorded at scattered locations elsewhere such as Birling Gap and Hastings Country Park. At Birling Gap in 2007, Steven Falk found several individuals foraging on Bell Heather within an area of chalk heath. It has not, however, been recorded from the larger area of chalk heath nearby at Lullington Heath.

Chalk heath with Bell Heather, Birling Gap [SF]

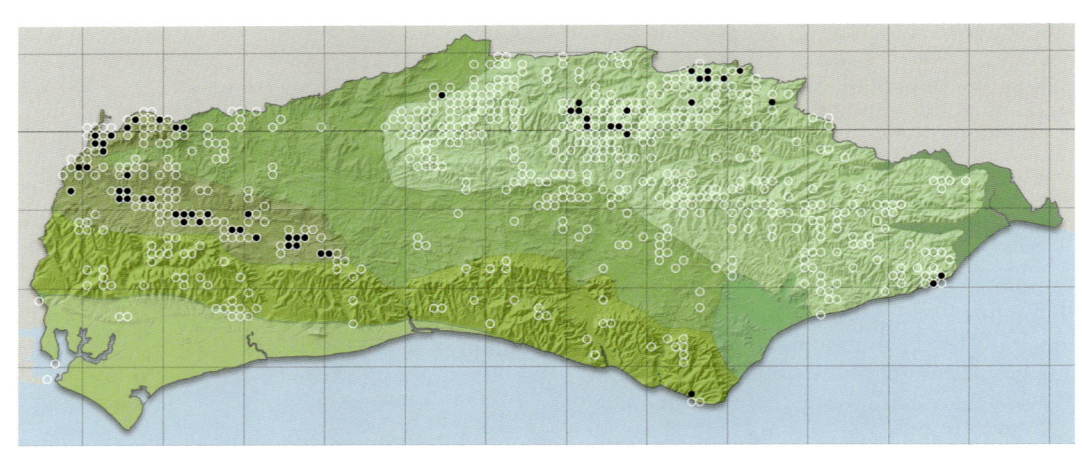

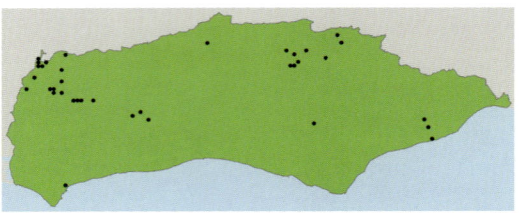

Large map: distribution 2000–2023, with white circles showing distribution of Heather and Bell Heather
Small map: distribution 1844–1999

432 *The Bees of Sussex*

Colletes succinctus ♀ [PC]

Colletes succinctus ♂ [PC]

Out in the field and under the microscope
Searching Heather flowers is an effective way of locating this bee, which is often seen flying with another Heather specialist, *Andrena fuscipes*. Both species can be present in large numbers, foraging for pollen.

Both males and females of *C. succinctus* can be confused with *C. hederae*, but fresh specimens of the two can be readily separated by the colour of the hair bands on the tergites. These are white in *C. succinctus* and yellow in *C. hederae*. Worn specimens should be separated by checking the galea— the outer portion is smooth and polished in *C. succinctus* but roughened in *C. hederae*.

C. succinctus does resemble a third species in the genus, *C. halophilus*, but this species is restricted to coastal saltmarshes.

Behaviour and interactions
Males adopt two mate-seeking strategies, one that involves patrolling and scent-marking a circuit and another that involves patrolling rapidly above a nest site and attempting to mate as females emerge from their brood cells. Males will often dig down to reach a female before she has emerged.

While in northern areas of Britain this bee often forms large nesting aggregations encompassing tens of thousands of nests, in southern counties smaller aggregations are established. These tend to be sited in areas of light sandy soil. Females are reported to forage up to 1.5 km from a nest site.

Brood cells are targeted by the cuckoo bee *Epeolus cruciger*.

On the wing
Sussex records run from late June until mid September.

Feeding and foraging
Pollen is gathered almost exclusively from plants in the Heather family, but also from ragwort and, later in the season, Ivy (compounding the risk of confusing this species with *C. hederae*). Mike Edwards has also observed this species foraging from Wild Mignonette, which in Austria is known as the main source of pollen.

Plant species visited at other times include Yarrow, Sheep's-bit, Weld, Creeping Thistle and Hemp-agrimony.

Hylaeus – Yellow-face Bees

These are virtually hairless black bees. All but one of the species have cream or yellow markings on the face. Most nest in dead stems or wood, generally establishing a linear series of brood cells, although one species uses the abandoned galls of a reed gall fly. This is the only genus of pollen-collecting bees that lacks specially adapted hairs with which to transport pollen, instead transporting pollen mixed with nectar internally. Brood cells are lined with a transparent film that is 'painted' onto the surface of the cell with the bee's mouthparts. Females provision each brood cell with a semi-liquid mix of pollen and nectar. There are no known species of cuckoo bee which target the brood cells of the species in this genus. Twelve species have been recorded in Britain, with eleven known from Sussex.

Hylaeus annularis Shingle Yellow-face Bee
FIRST RECORD 1896, Hastings (Edward Saunders) TOTAL RECORDS 13

Geography and history
This is a coastal species which, nationally, has been found at a handful of places along the east and south coasts of England between Suffolk and Dorset. It can be locally abundant, with any areas of vegetated shingle and sand dunes potentially supporting a population.

Hylaeus annularis ♀ [SF]

Hylaeus annularis ♂ [SF]

Large map: distribution 2000–2023, with white circles showing distribution of Sea-kale
Small map: distribution 1844–1999

434 *The Bees of Sussex*

In Sussex, there are very few records, with most from the coast between Winchelsea and Camber Sands. It has been recorded here on nine occasions, most recently in 2012 at Rye Harbour Nature Reserve. It has also been recorded once from Tide Mills, in 2011. There are two records from west Sussex, one from Shoreham where it was collected in 1905, and a second from Lancing in 1954.

Confusion over the correct scientific name for this species was only resolved recently when it was concluded that the species previously known as *Hylaeus spilotus* and *H. masoni* should both be referred to as *H. annularis*. To compound matters, the name *H. annularis* had been wrongly assigned to a species now known as *H. dilatatus*.

Out in the field and under the microscope

H. annularis is most likely to be found visiting plants such as Sea-kale and Wild Carrot.

Females have two almost circular yellowish markings on the face that do not reach the eye margins, and each mandible has three teeth. They closely resemble *H. dilatatus* but have a polished and smooth surface on a densely punctate scutum and on the marginal area of tergite one. In contrast, *H. dilatatus* has a slightly roughened surface on the scutum and on the marginal area of tergite one.

Males have flattened, shield-like yellow and black segments at the base of each of the antennae and have entirely black mandibles. In *H. dilatatus* males, the lowest antennal segment has a slightly more exaggerated downward bulge while the mandibles are yellowish white.

Behaviour and interactions

H. annularis is understood to nest in the ground in sandy areas. The only time that this behaviour has been noted in Britain was when Mike Edwards observed a female entering a nest at Camber Sands in 1981.

In France, there are reports of *H. annularis* nesting in dead plant stems.

On the wing

Typically, *H. annularis* is on the wing between late June and the end of August.

Feeding and foraging

H. annularis visits plants such as Sea-kale, bramble, Wild Carrot, Wild Parsnip, Creeping Thistle, Yarrow, Sea Mayweed and ragwort. Its pollen preferences are unknown.

Sea-kale on vegetated shingle, Rye Harbour Reserve [BY]

Hylaeus brevicornis Short-horned Yellow-face Bee
FIRST RECORD 1882, Hastings (Edward Saunders) TOTAL RECORDS 210

Geography and history
Hylaeus brevicornis is widely distributed across Sussex and is found in a variety of habitats where scrub and bramble are present. These include chalk grassland, open woodland, heathland and soft-rock cliffs. It has also been recorded from a scrubby area of neglected grassland in Lewes. It is rarely abundant in any of the places where it is found.

Out in the field and under the microscope
Because of its small size *H. brevicornis* is often overlooked. It is most likely to be found feeding or foraging on plants such as Wild Mignonette and Wild Carrot, although it can also be found by collecting dead bramble stems in late winter and rearing the adults in the warmth of a home.

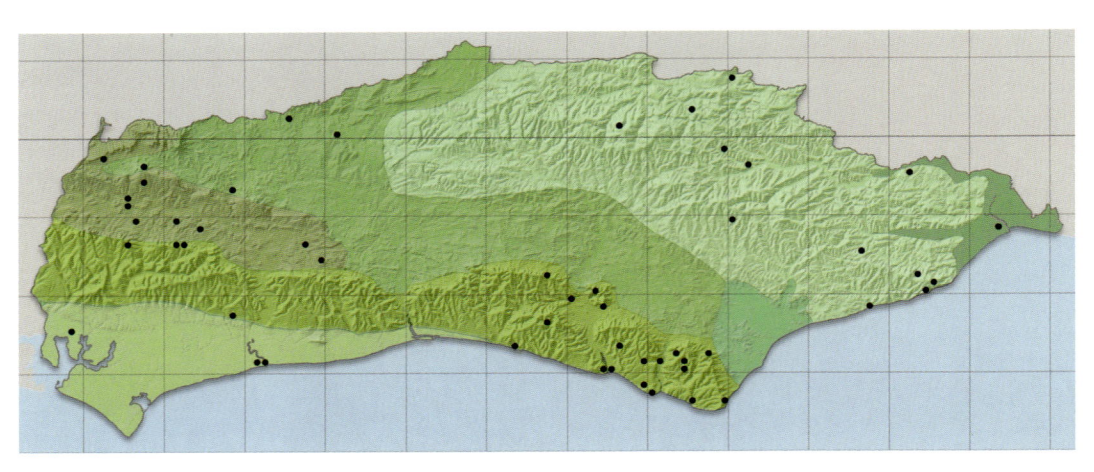

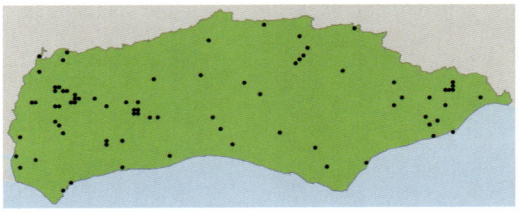

Large map: distribution 2000–2023
Small map: distribution 1844–1999

436 *The Bees of Sussex*

Females have a round face and three teeth on each of the mandibles. They can be separated from other similar species by the two narrow markings on the face and by the very slightly roughened surface to tergite one. The markings on the face just about touch the inner margins of the eyes but do not extend as far as the edge of the clypeus.

Males have extensive areas of white on the face and have a thickened black segment at the base of each of the antennae. This is rounded, rather than being angular. The first few antennal segments are very wide.

Behaviour and interactions
H. brevicornis nests in the dead stems of plants with a pithy centre such as bramble and elder, excavating a linear nest within the stem. The nest itself might be between 7 and 12.5 cm in length, with as many as 14 brood cells per nest. Males are generally towards the front of the nest, and females towards the rear.

The completed brood cell is provisioned with a semi-liquid mix of pollen and nectar. An egg is laid on top of the food store, the larvae floating on the provisions as it grows. *H. brevicornis* is often supplanted in the nest by other species, including by the wasp *Trypoxylon attenuatum* which frequently establishes its own nests within the same plant stem.

H. brevicornis has a long flight period, which suggests that it may have more than one brood, as has been reported from the continent.

Trypoxylon attenuatum ♀ [SF]

On the wing
Sussex records run from the beginning of May until September.

Feeding and foraging
This species collects pollen from a wide range of plant families, including Knotweed, Rose, Pea, Carrot and Bellflower.

Plants visited include Wild Mignonette, Bog Pimpernel, bramble, water-dropwort, Hogweed, Wild Carrot, Common Ragwort, Creeping Thistle and Beaked Hawk's-beard.

Hogweed [PC]

Hylaeus communis Common Yellow-face Bee
FIRST RECORD 1888, Bognor Regis (Henry Guermonprez) TOTAL RECORDS 612

Geography and history
Hylaeus communis is an extremely widespread bee and the most frequently recorded in the genus. It occurs in a very wide range of habitats and is frequently found in gardens. The very first Sussex record was made in a garden in Albert Road, Bognor Regis, in 1888.

Out in the field and under the microscope
This bee most closely resembles the reed specialist *H. pectoralis*, a much scarcer species. Like *H. pectoralis*, female *H. communis* have two teeth on each of the mandibles and a pair of facial markings that follow the margins of the eyes for a short distance and do not reach all the way across to the clypeus. The two species can be separated by the presence of very distinct ridges running vertically up the clypeus in *H. communis*.

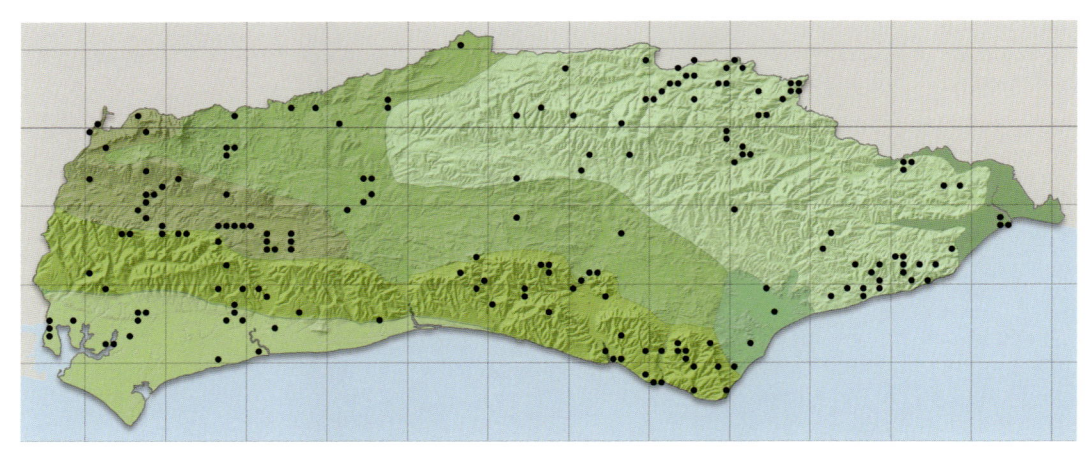

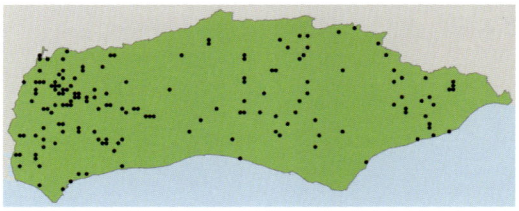

Large map: distribution 2000–2023
Small map: distribution 1844–1999

438 *The Bees of Sussex*

Males also have yellow markings on the face, and these are unique among the *Hylaeus* in that they curl around the bases of the antennae.

H. communis is generally found foraging or feeding from plants such as bramble and thistle, or at rest on sunlit fence posts or pieces of timber.

Behaviour and interactions

H. communis nests in a wider range of niches than other species of *Hylaeus*, including within dead bramble, elder and rose stems, within abandoned reed galls, and within holes in dead trees, branches, fence posts, sandy banks and walls. In 2023, *H. communis* bees (three males and one female) were reared from an empty Marble Gall collected near Midhurst, the first time that this behaviour has been recorded. The gall had been created by the tiny wasp *Andricus kollari*, and once a new generation of wasps had vacated the gall, the empty chamber was used by a female *H. communis* for egg laying.

Marble galls (left) created by the tiny wasp *Andricus kollari* (below) [both PB]

Hylaeus communis ♀ lining a cell [PW]

H. communis nests are generally linear and might contain as many as eight brood cells. As with other species of *Hylaeus*, *H. communis* is often supplanted in the nest by other hole-nesting aculeates.

It is known to often produce two (and sometimes three) broods per year on the continent, and the long flight period for this bee in Sussex may be evidence that two broods are produced here too, in at least some years. Further north, it produces just one brood per year.

On the wing

Sussex records run from early April until mid September.

Feeding and foraging

This bee collects pollen from a variety of plants, including members of the Carrot, Daisy, Borage, Cabbage and Dead-nettle families.

It has also been recorded visiting plants such as Greater Stitchwort, White Bryony, cotoneaster and Rosebay Willowherb.

Hylaeus confusus White-jawed Yellow-face Bee
FIRST RECORD 1903, Bognor Regis (Henry Guermonprez) TOTAL RECORDS 216

Geography and history
Hylaeus confusus is widely distributed across Sussex and can be locally common. It is associated with open broadleaved woodland, but also occurs in a wide range of other habitats where there is abundant scrub and bramble. These include grasslands and heathlands.

Until recently, *H. confusus* and the very similar *H. incongruus* were treated as one species.

Out in the field and under the microscope
This species is often found visiting flowering plants such as bramble and thistle, but it might also be found at rest on sunlit fence posts and foliage. It can also be reared from bramble stems collected in late winter.

Females belong to a large group in this genus that has two teeth on the mandibles and an elongated head. They are especially close to *H. incongruus*, and individuals of the two species can be tricky to separate because the differences between them are so subtle, and not all examples can be successfully identified.

One key character to check in the females is the distance between the bottom of each eye and the base of each of the mandibles. In the case of *H. confusus* this gap is very short, while in *H. incongruus* the gap is clearly longer.

H. confusus males are more straightforward. Two characters to look for are the white markings on the mandibles and the black labrum which sometimes has a white marking in the centre. By way of contrast, both the mandibles and labrum are white in *H. incongruus*.

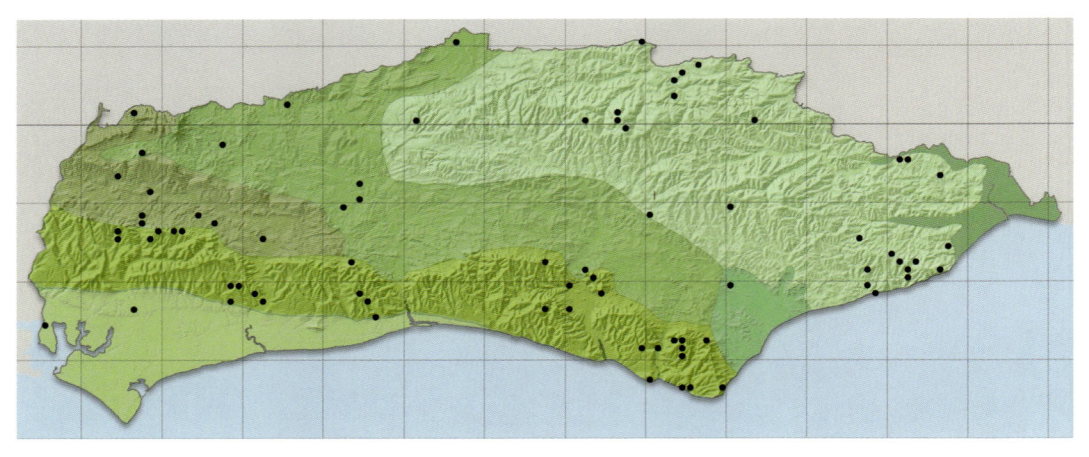

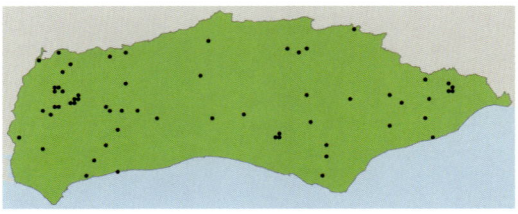

Large map: distribution 2000–2023
Small map: distribution 1844–1999

The Bees of Sussex

Hylaeus confusus ♂ [CKL]

Hylaeus confusus ♂ [IT]

Behaviour and interactions
H. confusus is known to nest in all sorts of different cavities, including dead bramble stems, holes in fence posts and dead timber, and also bee hotels. Detailed information on its behaviour is patchy as it was only recently recognised as a distinct species.

It is known to be at least partially bivoltine on the continent, and the long flight season in Sussex suggests this might also be the case here.

On the wing
Sussex records run from the beginning of May until late August.

Feeding and foraging
Pollen is collected from members of several different plant families, including Daisy, Bellflower, Pea, Rose and Mignonette.

There are also reports of *H. confusus* visiting a wide range of other species, including plants in the Buttercup, Speedwell, Carrot and Cabbage families.

Hylaeus cornutus Spined Hylaeus
FIRST RECORD 1879, Hollington Wood (Edward Saunders) TOTAL RECORDS 28

Geography and history
Hylaeus cornutus is restricted to southern England, reaching only as far north as Warwickshire and Norfolk. Most modern Sussex records are from areas of chalk grassland on the South Downs, while nationally it is also known from chalk heaths and open woodlands.

The very first county record for *H. cornutus* is from Hastings where it was recorded in 1879. It was then recorded in Haywards Heath and Hastings but subsequently went unrecorded until 1994. This is when it was recorded in Forest Row after a gap of 89 years by Alfred Jones.

Wild Carrot, Tide Mills [SF]

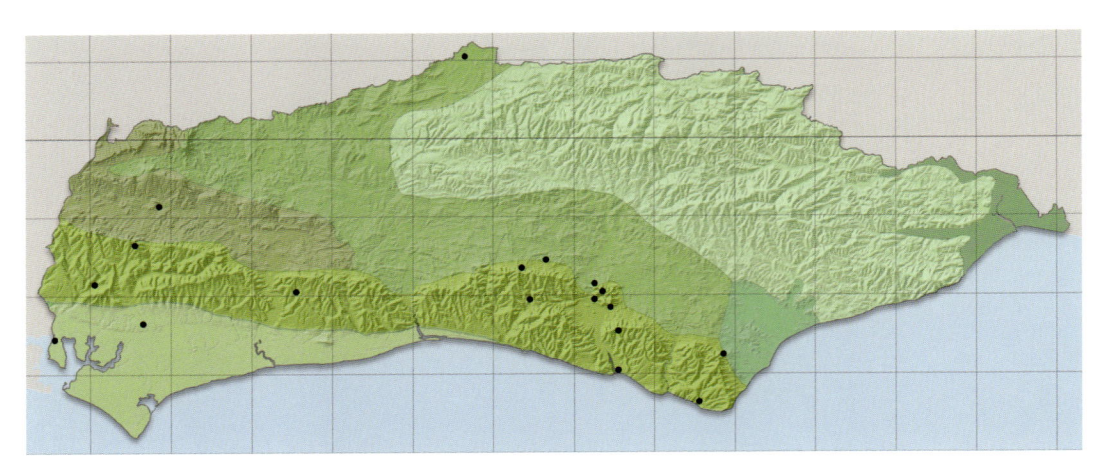

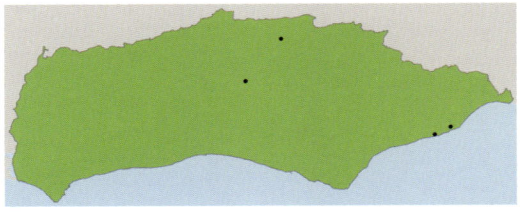

Large map: distribution 2000–2023
Small map: distribution 1844–1999

Since then, there have been a further 24 records, and it seems that this species may have increased in abundance recently. Recent records, all since 2014, are from Chichester, Stoughton, Willingdon, Falmer, Southease, Ditchling Beacon, Tide Mills, and gardens in Lewes and Midhurst.

Out in the field and under the microscope

H. cornutus is the only species of *Hylaeus* in which the males and females do not have yellow markings on the face—their faces are entirely black. Females are also unique in the genus in that they have a pair of triangular horns, one on each side of the lower face.

Males have bright yellowish-orange antennae (this colouring is just about visible to the naked eye) and largely yellow legs, with just black femora and a black marking on each of the tibiae.

This bee is most often found feeding or foraging on flower heads, particularly on plants from the Carrot family. Edward Saunders' first county record in 1879 was of "five ♂ off *Achillea* in Hollington Wood".

Hylaeus cornutus ♀ [SF]

Hylaeus cornutus ♂ [JE]

Behaviour and interactions

H. cornutus nests in hollow plant stems, using species such as bramble, dock, Wild Parsnip and teasel. The function of the horns on the front of a female's face had been assumed to be related to pollen transport, but this has now been shown to be unlikely. Instead, these may have a function in nest building.

On the wing

Sussex records run from mid June until mid August.

Feeding and foraging

This species collects pollen from plants in the Carrot and Daisy families, especially Wild Carrot and Wild Parsnip on the South Downs. Hogweed and Wild Angelica are also regularly visited for pollen.

H. cornutus has also been recorded visiting plants such as Lesser Stitchwort, Ox-eye Daisy, Yarrow, Wild Mignonette and Field Scabious.

Hylaeus dilatatus Chalk Yellow-face Bee
FIRST RECORD 1902, Linch (Henry Guermonprez) TOTAL RECORDS 200

Geography and history
In Sussex, most records for *Hylaeus dilatatus* are from the South Downs, where it has been found on chalk grasslands such as Malling Down and Lewes Cemetery and in open woodland such as West Dean Woods and Rewell Wood.

The scientific name *H. dilatatus* has only been used for this species since 2008. Before then, it was referred to as *H. annularis*, but this name has now been assigned to a separate species.

Malling Down and Lewes
[AKP]

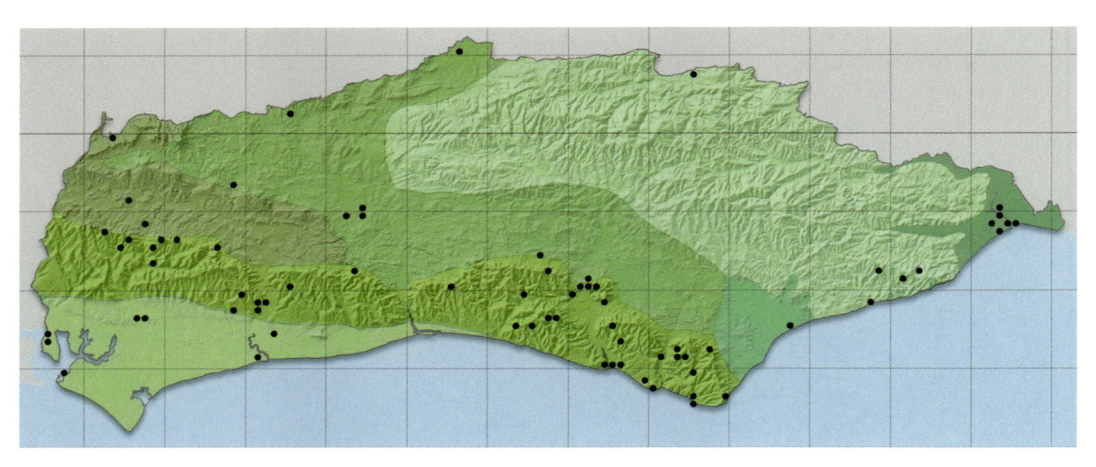

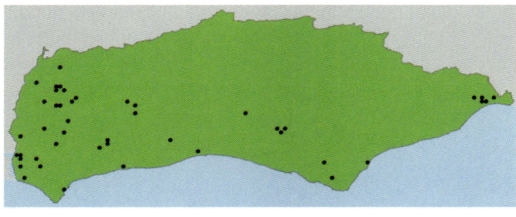

Large map: distribution 2000–2023
Small map: distribution 1844–1999

Hylaeus dilatatus ♀ [PC]

Hylaeus dilatatus ♂ [MB]

Out in the field and under the microscope
H. dilatatus is one of three species—the other two being *H. annularis* and *H. brevicornis*—in the genus in which females have three teeth on each of the mandibles and a roundish face. Of these other species, it most closely resembles *H. annularis* as both *H. dilatatus* and *H. annularis* females have small roundish yellow spots on the face that do not reach the margins of the eyes. In the third species, *H. brevicornis*, the triangular markings do just reach these margins.

Females of *H. dilatatus* and *H. annularis* can be separated by checking the surface of the scutum and the marginal area of tergite one. In *H. dilatatus* these areas have a slightly roughened surface, while in *H. annularis* the surfaces are completely smooth and shiny.

Male *H. dilatatus* have a very inflated yellow and black segment at the base of each of the antennae and extensive yellow markings on the face. The yellow markings on the mandibles help to separate this species from *H. annularis*, which is completely black here. As with the females, the males have a slightly roughened surface on the marginal area of tergite one.

H. dilatatus and *H. annularis* can sometimes be seen together on areas of vegetated shingle. As well as being found on flowers, *H. dilatatus* can be reared from bramble stems collected in late winter.

Behaviour and interactions
This *Hylaeus* is a stem-nester, usually establishing its nests within dead bramble and rose stems. However, it might also use a plant such as dock or decaying wood. The nest might be sited in a tunnel excavated by the female herself, or she may take advantage of an existing cavity such as an abandoned solitary wasp's nest.

On the wing
Sussex records run from the beginning of May until mid September.

Feeding and foraging
H. dilatatus collects pollen from a wide range of different families, including Carrot, Daisy, Bellflower and Dead-nettle.

Hylaeus hyalinatus Hairy Yellow-face Bee
FIRST RECORD 1901, Bognor Regis (Henry Guermonprez) TOTAL RECORDS 196

Geography and history
Hylaeus hyalinatus is widely distributed in Sussex but rarely abundant. There is a concentration of records along the coast, with Chichester Harbour, Pagham Harbour, Littlehampton, the Cuckmere Estuary, Glyne Gap and Rye Harbour all providing recent records.

This *Hylaeus* is most often associated with areas of vegetated shingle, sandpits, coastal grasslands, soft-rock cliffs and sand dunes, but it is also known from gardens. There have been recent records from gardens in Lewes, Hailsham and Midhurst.

Out in the field and under the microscope
This bee is usually encountered as singletons, with males and females most often found feeding or foraging on plants such as Wild Mignonette, bramble, Sea-kale and members of the Daisy family. It is, however, also possible to come across several in one place—in a garden in Tunbridge Wells, Ian

Vegetated shingle with Sea-kale, Glyne Gap [PC]

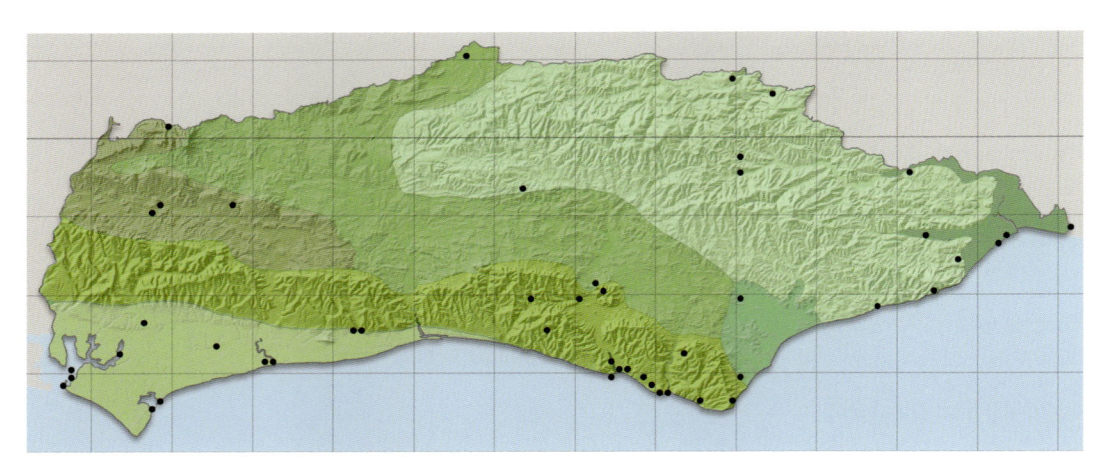

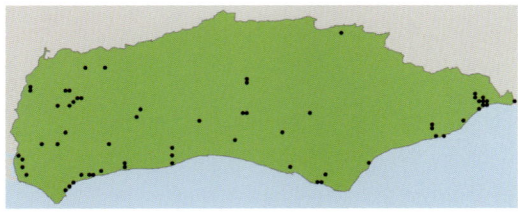

Large map: distribution 2000–2023
Small map: distribution 1844–1999

446 *The Bees of Sussex*

Hylaeus hyalinatus ♀ [PC]

Hylaeus hyalinatus ♂ [PC]

Beavis has observed "a large group (apparently all females) […] swarming around a small Hebe bush, settling briefly on its flowers". This behaviour has also been observed on individual Sea-kale plants on the shingle at Glyne Gap.

Females have triangular markings on the face that reach from the margins of the eyes to the margins of the clypeus, two teeth on the mandibles, and a relatively large gap between the base of each mandible and the bottom of each eye.

Males have a sparse covering of very long hairs on the face (which can rub off over time) and a dense covering of hairs on the underside of the thorax (which can also rub away), as well as a spoon-shaped sternite eight. Like the females, the males have a relatively large gap between the base of each mandible and each eye.

Behaviour and interactions

H. hyalinatus nests in a variety of different situations. Steven Falk has observed females nesting "in a sand face at Seaford Head Golf Course", while crumbling mortar joints in walls, beetle holes in deadwood and dead plant stems such as bramble and hawthorn are also reported.

This bee will often take advantage of nests created by *Megachile* bees and by *Trypoxylon* wasps.

On the wing

Sussex records run from the end of April until the end of August.

Feeding and foraging

H. hyalinatus collects pollen from several different plant families, including Cabbage, Carrot, Daisy, Bellflower, Rose and Mignonette.

It is also known to visit a wide range of other plants, including Sea-kale, White Bryony, sedum, spurge and geranium.

Hylaeus hyalinatus ♂ on Sea-kale at Glyne Gap [PC]

Hylaeus incongruus White-lipped Yellow-face Bee
FIRST RECORD 1879, Hollington Wood (Edward Saunders)   TOTAL RECORDS 38

Geography and history
Across Britain *Hylaeus incongruus* is a scarce bee that is restricted to Dorset, Hampshire, Surrey, Berkshire, Kent and Sussex. It is a species of heathlands with abundant scrub, but it is also known from open woodlands and chalk grasslands.

The earliest Sussex record for *H. incongruus* is from August 1879 when Edward Saunders "caught three specimens on bramble flowers in Hollington Wood near Hastings". This was also the first British record.

Iping Common [NS]

Large map: distribution 2000–2023
Small map: distribution 1844–1999

This first record is one of only two from Hastings, although there is also a series of records from just along the coast near Rye. These run from 1926 to 1930, since when *H. incongruus* has not been recorded in east Sussex.

The remaining county records are all from west Sussex, where it was first recorded at Ambersham Common in 1977. Most of the west Sussex records are from places like Stedham, Iping and Woolbeding Commons on the Lower Greensand, with just three from Goodwood, West Dean Woods and Bignor Hill on the South Downs.

Out in the field and under the microscope

H. incongruus is often found at the edge of woodland and close to areas of scrub within heathland. It is generally found visiting the flowers of plants such as thistle and bramble, but it might also be seen investigating potential nest sites in pieces of deadwood.

Females have two teeth on the mandibles and two triangular yellow markings on the face that extend from the margins of the eye to the edge of the clypeus. They are very similar to *H. confusus* but, among other differences, have more extensive white markings.

Males have a white labrum and white-marked mandibles.

Behaviour and interactions

This *Hylaeus* is believed to nest in dead and decaying timber, although on the continent there are also reports of *H. incongruus* nesting in reed and bramble stems.

On the wing

Typically, *H. incongruus* is on the wing from the beginning of June until the end of August.

Feeding and foraging

H. incongruus collects pollen from several different plant families, including Carrot, Daisy, Cabbage, Bellflower and Pea.

It has also been recorded visiting plant species such as bramble, Wild Parsnip, Hogweed and Tormentil.

Hylaeus pectoralis Reed Yellow-face Bee
FIRST RECORD 1976, Church Norton (Mike Edwards) TOTAL RECORDS 24

Geography and history
Hylaeus pectoralis is unusual among British bees in that it is very closely associated with Common Reed, a plant of fresh or brackish water. As such, it is restricted to reedbeds, reed-filled ditches, and stands of reed around the margins of brackish coastal sites. Along with *Macropis europaea* and *Colletes halophilus*, this is one of just three British species linked to wetlands.

For many years, this species was known in Britain only from the East Anglian fens. Most records are now concentrated along the Hampshire coast, the Sussex coast, the Thames Estuary and East Anglia. It is also known from scattered locations as far north as Warwickshire.

Reedbed in east Sussex [BY]

Large map: distribution 2000–2023, with white circles showing distribution of Common Reed
Small map: distribution 1844–1999

In Sussex, it was first recorded at Chichester Harbour in 1976 when Mike Edwards reared an adult from a nest on a reed stem. Elsewhere in west Sussex, it has since been recorded at Medmerry, Pagham Harbour, Littlehampton, Burton Mill Pond and Binsted.

In east Sussex, it has been recorded on the edge of farmland near Newhaven, on the Pevensey Levels, and from Filsham Reedbed.

Hylaeus pectoralis ♀ [CKL]

Hylaeus pectoralis ♂ [PH]

Out in the field and under the microscope
H. pectoralis is a tricky bee to find, but it can be seen by searching flower heads close to areas of reed, by sweeping a net over these areas, or by rearing adults from reed galls collected in late winter.

Females closely resemble the much more abundant *H. communis*. Like this other species, they lack patches of white hairs on the hind margin of tergite one. However, they can be separated by the absence of the strong vertical ridges found on the clypeus in *H. communis*.

Male *H. pectoralis* have swollen black segments at the base of the antennae and very dense fringes of white hairs on sternites two, three and four.

Behaviour and interactions
This species of *Hylaeus* usually establishes its nests within abandoned galls on reed stems created by the fly *Lipara lucens*, although it will also use the reed stems themselves. It might also use dead bramble stems.

Within a gall, each female will create up to eight brood cells, the number of cells being partly determined by the size of the gall. The cells are arranged linearly, and the entrance to the nest is sealed with a plug made from reed-leaf fragments. These nests appear to be able to withstand short periods of submersion.

Abandoned gall on reed stem [BY] created by *Lipara lucens* [PC]

On the wing
Sussex records run from the beginning of April until late August. Elsewhere, *H. pectoralis* is reported to be on the wing from the beginning of June until late September.

Feeding and foraging
Pollen is collected from plants such as Creeping Thistle, Marsh Thistle, Purple Loosestrife, bramble and cinquefoil.

Other plant species visited include Hogweed, Wild Carrot, bindweed and Wild Angelica.

Hylaeus pictipes Little Yellow-face Bee
FIRST RECORD 1896, Hastings (Frisby) **TOTAL RECORDS** 41

Geography and history
Hylaeus pictipes used to be widely distributed in southern England, and although it has apparently declined, there have been several recent records in Sussex that suggest its population may have recovered a little.

This species of *Hylaeus* is found in a variety of different habitats, including chalk grassland, woodland, coastal grasslands, sand dunes and vegetated shingle. It is frequently found in gardens.

Almost all Sussex records are thinly scattered across southern areas of the county, particularly along the eastern South Downs, and locations include Birling Gap, Cuckmere Haven, Cow Gap, Seaford Head and Eastbourne. Great Dixter and a garden in Lewes have both been regular sources of records since 2017.

Thatched barn and log pile at Great Dixter, nesting habitats for *Hylaeus pictipes* [AP]

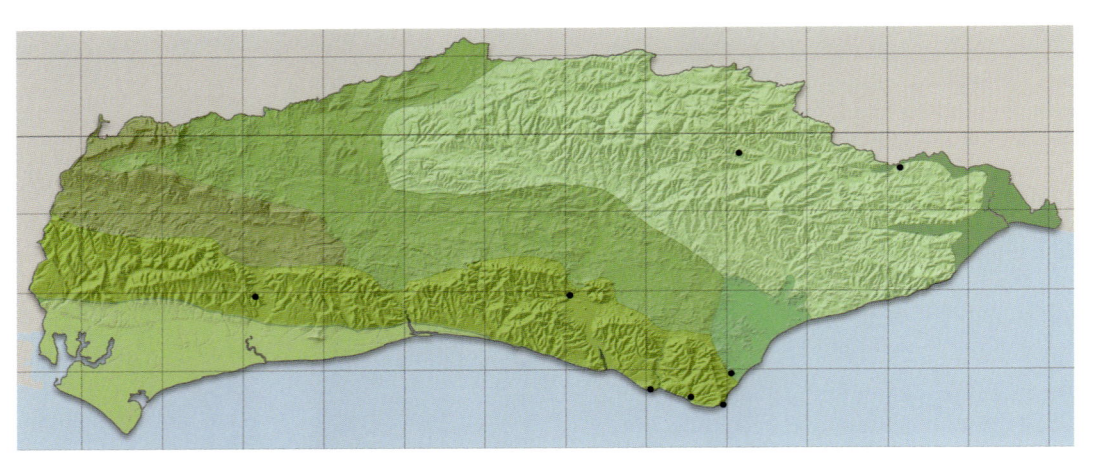

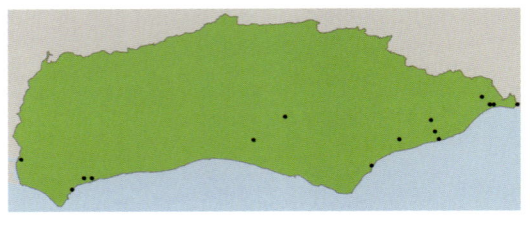

Large map: distribution 2000–2023
Small map: distribution 1844–1999

Out in the field and under the microscope
H. pictipes is frequently found on Wild Mignonette, but other foodplants such as Weld and bramble are also worth checking. Andy Phillips has also observed that this bee is "particularly numerous at Great Dixter, especially during July, foraging from Greater Masterwort".

Males and females are among the smaller *Hylaeus*. Females have two teeth on the mandibles, a face that is longer than it is wide, and broad yellowish bands on the face that extend from the margins of the clypeus to the inner eye margins. One definitive character to look for is the groove running up each side of the face, close to the innermost eye margins—this extends to a point in line with the rearmost ocelli, which is much further than for any other species of *Hylaeus*.

Males are more straightforward and have relatively narrow, parallel-sided segments at the base of each of the antennae and a completely smooth and densely punctate surface to tergite one.

Hylaeus pictipes ♀ [PB]

Hylaeus pictipes ♂ [PB]

Behaviour and interactions
Andy Phillips has reported finding *H. pictipes* nesting in the thatch at Great Dixter, while elsewhere it has been observed nesting in dead bramble and rose stems, as well as in abandoned beetle burrows in old fence posts and gorse stems. George Else and Mike Edwards have reared adults from nests within gorse stems established by the beetle *Anobium punctatum*.

On the continent, it is also reported to nest in the mortar joints of walls.

On the wing
Sussex records run from mid June until late August.

Feeding and foraging
Pollen is collected from several different plant families, including the Mignonette, Carrot, Daisy, Bellflower, Borage and Rose families.

Hylaeus signatus Large Yellow-face Bee
FIRST RECORD 1900, Eastbourne (Charles Nurse)　TOTAL RECORDS 47

Geography and history
In Britain *Hylaeus signatus* is a widely distributed bee with records recently extending as far north as Edinburgh. Most records, however, are concentrated in central and southern England.

All but two of the Sussex records since 2000 are from the eastern South Downs between Falmer and Birling Gap. During this period, *H. signatus* has been found at several sites here, including Southerham Farm, Deep Dean, Malling Down, Denton Hill and Castle Hill. The two remaining records are from farmland near Easebourne and from Knepp.

In Sussex, *H. signatus* has been most frequently found on chalk grasslands, but elsewhere in the country it is also known from gardens, quarries and coastal sites.

Denton Hill [SF]

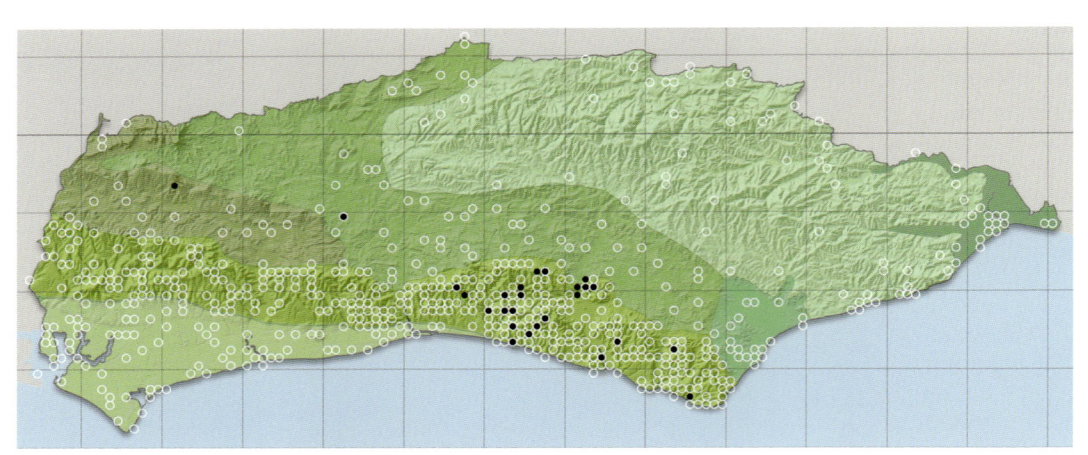

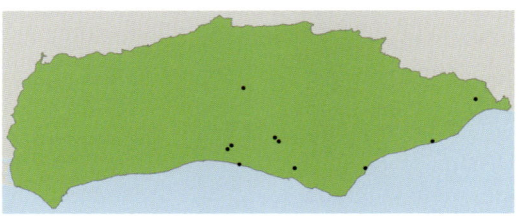

Large map: distribution 2000–2023, with white circles showing distribution of Weld and Wild Mignonette
Small map: distribution 1844–1999

454　*The Bees of Sussex*

Hylaeus signatus ♀ [PC]

Hylaeus signatus ♂ [PC]

Out in the field and under the microscope
The most successful way of locating this species is to check flowering Weld and Wild Mignonette plants where several females might be seen foraging on groups of flowers. Males are also often found on these plants.

This is the largest of the *Hylaeus*, with males frequently larger than the females. These have two teeth on each of the mandibles and yellow markings on the face that hug the margins of the eyes, just reaching the edges of the clypeus. They have black hind tibiae and a small fringe of white hairs at the sides of the marginal area of tergite one.

Males have obviously swollen, shiny areas on sternites three and four, a character shared only with *H. confusus* and *H. punctulatissimus* (the last of these has been confirmed only once in Britain). In *H. confusus* males, these swellings are much less apparent.

Behaviour and interactions
This species of *Hylaeus* nests in a variety of situations, including in dead bramble or rose stems and in earth banks and crumbling mortar in walls. It can sometimes form large aggregations.

Females generally excavate their own nest chamber but might take advantage of an existing abandoned nest, including those of *Colletes daviesanus*.

Hylaeus signatus ♂'s on Wild Mignonette [PC]

On the wing
Sussex records run from the beginning of June until early August. Elsewhere, *H. signatus* is reported to be active into September.

Feeding and foraging
Pollen is collected only from plants in the Mignonette family, particularly Weld and Wild Mignonette. This family is also the preferred nectar source.

Other plant families visited include Rose, Cabbage, Bellflower and Daisy.

The Bees of Sussex

Extinct species

While many new species are colonising Sussex, ten have completely disappeared from the county since the late nineteenth century. There is also an eleventh species that was understood to be here, but there are now questions about its status in Sussex (and in Britain).

The following brief accounts summarise the history and geography of these eleven species and include maps that show their historic range in the county between the years 1844 and 1999.

Bombus cullumanus Cullum's Bumblebee

FIRST RECORD 1844, Brighton Downs (W. Walcott) **TOTAL RECORDS** 6

Bombus cullumanus was previously widely distributed across western Europe but is now 'Critically Endangered', although it remains common within its range in Asia. Here in Europe, it is probably heading towards extinction, and this follows a significant population decline triggered by the destruction of species-rich grasslands, particularly those supporting abundant clovers and vetches.

The last British record is from Berkshire (possibly 1941). In Britain it was a species of extensive chalk grasslands but, despite extensive searching in locations where it was formerly found, it has not been seen again. Given the distinctiveness of the males, it seems unlikely that it has been overlooked.

The final Sussex record is from September 1923 when 40 males were seen in two localities by Edward Nevinson and his son while searching the downs near Seaford. The other five Sussex records are all from a small area of the South Downs between Brighton, Newhaven and Seaford. The earliest is of a male collected in 1844 on the Brighton Downs and is one of the very first aculeate records for Sussex. The remaining four records are from the Newhaven and Seaford areas and were collected during 1921 and 1922.

Females are very similar in appearance to *B. lapidarius* and were not associated with the males until 1926. In fact, all but one of the Sussex records are of males, reflecting the difficulty that early naturalists had in linking the females to the very different-looking males.

Bombus cullumanus queen from Berkshire, originally misidentified as *B. lapidarius* [SF]

Bombus cullumanus ♂ caught by Edward Nevinson near Seaford in September 1923, the final time this bumblebee was recorded in Sussex [SF]

The Bees of Sussex

Bombus distinguendus Great Yellow Bumblebee
FIRST RECORD 1971, Findon Valley (G.V. Geiger) **TOTAL RECORDS** 1

Bombus distinguendus is in decline across western Europe and is regarded as 'Vulnerable' on the Red List for European bees. As well as being dependent on extensive areas of species-rich grassland, it is very vulnerable to climate change.

This large and impressive bumblebee was formerly found across Britain, but even in the early twentieth century it was described as "rare in the south of England" by Frederick Sladen.

The destruction of flower-rich grasslands was the principal cause of its decline, with populations now confined to landscapes where extensive areas of semi-natural grassland survive. These include the machair grasslands of the southern Outer Hebrides, as well as the Inner Hebrides, Orkney, and the north coast of Scotland. It also occurs on the west coast of Ireland.

There is only one record for *B. distinguendus* in Sussex. This specimen was collected near Findon Valley, Worthing, in September 1971 and confirmed by David Alford as part of the Bumblebee Distribution Maps Scheme.

Bombus distinguendus queen, Scotland [NO]

Bombus subterraneus Short-haired Bumblebee
FIRST RECORD 1844, Brighton Downs (W. Walcott) **TOTAL RECORDS** 59

In the early twentieth century this was a widely distributed bumblebee across scattered locations in Wales and southern England, including in Sussex. It then went into decline, largely because of the destruction of large areas of flower-rich grassland. *Bombus subterraneus* was last recorded in Britain from Dungeness in Kent in 1988 and was declared extinct in 2000.

Here in Sussex, *B. subterraneus* was found on flower-rich grassland as well as on vegetated shingle and saltmarsh along the coast. The final Sussex record is from 1986 and was made at Rye Harbour Nature Reserve by Gerald Dicker.

It was also a regular visitor to gardens, with Portsmouth Museum holding several specimens that were collected from a garden in Albert Road, Bognor Regis, in the late nineteenth and early twentieth centuries.

Between 2009 and 2016, *B. subterraneus* was reintroduced at Dungeness in Kent through the release of 204 queens that had been collected from southern Sweden. Although the reintroduction does not appear to have succeeded, other species of bumblebee such as *B. ruderatus* have benefitted significantly from the associated land management changes instigated by local landowners.

Bombus subterraneus queen shortly after release at Dungeness in Kent, 2016 [PG]

The Bees of Sussex 457

Bombus sylvarum Shrill Carder Bee
FIRST RECORD 1879, Hastings (Edward Saunders) TOTAL RECORDS 80

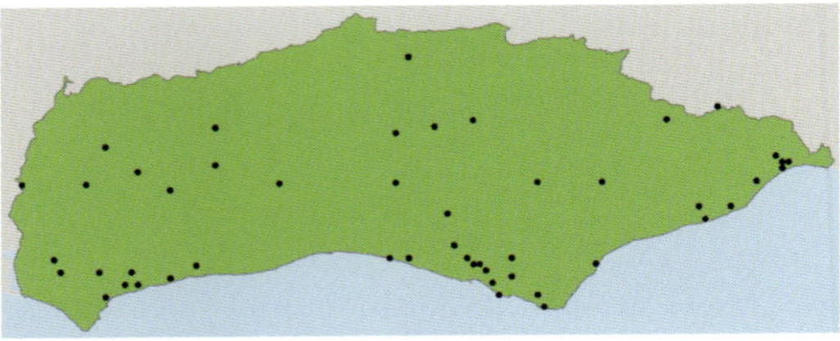

In some parts of Europe *Bombus sylvarum* is still a common bee. Until the second half of the twentieth century it was widely distributed throughout lowland and coastal areas of England and Wales, becoming more localised in central and northern areas of England.

Because of the destruction of large areas of flower-rich grassland from the early twentieth century onwards, known breeding populations are now restricted to five landscapes in Britain: the Thames Estuary, Somerset Levels, Gwent Levels, Kenfig-Port Talbot and South Pembrokeshire. A sixth population on Salisbury Plain, Wiltshire, last recorded in 2008, is now thought to be locally extinct, but this species could still be present here—and in other locations—but overlooked.

Up until the second half of the twentieth century, *B. sylvarum* was still found across Sussex. Its final strongholds were the countryside around Rye and between Brighton and Beachy Head, and from 1965 to 1970 it was recorded by Alfred Jones in Southease, South Heighton, Tarring Neville, Lewes, Alfriston and East Dean. It was last recorded in the county at Rye Harbour Nature Reserve by Gerald Dicker in 1986.

In parts of its range on the continent (the west coast of France, for example), an all-black form with a red tail known as *nigrescens* is common. Six males of this form were collected in the Seaford area in August 1921 by Charles Mortimer, this date suggesting that they had bred here. This form closely resembles *B. ruderarius*.

Bombus sylvarum queen [PB]

Eucera nigrescens Scarce Long-horned Bee
FIRST RECORD 1906, Hastings (unknown) TOTAL RECORDS 3

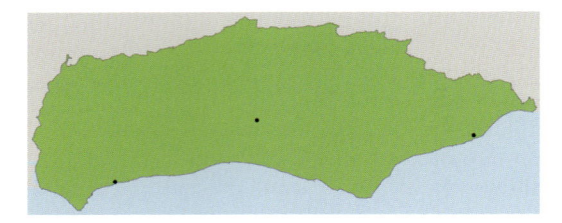

Eucera nigrescens is believed to be extinct in Britain and was last recorded in Kent in 1970. In Britain, it was associated with open habitats rich in clovers and vetches, particularly grassland and woodland.

While *E. nigrescens* was most frequently recorded in Kent, it used to be widely distributed across southern England, its range extending as far west as Devon and as far north as Gloucestershire and Berkshire. It is not thought to have been as abundant in Britain as its close relative *E. longicornis*, but, like this other species, it is still widely distributed in parts of Europe.

There are just three records for *E. nigrescens* in Sussex, from 1906, 1907 and 1941. The first two are from Hastings and Felpham respectively, and the last record—of three males swept from Tufted Vetch near Hassocks by Donald Baker—was not published until 1964 and was the first account of E. nigrescens as a British species. By this time, it had all but disappeared from the country.

Eucera nigrescens ♂ [PW]

Melecta luctuosa Square-spotted Mourning Bee
FIRST RECORD 1905, Brighton (William Unwin) TOTAL RECORDS 1

There are just a handful of British records for *Melecta luctuosa*, all from southern England, but it has not been seen here since the 1920s. It is still found in some parts of continental Europe, but even there it is in decline.

There is just one record from Sussex, cited by Edward Saunders in *The Victoria History of the County of Sussex*, published in 1905. This record is from Brighton.

The distribution of *M. luctuosa* in Britain was tied to locations occupied by its host *Anthophora retusa*. This last species has undergone a significant decline but still occurs in places with light soils such as heathlands, coastal dunes and soft-rock cliffs. The only surviving Sussex colony for *A. retusa* is at Seaford Head, but searches here have found just *M. albifrons*, and not *M. luctuosa*.

Melecta luctuosa ♂ [PW]

Nomada sexfasciata Six-banded Nomad Bee
FIRST RECORD 1879, Fairlight (Edward Saunders) TOTAL RECORDS 3

Nomada sexfasciata is long extinct in Sussex and is teetering on the edge of extinction in Britain, although it is still widely distributed in some parts of continental Europe. Like its hosts *Eucera longicornis* and *E. nigrescens*, this *Nomada* is associated with grasslands that are rich in plants from the Pea family such as Common Bird's-foot-trefoil and Kidney Vetch. It was also known from heathland and woodland.

In Britain, it is now known from just a single site on the south Devon coast. Here, it has not been seen in good numbers for some years, and its future seems precarious. By way of contrast, in 1896 Edward Saunders described it as "not rare in many localities in May and June".

The earliest reference to *N. sexfasciata* in Sussex is of a female collected from Fairlight by Edward Saunders in August 1879. Both this specimen and one collected in 1920 from Hastings by Rosse Butterfield are in the Natural History Museum. The Cliffe Castle Museum, Keighley, holds six additional specimens—two of which are also labelled "Hastings"—in a collection that belonged to Butterfield. Sadly, none of these has a date attached, although we know he lived from 1874 to 1938.

There is also a record of *N. sexfasciata* from Brighton by William Unwin, cited in *The Victoria History of the County of Sussex*, published in 1905. Two of his specimens are in the Booth Museum, although neither has a date or location label attached.

Nomada sexfasciata ♀ [PB]

Osmia xanthomelana Cliff Mason Bee
FIRST RECORD 1876, Eastbourne (Frederick Smith) **TOTAL RECORDS** 1

Osmia xanthomelana is one of Britain's most threatened bees. It is long extinct in Sussex and nationally now survives at just two small and vulnerable sites close to each other on the Llyn Peninsula in north Wales. It was once a widely distributed but very local species found at a handful of scattered locations in southern England as well as on the coasts of west and north Wales.

Osmia xanthomelana ♀ [SF]

The only evidence for the occurrence of *O. xanthomelana* in Sussex is from Frederick Smith, who gives Eastbourne as the location in *A Catalogue of British Hymenoptera in the British Museum*, published in 1876. He gives an early description of its nesting behaviour, reporting that the cells are sited at the base of grass stems and that these are "pitcher-shaped and constructed of mud mixed with small pebbles, the cells are rounded at the bottom, but flattened at the top, and closed by a lid; a nest, when completed, usually contains five or six cells". Although Smith does not give any further information as to location, Cow Gap below Beachy Head is thought to be the likeliest spot. This is an area of slumping cliffs and seepages, with the south-facing slopes supporting very species-rich chalk grassland. Although Cow Gap is a good site for other species of *Osmia*, it is unlikely to now be suitable for *O. xanthomelana*—the seepages are almost entirely occupied by thick bands of scrub.

Andrena rosae Perkins' Mining Bee
FIRST RECORD 1853, Lewes (William Unwin) **TOTAL RECORDS** 5

Andrena rosae is an extremely scarce bee with very few modern records. Since 2000 it has been recorded only from Cornwall, north Devon, the south Midlands, Pembrokeshire, south Glamorgan, and Kent—where it was recorded in 2000 in Tunbridge Wells, close to the Kent/Sussex County boundary. Here, Ian Beavis found a male on a cultivated plant from the Speedwell family. Most recent records are from Cornwall.

Andrena rosae ♀ [SF]

In Sussex, there are five records for *A. rosae*, two from Lewes in 1853 and 1855, two from Hastings in 1885 and 1887, and a fifth from Chichester in 1919. William Unwin reported that he had swept the first of these, a female, "in April 1853, near Landport, Lewes, from the blossoms of the blackthorn". The 1855 record is of two females swept from Sallow catkins from the same location.

The six specimens supporting the two Hastings records were discovered in the collections of the Natural History Museum, while the Chichester specimen was located by Thomas Wood in the National Museum of Ireland in Dublin.

Intriguingly, there is a specimen of *A. rosae* in the Booth Museum, with the collector's name given as Unwin. Unfortunately, the specimen does not have a date or location label, but it is possible that this is one of the specimens that William Unwin reported finding in Lewes in 1853 and 1855.

Halictus maculatus Box-headed Furrow Bee
FIRST RECORD 1879, Fairlight Glen (Edward Saunders) **TOTAL RECORDS** 1

Halictus maculatus is long extinct in mainland Britain—although, interestingly, it was found as recently as 2007 on Sark in the Channel Islands.

Before becoming extinct, it was recorded on 35 occasions in Britain from half a dozen locations across southern England. Most records are from south Devon where there were two thriving nest sites close to each other. This is where the last British record was made in 1930.

There is a single record for this bee in Sussex, made by Edward Saunders on a visit to Hastings in August 1879. Writing in *The Entomologist's Monthly Magazine*, he reported taking "one ♀ off [Smooth Hawk's-beard] in Fairlight Glen. This is one of our rarest bees, only two specimens, so far as I know, having occurred in England before."

Halictus maculatus ♀ [PW]

Rophites quinquespinosus Five-spined Rophites
FIRST RECORD 1877, Guesting (Edwin Bloomfield) **TOTAL RECORDS** 2

Rophites quinquespinosus has been recorded twice in Britain, both times in Sussex and both times by Edwin Bloomfield, a vicar living in Guestling. He reported that he had found this species in woodland near Guestling, with the dates given as 1877 and 1878 and with the last specimen surviving in the Natural History Museum. Records from successive years suggest that a breeding population had been established.

Bloomfield's second sighting was reported in *The Entomologist's Monthly Magazine* in 1878: "I have been fortunate enough to meet with a second specimen of *Rophites quinquespinosus*, having taken a specimen […] on the afternoon of the 4th of August."

Rophites quinquespinosus ♀. The label suggests that it was collected from Guestling in July 1878. [SF]

There are reasons to be sceptical about the validity of these records. Not only does the month reported above not match the month given on the label attached to the specimen, but on the continent *R. quinquespinosus* occurs in habitats that are very different to the woodland found near Guestling—in Germany, some populations are found in warm, dry, upland meadows and nutrient-poor grasslands.

Woodland near Guestling on the left [PG] and upland meadows in southern Germany on the right [PW]

The Bees of Sussex 461

Making bees welcome—a personal journey

Gardens can be magnets for bees and frequently offer more opportunities to feed, forage or nest than the surrounding countryside. My garden in Lewes, to start close to home, attracts 75 species, almost a third of all those found in the county. But this isn't unusual, and you could achieve a similar tally by following a few simple principles to make bees more welcome.

As well as supporting a thriving bee population, a bee-friendly garden, allotment or balcony is the perfect place to observe how bees behave up close, or simply sit back and watch—a win for the bees and a win for you.

Here are ten things I've done and you could do too:

1. **Read the label.** Garden centres and nurseries make it easier for us by labelling many plants and seeds "pollinator-friendly". If you're thinking through planting options for a particular spot, make this the deciding factor. You'll almost always find one that fits your criteria as well as those of the bees.

2. **Cater for all shapes and sizes.** Not all bees are the same, and a flower that's perfect for one will be a non-starter for another. A bee with a short tongue—*Lasioglossum morio*, say—will head straight for a flat, open flower like forget-me-not, but a species with a long tongue—*Bombus hortorum*, perhaps—will make a beeline for a tube-like flower such as foxglove. Be inclusive by choosing plants with different flower shapes.

3. **Keep supply lines open.** Different bees need pollen and nectar at different times—spring, summer, autumn, and winter too—so extend your flowering season with plants which bloom one after the other, from one season to the next.

Bombus hortorum worker, a long-tongued bee [NO]

Gardens can be magnets for bees—my garden in Lewes has attracted 75 species [WP]

The Bees of Sussex

4. **Plan for fussy eaters.** If there's a particular species you want to see, don't leave it to chance but give it to them on a plate. So, *Melitta leporina* loves a White Clover in the lawn, while *Chelostoma campanularum* won't be able to resist a bellflower. In other words, plant it and they should come.
5. **Go wild.** Add some native wildflowers to your lawn and let some of your grass grow long. British bees have evolved alongside native wildflowers, so by making space for one you make space for the other. And—a bonus—some species will take advantage of the longer grass for nesting.

Melitta leporina ♂ visiting White Clover [PC]

6. **Mix it up.** Don't throw your lawn mower away quite yet, and keep some areas short. Mown grass, especially if it's south-facing and sheltered, could be a nesting hotspot for any bee partial to a bit of warmth.
7. **Embrace the ageing process.** As they grow old, trees and shrubs develop increasing amounts of dead and decaying wood, which makes them a perfect place for bees to nest. Let things take their natural course, and don't prune too hard.
8. **Give a helping hand.** Artificial nests in sunny spots are bound to find willing residents. It doesn't need to be fancy—holes of different sizes in wooden posts, a few bamboo canes, a sandy patch or anything else which mimics a bee's natural preferences will do the trick. Look for things you can adapt around the garden, or—if you want to go high-end—build a full-blown bee hotel.
9. **Change your mindset.** 'Weeds' are all in the mind, and many we've traditionally shunned—dandelion and buttercup, for example—are sources of pollen and nectar too.
10. **Put the spray away.** No one wants to hang around when pesticides are in the air, least of all a bee. So avoid using these wherever and whenever possible, especially where plants are in flower and when bees are active. It's harder work but kinder to the bees.

Megachile bees at a bee hotel [CG]

Gazetteer

Below is a list of sites mentioned in the text, plus grid references and summaries of key information. Each location is numbered and shown on the map below. Most are open access but a few are not, and permission to visit must be obtained if you're hoping to explore beyond public rights of way.

Site name	Grid Ref	Notes
Abbot's Wood • 57	TQ5607	Woodland managed by Forestry England.
Amberley Wildbrooks • 23	TQ0314	Extensive wetland part-owned by RSPB and Sussex Wildlife Trust.
Ambersham and Heyshott Commons • 14	SU9119	Open access complex of heathland, scrub and secondary woodland.
Ashcombe Bottom • 32	TQ3711	Chalk grassland, secondary woodland and scrub managed by National Trust.
Ashdown Forest • 42	TQ4529	2,600 ha heathland and woodland landscape managed by Conservators of Ashdown Forest.
Batemans • 61	TQ6723	National Trust house, garden and farmland.
Beachy Head • 59	TV5895	Chalk grassland and scrub owned by Eastbourne Borough Council.
Birling Gap • 54	TV5596	Chalk grassland and chalk heath managed by National Trust.
Blackcap • 33	TQ3712	Chalk grassland, secondary woodland and scrub managed by National Trust.
Broadwater Warren • 55	TQ5537	A matrix of heathland and woodland managed by RSPB.
Burton Mill Pond • 17	SU9717	A mosaic of heathland, wet woodland and a freshwater lake part-managed by Sussex Wildlife Trust.
Camber Sands • 73	TQ9518	Sand-dune system managed by East Sussex County Council.
Castle Hill • 34	TQ3707	Chalk grassland managed by Natural England.
Chailey Common • 35	TQ3820	Heathland, scrub and woodland complex managed by East Sussex County Council.
Chichester Harbour, incl. Thorney Island • 1	SU7500	Estuary with mudflats, salt marsh, shingle and reedbeds plus coastal woodland and grassland.
Combe Haven • 65	TQ7709	Reedbed, grazing marsh and alluvial meadow and including Filsham Reedbed.
Cow Gap • 60	TV5995	Coastal chalk grassland, seepages and scrub managed by Eastbourne Borough Council.
Cuckmere Haven and Seven Sisters • 48	TV5197	Saltmarsh, vegetated shingle and chalk grassland managed by South Downs National Park Authority and National Trust.
Deep Dean • 51	TQ5302	Chalk grassland and scrub managed by South East Water.
Devil's Dyke • 28	TQ2611	Chalk grassland and scrub managed by National Trust.
Devil's Jumps • 5	SU8217	Chalk grassland managed by Murray Downland Trust.
Ditchling Beacon • 30	TQ3313	Chalk grassland, scrub and secondary woodland managed by Sussex Wildlife Trust.
Ditchling Common • 31	TQ3318	Species-rich grassland, woodland, scrub and lakes managed by East Sussex County Council.
East Head • 2	SZ7699	Sand dune, salt marsh and tidal estuary managed by National Trust.
Ebernoe Common • 18	SU9726	Wood pasture, woodland, grassland and scrub mosaic managed by Sussex Wildlife Trust.
Fairmile Bottom • 19	SU9809	Chalk grassland and scrub managed by West Sussex County Council.
Filsham Reedbed • 66	TQ7709	Reedbed, scrub and grassland managed by Sussex Wildlife Trust and forming part of Combe Haven.
Flatropers Wood • 71	TQ8623	Woodland with clearings managed by Sussex Wildlife Trust.
Fore Wood • 62	TQ7513	Coppiced woodland managed by RSPB.
Friston Forest • 52	TV5499	Plantation woodland with pockets of grassland and scrub managed by Forestry England.
Galley Hill & Glyne Gap (Bulverhythe) • 63	TQ7607	Low, sandstone cliffs between Bexhill and Bulverhythe.
Gills Lap • 44	TQ4631	Heathland, scrub and woodland mosaic managed by Conservators of Ashdown Forest.
Graffham Common • 15	SU9319	Wooded heath managed by Sussex Wildlife Trust.
Great Dixter • 68	TQ8125	Gardens, house and wider countryside managed by Great Dixter Charitable Trust.
Great Wood • 64	TQ7615	Woodland with heathy clearings and wet flushes managed by Forestry England.
Hargate Forest • 58	TQ5736	Woodland with areas of heathland managed by Woodland Trust.
Harting Down • 3	SU7918	Chalk grassland, scrub and woodland managed by National Trust.
Hastings Country Park • 70	TQ8510	Gill woodland, soft-rock cliffs, grassland and farmland managed by Hastings Borough Council. Includes Fairlight Glen.
Heyshott Down and Escarpment • 13	SU8916	Steep-scarp chalk grassland, scrub and secondary woodland managed by Murray Downland Trust.
High & Over and Cradle Hill • 49	TQ5001	Chalk grassland and scrub managed by National Trust.

The Bees of Sussex

Site name	Grid Ref	Notes
Hope Gap ● 47	TV5097	Dry valley chalk grassland, scrub and cliffs at Seaford Head managed by Sussex Wildlife Trust.
Iping and Stedham Commons ● 8	SU8521	Heathland, scrub and woodland managed by Sussex Wildlife Trust.
Isle of Thorns ● 38	TQ4230	Heathland, scrub and woodland mosaic managed by Conservators of Ashdown Forest.
Kingley Vale ● 6	SU8210	Chalk grassland and Yew woodland managed by Natural England.
Knepp ● 26	TQ1521	Rewilded grassland, scrub, woodland and wetland mosaics managed by Knepp Estate.
Lavington Common ● 16	SU9418	Heathland, scrub and woodland complex managed by National Trust.
Levin Down ● 12	SU8813	Chalk grassland, chalk heath and woodland managed by Sussex Wildlife Trust.
Lord's Piece ● 21	SU9917	Open access heathland and acid grassland mosaic.
Lullington Heath ● 53	TQ5401	Chalk heath, chalk grassland and scrub managed by Natural England.
Lydd Ranges (Midrips) ● 74	TR0018	Vegetated shingle within live-firing range managed by the Ministry of Defence.
Lynchmere Common ● 10	SU8631	Heathland, scrub and woodland managed by Lynchmere Society.
Malling Down ● 39	TQ4210	Chalk grassland, scrub and woodland managed by Sussex Wildlife Trust.
Markstakes Common ● 36	TQ3918	Wooded common owned by Lewes District Council.
Marline ● 67	TQ7711	Complex of species-rich meadows, pasture and woodland managed by Sussex Wildlife Trust.
Medmerry ● 7	SZ8295	Intertidal mudflats, saltmarsh, grassland and farmland managed by RSPB.
Midhurst Common ● 11	SU8721	Open access heathland and woodland.
Mount Caburn ● 41	TQ4408	Open access chalk grassland.
Nymans ● 29	TQ2629	National Trust garden, woodland and farmland.
Old Lodge ● 45	TQ4630	Heathland and woodland within Ashdown Forest managed by Sussex Wildlife Trust.
Rewell Wood ● 20	SU9807	Privately owned woodland with wide rides and clearings.
Rowland Wood and Park Corner Heath ● 50	TQ5114	Woodland complex managed by Butterfly Conservation.
Rye Harbour Nature Reserve ● 72	TQ9317	Coastal habitats including vegetated shingle and saltmarsh managed by Sussex Wildlife Trust.
Seaford Head ● 46	TV4997	Chalk grassland and scrub managed by Sussex Wildlife Trust.
Selwyns Wood ● 56	TQ5520	Woodland managed by Sussex Wildlife Trust.
Sheffield Park ● 37	TQ4123	National Trust house, garden and parkland.
Southerham Farm ● 40	TQ4310	Chalk grassland, scrub and woodland managed by Sussex Wildlife Trust.
St Helen's Wood ● 69	TQ8111	Complex of woodland, gill streams and grassland managed by Hastings Borough Council.
Sullington Warren ● 25	TQ0914	Heathland and woodland managed by National Trust.
Waltham Brooks ● 22	TQ0215	Grazing marsh managed by Sussex Wildlife Trust.
Weavers Down ● 4	SU8130	Open access acid grassland and heathland.
Well Bottom ● 43	TQ4505	Open access chalk grassland and scrub.
West Dean Woods ● 9	SU8415	Coppiced woodland managed by Sussex Wildlife Trust.
Wiggonholt Common ● 24	TQ0516	Heathland, acid grassland and woodland managed by RSPB.
Woods Mill ● 27	TQ2113	Complex of grassland, hedgerows, woodland and wetland managed by Sussex Wildlife Trust.

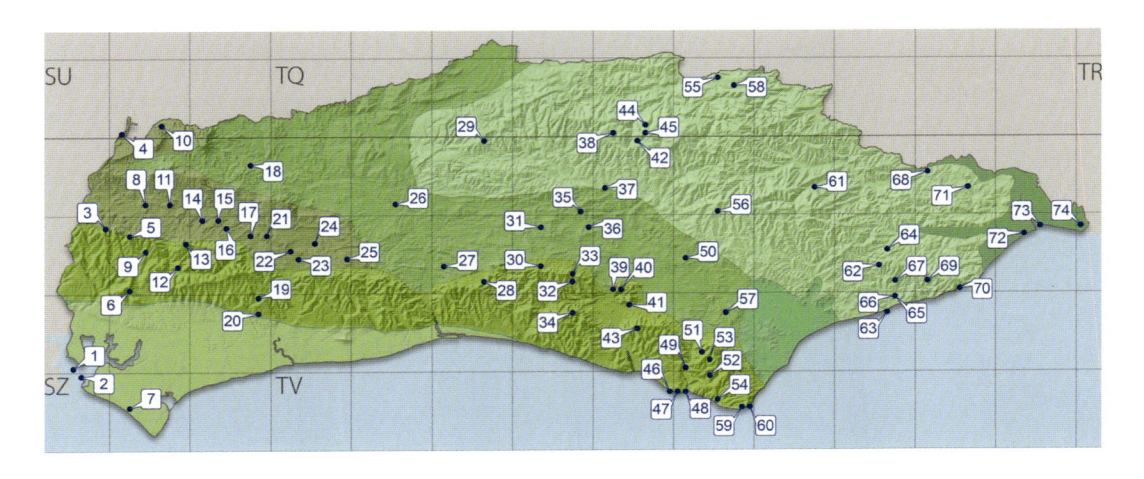

How to make a record

Good data underpins good decision-making, whether this relates to how to manage a parcel of land or how to respond to pollinator declines. Nationally, BWARS hold 834,000 unique records–a powerful dataset that has been used for many different purposes, including most recently on-going work to assess the conservation status of each species.

With experience, it is possible to identify a number of bee species just from photographs or in the field. These include *Anthidium manicatum*, *Bombus lapidarius* and even some species of *Andrena*, including *A. cineraria* and *A. fulva*. However, identification is simply not possible in this way for the majority of species because distinguishing features are frequently only visible with the aid of a microscope.

This means that a selection of bees encountered will need to be collected, pinned and examined with the aid of a microscope before they can be identified. Collecting–and therefore killing–small numbers has a negligible impact on populations, with any negatives vastly outweighed by the value of the data that this work generates.

To illustrate this, some 5,000 or so new records have been sourced for *The Bees of Sussex* from pinned and labelled specimens held in museum collections, data that would have been lost if the recorders had not collected and stored these specimens. This data includes two species, *Andrena rosae* and *Stelis phaeoptera*, which had not otherwise been known from Sussex.

A 'good' record will include the name of the species, the date on which it was seen, the grid reference and site name of the location, and the names of the recorder and the determiner (this is the person who confirms the identification, and may be the same as the recorder).

If you are interested in collecting specimens, you should follow the code of conduct agreed by the Joint Committee for the Conservation of British Invertebrates (available through the British Entomological and Natural History Society website). Your data should be submitted to both the Bees, Wasps and Ants Recording Society and to the Sussex Biodiversity Record Centre (both maintain their own databases, and you can find their contact details online). Many recorders use an Excel spreadsheet for this purpose. The fictitious example below illustrates a typical format:

Species name	Recorder	Determiner	Location name	Vice county	Sample date	Grid ref	Sample type
Andrena subopaca	James Power	James Power	Malling Down	East Sussex	07/06/2023	TQ429112	Pinned specimen
Bombus hortorum	James Power	James Power	Malling Down	East Sussex	07/06/2023	TQ429112	Field observation
Bombus hypnorum	James Power	James Power	Malling Down	East Sussex	07/06/2023	TQ429112	Field observation
Lasioglossum parvulum	James Power	James Power	Malling Down	East Sussex	07/06/2023	TQ429112	Pinned specimen
Andrena flavipes	James Power	James Power	Malling Down	East Sussex	07/06/2023	TQ429112	Field observation
Lasioglossum pauxillum	James Power	James Power	Malling Down	East Sussex	07/06/2023	TQ429112	Pinned specimen
Lasioglossum calceatum	James Power	James Power	Malling Down	East Sussex	07/06/2023	TQ429112	Pinned specimen

An alternative approach is to use iRecord, a simple-to-use app that allows you to upload a photograph and generate a grid reference. The number of species which can be identified through this channel is limited, but it is a significant source of data for many of the species that are easier to identify, including *Anthophora plumipes*, *B. lapidarius*, *B. pascuorum* and *Osmia bicornis*. Even if you just use iRecord and limit yourself to these species, then, you'll still be contributing to the growing understanding of bees in Sussex.

References

Abrahams, F. et al. (2018) *The Flora of Sussex*. Pisces Publications.

Alford, D.V. (1975) *Bumblebees*. Davis-Poynter Limited.

Allen, G. (2018) *Nomada fusca* Schwarz, 1986 found in the UK? *BWARS Newsletter*, Autumn 2018, 31.

Allen, G. (2020) *Bees, wasps and ants of Kent*, 2nd Edition. Kent Field Club.

Andrewes, C.H. (1946) *Andrena vaga* Panz. (Hym., Apidae) in Sussex. *The Entomologist's Monthly Magazine*, 82:39.

Anon. Brazenor Brothers (1863/8 – 1937) Unpublished note.

Anon. (1887) Unwin obituary. *The Entomologist's Monthly Magazine*, 24: 47.

Anon. (1938) Obituaries. *Nature*, 142: 986.

Anon. (1980) *Atlas of the Bumblebees of the British Isles*. Institute of Terrestrial Ecology.

Baker, D.B. (1964) Two bees new to Britain. *The Entomologist's Monthly Magazine*, 100: 279–286.

Baldock, D. (2008) *Bees of Surrey*. Surrey Wildlife Trust.

Beavis, I.C. (2002) Aculeate Hymenoptera of Tunbridge Wells and the Central High Weald. *Entomologist's Gazette*, 53: 97–129.

Beavis, I.C. (2007) Aculeate Hymenoptera of Tunbridge Wells and the Central High Weald: first supplement. *Entomologist's Gazette*, 58: 85–118.

Beavis, I.C. (2012) *Osmia pilicornis* Smith and other rarities in the High Weald near Tunbridge Wells. *BWARS newsletter*, Spring 2012: 26–27.

Bennett, W.H. (1888) Coleoptera and Hymenoptera in the Hastings District. *The Entomologist's Monthly Magazine*, 25: 164–165.

Benton, T. & Owens, N. (2023) *Solitary Bees*. New Naturalist 146. William Collins, London.

Bogusch, P. (2003) Hosts, foraging behaviour and distribution of six species of cleptoparasitic bees of the subfamily Anthophorinae (Hymenoptera: Apidae). *Acta Societatis Zoologicae Bohemicae*, 67: 65–70.

Bogusch, P. & Straka, J. (2012) Review and identification of the cuckoo bees of Central Europe (Hymenoptera: Halictidiae: Sphecodes). *Zootaxa*, 3311: 1–41.

Bloomfield, E.N. (1878) *Rophites quinque-spinosus* and *Acronycta alni* near Hastings. *The Entomologist's Monthly Magazine*, 15: 113.

Brandon, P. (1998) *The South Downs*. The History Press.

Brock R.E. et al. (2021) No severe genetic bottleneck in a rapidly range-expanding bumblebee pollinator. *Proceedings of the Royal Society* B, 288: 20202639.

Butterfield, W. R. (1920) Notes on the local fauna and flora for 1919. *The Hastings and East Sussex Naturalist*, 3: 125–142.

Butterfield, R. (1921) *Sapyga clavicornis* Linn. and *Nomada guttulata* Schenck at Hastings. *The Entomologist's Monthly Magazine*, 57: 261.

Butterfield, W. R. (1922) Notes on the local fauna and flora for 1920. *The Hastings and East Sussex Naturalist*, 3: 211–224.

BWARS. https://w.w.w.bwars.com/

Chambers, V.H. (1949) The Hymenoptera aculeata of Bedfordshire. *Transactions of the Society for British Entomology*, 9: 197–252.

Crowther, L. et al. (2019) Spatial ecology of a range-expanding bumble bee pollinator. *Ecology and Evolution*: 9, 986–997.

Davison, P.J. & Field, J. (2016) Social polymorphism in the sweat bee *Lasioglossum* (*Evylaeus*) *calceatum*. *Insectes Sociaux*: 63, 327–338.

Dawson, R. (2018) *Anthophora furcata* nest building at Dovestone Reservoir. *BWARS newsletter*, Autumn 2018: 23–24.

Dellicours, S. & Michez, D. (2010) Biologie, observations et collectes de trois espèces soeurs du genre *Melitta* Kirby 1802 (Hymenoptera, Melittidae). *Osmia*, 4: 29–34.

Early, J. (2006) Notes on *Nomada signata* clepto-parasiting *Andrena fulva* nesting in a garden lawn. *BWARS newsletter*, Autumn 2006: 24–25.

Early, J. (2016) Observations on foraging by *Hylaeus* (*Abrupta*) *cornutus*. *BWARS newsletter*, Autumn 2016: 21–24.

Early, J. (2019) Classic opportunism: *Megachile willughbiella* (Kirby, 1802): iso. Perkins: 1925 nesting in grow bags. *BWARS newsletter*, Spring 2019: 10.

Early, J. (2023) Seek and you will find – *Pemphredon austriaca*: ISO. Bleet & Early: 2022 reared from galls collected in West Sussex (VC13), East Sussex (VC14), Middlesex (VC21) and East Suffolk (VC25). *BWARS newsletter*, Autumn 2023: 8–11.

Earwaker, R. (2013) Hunting *Osmia pilicornis* (Smith) in the south-east. *BWARS newsletter*, Autumn 2013: 15–16.

Edwards, M. (1979) Further Aculeate Hymenoptera from Sussex. *The Bulletin of the Amateur Entomological Society*, 38: 69–74.

Edwards, M. (2019) *Stelis odontopyga* Noskiwiecz, 1926: iso. Edwards: 2019 (Hymenoptera: Megachilinae) recognised in Britain. *BWARS Newsletter*, Spring 2019: 34.

Edwards, M. & Jenner, M. (2005) *Field guide to the bumblebees of Great Britain and Ireland*. Ocelli.

Edwards, M. & Jenner, M. (2008) Investigation of the autecology of the bee *Anthophora retusa* (Hymenoptera: Apidae) in 2008. *Hymettus Research Report*.

Edwards, M. & Jenner, M. (2009) Investigation of the autecology of the bee *Anthophora retusa* (Hymenoptera: Apidae) part 2, 2009. *Hymettus Research Report*.

Edwards, M. et al. (2019) *Stelis odontopyga* Noskiwiecz (Hymenoptera: Megachilinae) new to Britain. *British Journal of Entomology and Natural History*, 32: 43–47.

Else, G. (1995) The distribution and habits of the bee *Hylaeus pectoralis* Förster, 1871, (Hymenoptera: Apidae) in Britain. *British Journal of Entomology and Natural History*, 8: 43–47.

Else, G. (1997) The status of *Stelis breviuscula* (Nylander) (Hymenoptera: Apidae) in Britain, with a key to the identification of the British species of *Stelis*. *British Journal of Entomology and Natural History*, 10: 214–216.

Else, G.R. (2012) Observations on *Andrena ferox* (Smith) (Hymenoptera: Apidae) in the New Forest woodlands. *BWARS Newsletter*, Spring 2012: 13–18.

Else, G.R. & Edwards, M. (2018) *Handbook of the Bees of the British Isles* – vols 1 and 2. Ray Society.

Else, G.R. (2018) *Melitta* sleeping clusters. *BWARS Newsletter*, Autumn 2018: 7–10.

Falk, S. Flickr collection, Apoidea (bees). https://www.flickr.com/photos/63075200@N07/collections/72157629294459686/.

Falk, S. (2004) The form of *Nomada fulvicornis* F. (Hymenoptera Apidae) associated with the mining bee *Andrena nigrospina* Thomson. *British Journal of Entomology and Natural History*, 17: 229–235.

Falk, S. (2011) A survey of the bees and wasps of fifteen chalk grassland and chalk heath sites within the East Sussex South Downs. Sussex Biodiversity Record Centre.

Falk, S. (2019) Short Communications: *Nomada zonata* Panzer (Hymenoptera: Apidae) in Britain. *British Journal of Entomology and Natural History*, 32: 175.

Falk, S. & Earwaker, R (2019). Dusky-horned Nomad Bee, *Nomada bifasciata*, new to Britain (Hymenoptera: Apidae). *British Journal of Entomology and Natural History*, 32: 170–175.

Falk, S. & Lewington, R. (2015) *Field Guide to the Bees of Great Britain and Ireland*. British Wildlife Field Guides, Bloomsbury.

Falk, S. et al. (2019) The Water-dropwort Mining Bee *Andrena ampla* Warncke (Hymenoptera: Apidae), new to Britain. *British Journal of Entomology and Natural History*, 32: 273–285.

Falk, S. et al. (2022) DNA and morphological characterisation of the Bilberry Nomad Bee *Nomada glabella sensu* Stöckhert nec Thomson in Britain with discussion of the remaining variation within *N. panzeri*. *British Journal of Entomology and Natural History*, 35, 91–111.

Gammans, N. & Allen, G. (2014) *The Bumblebees of Kent*. Kent Field Club.

Gammans, N. (2019) *The Short-haired Bumblebee Reintroduction Project: 10-year report*. Bumblebee Conservation Trust.

Guichard, K.M. (1938) *Andrena humilis* and *Picris hieracioides*. *The Entomologist's Monthly Magazine*, 74: 233–234.

Hallett, H.M. (1929) *Hylaeus gibbus* S. Saund. in Sussex. *The Entomologist's Monthly Magazine*, 65: 263.

Hallett, H.M. (1952) *Microdynerus exilis*. -S., *Sphecodes scabricollis* Wesm. and other Aculeate Hymenoptera in West Sussex. *The Entomologist's Monthly Magazine*, 88: 44.

Hargreaves, B. & White, S. (2021) *The Aculeate Hymenoptera (Bees, Wasps and Ants) of Lancashire and North Merseyside*. Lancashire & Cheshire Fauna Society.

Hawkins, R.D. (2011) *Lasioglossum sexstrigatum* (Hymenoptera: Apidae, Halictinae) new to Britain. *British Journal of Entomology and Natural History*, 24: 90–92.

Heal, N. (1998) Obituary: Gerald Henry Lethbridge Dicker, 1913–1997. *Transactions of the Kent Field Club*, 15 (2): 119–121.

Hennessy, G. et al. (2021) Phenology of the specialist bee *Colletes hederae* and its dependence on *Hedera helix* L. in comparison to a generalist, *Apis mellifera*. *Arthropod-plant Interactions*, 15: 183–195.

Hodge, P. (1996) The Brede Valley Entomological Assessment: a survey of insects within the River Brede floodplain between Westfield and Rye. Unpublished report for English Nature.

Jacquemyn, H. & Hutchings, M.J. (2015) Biological flora of the British Isles: *Ophrys sphegodes*. *Journal of Ecology*, 103: 1680–1696.

Jones, N. & Cheeseborough, I. (2014) *A Provisional Atlas of the Bees, Wasps and Ants of Shropshire*. Field Studies Council.

Kasparek, M. (2015) The cuckoo bees of the genus *Stelis* Panzer, 1806 in Europe, North Africa and the Middle East. *Entomofauna, Supplement*, 18.

Kirby-Lambert, C. (2016) *Nomada alboguttata* Herrich-Schäffer, 1839 new to the British Isles and *Nomada zonata* Panzer, 1798 first record for mainland Britain. *BWARS newsletter*, Autumn 2016: 29–30.

Knowles, A. (2018) Wildlife Reports. Bees, wasps and ants. *British Wildlife*, October 2018: 58.

Knowles, A. (2020) Wildlife Reports. Bees, wasps and ants. *British Wildlife*, February 2020: 213–215.

Knowles, A. (2021) Wildlife Reports: Bees, wasps and ants. *British Wildlife*, December 2021: 215–217.

Knowles, A. (2022) Wildlife Reports: Bees, wasps and ants. *British Wildlife*, December 2022: 214–216.

Knowles, A. (2023) Wildlife Reports: Bees, wasps and ants. *British Wildlife*, October 2023: 58–60.

Little, B. & Jarman, J. (2011) Observations of *Andrena tarsata* Nylander and *Nomada obtusifrons* Nylander in Central Scotland. *BWARS Newsletter*, Spring 2011.

Lyons, G. (2020) Brighton B Banks: invertebrate, botanical, and management survey 2020. Unpublished report for The LIVING Coast UNESCO Biosphere and Brighton and Hove City Council.

Macdonald, M. & Nisbet, G. (2006) *Highland Bumblebees: Distribution, Ecology and Conservation*. Highland Biological Recording Group, Inverness.

Michener, C.D. (2007) *The Bees of the World*, second edition. The John Hopkins University Press, Baltimore.

Morice, F.D. (1889) Rare aculeate hymenoptera in 1889. *The Entomologist's Monthly Magazine*, 25: 434–435.

Morice, F.D. (1892) Aculeate Hymenoptera in 1892. *The Entomologist's Monthly Magazine*, 29: 10–12; 90.

Morice, F.D. (1902) Hymenoptera near Haywards Heath, Sussex. *The Entomologist's Monthly Magazine*, 38: 223.

Mortimer, C.H. (1922a) A new British *Bombus*, *nigrescens* (Pérez), from Sussex. *The Entomologist's Monthly Magazine*, 58: 16–17.

Mortimer, C.H. (1922b) Occurrence of *Bombus cullumanus* (Kirby) in Sussex. *The Entomologist's Monthly Magazine*, 58: 19.

Moyse, R. (2021) Studying a nesting aggregation of *Andrena ferox*: iso. Perkins: 1919 on the North Downs in Kent in 2021. *BWARS Newsletter*, Autumn 2021: 14–15.

Nevinson, E.B. (1923) The survival of *Bombus cullumanus*. *The Entomologist's Monthly Magazine*, 59: 277–278

Nieto, A. *et al*. (2014) *European Red List of Bees*. Luxembourg: Publication Office of the European Union.

Notton, D.G. & Dathe, H.H. (2008) William Kirby's types of *Hylaeus* Fabricius (Hymenoptera, Colletidae) in the collection of the Natural History Museum, London. *Journal of Natural History*, 42: 1861–1865.

Notton, D.G. & Norman, H. (2017) Hawk's-beard Nomad Bee, *Nomada facilis*, new to Britain (Hymenoptera: Apidae). *British Journal of Natural History and Entomology*, 30: 201–214.

Ollerton, J. *et al*. (2014) Extinctions of aculeate pollinators in Britain and the role of large-scale agricultural changes. *Science*, 346: 1360–1362.

Owens, N. (2017) *The Bees of Norfolk*. Pisces Publications.

Owens, N. (2020) Two additional hosts confirmed for *Coelioxys inermis*: iso. Amiet *et al*.: 2004 in Norfolk. *BWARS Newsletter*, Autumn 2020: 18–21.

Page, S. *et al*. (2020) A conservation strategy for the Shrill Carder Bee *Bombus sylvarum* in England and Wales, 2020–2030. Bumblebee Conservation Trust.

Paxton, R.J. & Pohl, H. (1999) The Tawny Mining Bee *Andrena fulva* (Müller) (Hymenoptera, Andreninae), at a South Wales field site and its associated organisms: Hymenoptera, Diptera, Nematoda and Strepsiptera. *British Journal of Natural History and Entomology*, 12: 165–178.

Perkins, R.C.L. (1917) Notes on the collection of British Hymenoptera Aculeata formed by F. Smith. *The Entomologist's Monthly Magazine*, vol 53; 72.

Perkins, R.C.L. (1917) *Andrena bucephala* Steph. and *Nomada bucephalae* Perk. In Devonshire and notes on their habits. *The Entomologist's Monthly Magazine*, 53: 198–199.

Perkins, R.C.L. (1919) The British series of *Andrena* and *Nomada*. *The Transactions of the Entomological Society of London*, 1919: 218–319.

Pesenko, Y. *et al*. (2000) *Bees of the family Halictidae (excluding* Sphecodes*) of Poland: taxonomy, ecology, bionomics*. University Publishing House, Bydgoszcz.

Phillips, A. (2020) The Long-horned Bee (*Eucera longicornis*) at Great Dixter, East Sussex, UK. Unpublished report for Great Dixter Charitable Trust Ltd.

Power, J. (2023) The current status of *Osmia cornuta* iso: Amiet *et al*.: 2004 in Great Britain. *BWARS newsletter*, Spring 2023: 36–37.

Praz, C. *et al*. (2022) Unexpected levels of cryptic diversity in European bees of the genus *Andrena* subgenus *Taeniandrena* (Hymenoptera, Andrenidae): implications for conservation. *Journal of Hymenoptera Research*, 91: 375–428.

Prosi, R. *et al*. (2016) Distribution, biology and habitat of the rare European osmiine bee species *Osmia (Melanosmia) pilicornis* (Hymenoptera, Megachilidae, Osmiini). *Journal of Hymenoptera Research*, 52: 1–36.

Raw, A. (1974) Pollen preferences of three *Osmia* species (Hymenoptera). *Oikos*, 25: 54–60.

Richards, O.W. (1926) Capture in England of female and worker of *Bombus cullumanus* K. (Hym.). *The Entomologist's Monthly Magazine*, 62: 267–268.

Rowson, R. & Pavett, M. (2008) A visual guide for the identification of British *Coelioxys* bees. Privately published, Cardiff, UK.

Saunders, E. (1877) Notes on Hymenoptera captured in 1877. *The Entomologist's Monthly Magazine*, 14: 163–164.

Saunders, E. (1879) Descriptions of new species of British Hymenoptera. *The Entomologist's Monthly Magazine*, 15: 199–201.

Saunders, E. (1879) Notes on rare and other species of Hymenoptera taken in the neighbourhood of Hastings in 1879. *The Entomologist's Monthly Magazine*, 16: 97–98.

Saunders, E. (1881) Notes on Spring Hymenoptera in 1881. *The Entomologist's Monthly Magazine*, 18: 42–43; 114; 160–161; 199.

Saunders, E. (1882) Notes on Spring Hymenoptera at Hastings Spring 1882. *The Entomologist's Monthly Magazine*, 19: 20; 280.

Saunders, E. (1882) Synopsis of British hymenoptera. Diplotera and Anthophila, Part 1 to end of Andrenidae. *The Transactions of the Entomological Society of London*, 1882: 165–290.

Saunders, E. (1884) Synopsis of British hymenoptera part II. *The Transactions of the Entomological Society of London*, 1884: 159–250.

Saunders, E. (1887) Notes on British Hymenoptera. *The Entomologist's Monthly Magazine*, 24: 124–125.

Saunders, E. (1888) *Andrena* and *Stylops*. *The Entomologist's Monthly Magazine*, 25: 293–295.

Saunders, E. (1895) *Sphecodes rubicundus*. *The Entomologist's Monthly Magazine*, 31: 258–259.

Saunders, E. (1896) *The Hymenoptera Aculeata of the British Isles*. Reeve & Co.

Saunders, E. (1898) Aculeate Hymenoptera at Littlehampton. *The Entomologist's Monthly Magazine*, 34: 213–214.

Saunders, E. (1899) Two additional British species of *Andrena*. *The Entomologist's Monthly Magazine*, 35: 154–155.

Saunders, E. (1905) In Page, W. *The Victoria History of the County of Sussex*, volume 1. Archibald Constable and Co.

Saunders, E. (1906) Additions and corrections to the British list of hymenoptera since 1896. *The Entomologist's Monthly Magazine*; 42: 204.

Saunders, E. (1907) *Halictus brevicornis*, Schrank, as addition to the list of British Hymenoptera. *The Entomologist's Monthly Magazine*, 43: 40–41.

Saunders, E. (1907) Aculeate Hymenoptera in West Suffolk and at Eastbourne. *The Entomologist's Monthly Magazine*, 43: 67.

Saunders, P. (2020) Review of *Andrena rosae*: iso. Else & Edwards: 2018 in Cornwall. *BWARS Newsletter*, Autumn 2020: 12–18.

Saunders, P. (2023) Pollen preference, ecology and conservation of *Andrena simillima* (Hymenoptera: Andrenidae) in Cornwall. *British Journal of Entomology and Natural History*: 36, 101–112.

Saunders, S.S. (1881) Capture of rare Hymenoptera on the south coast. *The Entomologist's Monthly Magazine*, 18: 160–161.

Schindler, M. *et al.* (2018) Courtship behaviour in the genus *Nomada* – antennal grabbing and possible transfer of male secretions. *Journal of Hymenoptera Research*, 47–59.

Sladen, F.W.L. (1898) *Bombus smithianus* near Rye. *The Entomologist's Monthly Magazine,* 34: 254.

Sladen, F.W.L. (1912) *The Humble-bee. Its life-history and how to domesticate it.* Macmillan and Co., London.

Smit, J. (2018) Identification key to the European species of the bee genus *Nomada* Scopoli, 1770 (Hymenoptera: Apidae), including 23 new species. *Entomofauna Monographie* 3: 1-253.

Smith, F. (1844) Notes on the British Humble Bees. *The Zoologist*, 2: 548.

Smith, F. (1845) Descriptions of British species of bees belonging to the genera *Melecta* (Latreille), *Coelioxys* (Latreille), and *Stelis* (Panzer). *The Zoologist*, 3: 1146–1155.

Smith, F. (1846) Descriptions of the British species of bees comprised in the genera *Colletes* of Latreille and *Macropis* of Klug; with observations on their economy. *The Zoologist*, 4: 1277–1278.

Smith, F. (1846) A supplementary paper containing descriptions of a few species of bees recently discovered or omitted in the descriptions of the genera to which they belong. *The Zoologist*, 4: 1566–1568.

Smith, F. (1847) Descriptions of the British species of bees comprised in the genera *Andrena* of Fabricius. *The Zoologist*, 5: 1662–1671; 1732–1753.

Smith, F. (1848) Descriptions of the British species of bees belonging to the genus *Halictus* of Latreille. *The Zoologist*, 6: 2200–2209.

Smith, F. (1855) Notes on the new species of British aculeate hymenoptera. *The Entomologist's Annual* 1855–1857: 87–96.

Smith, F. (1876) *Catalogue of British hymenoptera in the British Museum. 1. Andrenidae and Apidae*. London.

Smith, F. (1891) *Catalogue of the British bees in the collection of the British Museum*. British Museum (Natural History), London.

Stace, C.A. (2019) *New Flora of the British Isles*, 4th Edition. C & M Floristics.

Unwin, W.C. (1858) A list of the insects observed in the southern parts of the county of Sussex. *The Naturalist*, 8: 18–20, 39–41, 91–93, 158–160, 208–210, 255–257; 276.

Westrich, P. (2019) *Die Wildbienen Deutchlands*, 2. Auflage. Eugen Ulmer KG, Stuttgart.

Westrich, P. (2020) Die Filzbiene *Epeolus variegatus* (Linnaeus 1758), ein weiterer Brutparasit der Seidenbiene *Colletes hederae* Schmidt & Westrich 1993 (Hymenoptera: Anthophila)? *Eucera* 14, 24–26.

Williams, P.H., Byvaltsev, A., Sheffield, C. *et al.* (2013) *Bombus cullumanus*—an extinct European bumblebee species? *Apidologie* 44, 121–132.

Wood, T.J. *et al.* (2022) *Andrena scotica* Perkins is the valid name for the widespread European taxon previously referred to as *Andrena carantonica* Pérez (Hymenoptera: Andrenidae). *British Journal of Entomology and Natural History*, 35: 393–408.

Wood, T.J. (2022) Nomenclatural changes to the British *Andrena* fauna and species to watch out for. *BWARS Newsletter*, Autumn 2022: 12–15.

Yarrow, I.H.H. & Guichard, K.M. (1941) Some rare hymenoptera aculeata, with two species new to Britain. *The Entomologist's Monthly Magazine*, 77: 2–13.

Index

Page numbers in **bold** refer to the main species accounts.

Andrena afzeliella 11, **206**, 286, 287, 295, 319
Andrena alfkenella 105, **208**, 239
Andrena ampla 97, 292, 293
Andrena angustior 99, **210**
Andrena apicata 125, **212**, 291
Andrena argentata 7, 91, 95, **214**, 411
Andrena barbilabris 90, 91, **216**, 406, 407, 411
Andrena bicolor 99, 211, **218**, 233, 279
Andrena bimaculata 111, **220**, 257, 311
Andrena bucephala 119, **222**, 245
Andrena chrysosceles 99, **224**
Andrena cineraria 115, 122, 123, **226**, 315, 466
Andrena clarkella 14, 125, **228**
Andrena coitana 128, 129, **230**, 307
Andrena confinis 143, **232**
Andrena congruens 232
Andrena denticulata 135, **234**
Andrena dorsata 142, 143, 232, 233, **236**, 255, 411
Andrena falsifica 8, 105, 209, **238**
Andrena ferox 127, 223, **240**
Andrena flavipes 99, 108, 109, 121, 235, **242**, 254, 255, 263, 314, 413
Andrena florea **244**
Andrena fucata 130, 131, 141, 231, **246**, 261
Andrena fulva 130, 131, 138, 139, **248**, 466
Andrena fulvago 9, 224, **250**, 263
Andrena fuscipes 7, 135, **252**, 433
Andrena gravida 4, 9, 243, **254**
Andrena haemorrhoa 127, 133, **256**
Andrena hattorfiana 17, **258**, 271
Andrena helvola 8, 131, 231, 247, **260**
Andrena humilis 121, **262**
Andrena labialis **264**, 413
Andrena labiata 2, 9, 116, 117, **266**
Andrena lapponica 112, 113, 131, 261, **268**
Andrena marginata 6, 92, 93, 259, **270**
Andrena minutula 105, **272**, 274, 275, 285, 303
Andrena minutuloides 105, 272, **274**, 285, 303
Andrena nigriceps 135, **276**, 301
Andrena nigroaenea 99, 103, 115, 265, **278**
Andrena nigrospina 110, 288, 289
Andrena nitida 115, **280**
Andrena nitidiuscula 135, 230, **282**
Andrena niveata 6, **284**
Andrena ovatula 206, 207, 237, **286**, 295, 319
Andrena ovatula agg. 206, 286
Andrena pilipes 111, 115, 257, **288**
Andrena parvula 273
Andrena praecox 100, 101, 212, 213, **290**
Andrena proxima 8, 97, **292**
Andrena rosae 2, 11, 127, 245, **460**, 466

Andrena russula 141, 207, 237, 287, **294**, 319
Andrena scotica 11, 103, 115, 127, 223, **296**, 312, 313
Andrena scotica agg. 296, 312
Andrena semilaevis 105, 230, **298**
Andrena similis 294
Andrena simillima 11, 135, 277, **300**
Andrena subopaca 105, 272, **302**
Andrena synadelpha 8, 131, **304**, 317
Andrena tarsata 2, 129, **306**
Andrena thoracica 9, 115, **308**
Andrena tibialis 111, 257, **310**
Andrena tridentata 235
Andrena trimmerana 11, 115, 127, 245, 296, 297, **312**
Andrena vaga 4, 9, 123, 227, **314**
Andrena varians 99, 101, 131, 269, 305, **316**
Andrena ventralis 4, 206
Andrena wilkella 140, 141, 155, 207, 237, 287, 295, **318**
Anthidium manicatum **146**, 204, 205, 466
Anthophora bimaculata **30**, 32, 36, 155, 161, 163
Anthophora furcata **32**, 36, 160, 161, 163
Anthophora plumipes 30, **34**, 39, 88, 89, 466
Anthophora quadrimaculata 32, **36**, 161, 163
Anthophora retusa 2, 6, 34, 35, **38**, 89, 459
Apis mellifera **40**, 248, 278
Blood Bee, Bare-saddled **388**
Blood Bee, Box-headed **402**
Blood Bee, Dark **404**
Blood Bee, Dark-winged **394**
Blood Bee, Dull-headed **390**
Blood Bee, False-margined **400**
Blood Bee, Furry-bellied **396**
Blood Bee, Geoffroy's **392**
Blood Bee, Little Sickle-jawed **398**
Blood Bee, Red-tailed **412**
Blood Bee, Reticulate **410**
Blood Bee, Rough-backed **414**
Blood Bee, Sandpit **406**
Blood Bee, Sickle-jawed **408**
Blood Bee, Spined **416**
Blood Bee, Swollen-thighed **386**
Bombus barbutellus 2, **42**, 44, 49, 53, 69, 78
Bombus bohemicus 2, **44**, 59
Bombus campestris 2, 44, **46**, 51, 61, 63, 67, 78
Bombus cryptarum 45, 58
Bombus cullumanus 2, 6, 11, 56, **456**
Bombus distinguendus 2, 6, **457**
Bombus hortorum 42, 43, **48**, 55, 68, 69, 462
Bombus humilis 2, 6, 9, 47, **50**, 61, 62
Bombus hypnorum 4, **52**, 62, 75
Bombus jonellus 7, 48, **54**, 75

Bombus lapidarius **56**, 67, 70, 456, 466
Bombus lucorum 45, 58
Bombus lucorum agg. 42, 45, **58**, 73, 76
Bombus magnus 45, 58
Bombus monticola 75
Bombus muscorum 2, 6, 8, 9, 13, 50, 51, **60**, 62
Bombus norvegicus 53
Bombus pascuorum 32, 46, 47, 51, 61, **62**, 71, 466
Bombus pomorum 42, 56
Bombus pratorum 47, **64**, 74, 75
Bombus ruderarius 2, 9, 47, 56, **66**, 70, 458
Bombus ruderatus 2, 8, 9, 43, 48, 49, 55, **68**, 457
Bombus rupestris 56, 57, **70**
Bombus soroeensis 2, 6, 9, **72**
Bombus subterraneus 2, 6, 9, 13, 48, 55, 68, **457**
Bombus sylvarum 2, 9, 13, 71, **458**
Bombus sylvestris 53, 55, 65, **74**
Bombus terrestris 59, 64, 65, 73, **76**, 78, 79
Bombus vestalis 44, 77, **78**
Bumblebee, Broken-belted **72**
Bumblebee, Buff-tailed **76**
Bumblebee, Cullum's **456**
Bumblebee, Early **64**
Bumblebee, Garden **48**
Bumblebee, Great Yellow **457**
Bumblebee, Heath **54**
Bumblebee, Large Garden **68**
Bumblebee, Red-tailed **56**
Bumblebee, Short-haired **457**
Bumblebee, Tree **52**
Bumblebee, White-tailed **58**
Carder Bee, Brown-banded **50**
Carder Bee, Common **62**
Carder Bee, Moss **60**
Carder Bee, Red-shanked **66**
Carder Bee, Shrill **458**
Carpenter Bee, Little Blue **80**
Carpenter Bee, Violet **144**
Ceratina cyanea **80**
Chelostoma campanularum **148**, 463
Chelostoma florisomne 148, **150**
Coelioxys conoidea 7, 17, **152**, 177
Coelioxys elongata 31, **154**, 156, 157, 169, 171, 175, 181
Coelioxys inermis 155, **156**, 169, 173, 175, 179
Coelioxys mandibularis 155, 156, 157, **158**, 169, 171, 173, 177, 179
Coelioxys quadridentata 33, 36, **160**, 163, 171, 181
Coelioxys rufescens 31, 33, 37, 161, **162**, 169, 171
Colletes cunicularius 4, 5, 6, 8, 9, 11, 315, **418**, 431
Colletes daviesanus 84, 85, **420**, 455
Colletes floralis 421
Colletes fodiens 83, 85, **422**, 429, 431

The Bees of Sussex 471

Colletes halophilus 6, 8, 9, 22, 84, 85, **424**, 427, 433, 450
Colletes hederae 4, 83, 85, 405, 425, **426**, 433
Colletes marginatus 5, 9, 83, **428**
Colletes similis 85, 423, 429, **430**
Colletes succinctus 7, 83, 427, **432**
Colletes, Bare-saddled **430**
Colletes, Davies' **420**
Colletes, Early **418**
Colletes, Hairy-saddled **422**
Colletes, Heather **432**
Colletes, Margined **428**
Cuckoo Bee, Barbut's **42**
Cuckoo Bee, Field **46**
Cuckoo Bee, Forest **74**
Cuckoo Bee, Gypsy **44**
Cuckoo Bee, Red-tailed **70**
Cuckoo Bee, Vestal **78**
Dark Bee, Banded **204**
Dark Bee, Little **196**
Dark Bee, Plain **202**
Dark Bee, Smooth-saddled **198**
Dark Bee, Spotted **200**
Dasypoda hirtipes 7, 8, 9, **20**
Epeoloides coecutiens 23
Epeolus cruciger 7, 11, **82**, 84, 85, 423, 427, 429, 433
Epeolus fallax 427
Epeolus variegatus 11, 82, 83, **84**, 421, 423, 425, 427, 429, 431
Epeolus, Black-thighed **84**
Epeolus, Red-thighed **82**
Eucera longicornis 4, 8, 9, **86**, 265, 458, 459
Eucera nigriscens 2, 86, 87, **458**, 459
Flower Bee, Fork-tailed **32**
Flower Bee, Four-banded **36**
Flower Bee, Green-eyed **30**
Flower Bee, Hairy-footed **34**
Flower Bee, Potter **38**
Furrow Bee, Bloomed **332**
Furrow Bee, Box-headed **461**
Furrow Bee, Bronze **330**
Furrow Bee, Bull-headed **384**
Furrow Bee, Chalk **342**
Furrow Bee, Common **336**
Furrow Bee, Downland **326**
Furrow Bee, Four-spotted **372**
Furrow Bee, Fringed **376**
Furrow Bee, Furry-claspered **346**
Furrow Bee, Green **356**
Furrow Bee, Grey-tailed **366**
Furrow Bee, Least **354**
Furrow Bee, Lobe-spurred **364**
Furrow Bee, Long-faced **368**
Furrow Bee, Orange-footed **382**
Furrow Bee, Orange-legged **328**
Furrow Bee, Red-backed **344**
Furrow Bee, Ridge-cheeked **370**
Furrow Bee, Shaggy **380**
Furrow Bee, Sharp-collared **352**
Furrow Bee, Short-horned **334**
Furrow Bee, Small Shiny **374**
Furrow Bee, Smeathman's **378**
Furrow Bee, Smooth-faced **340**
Furrow Bee, Smooth-gastered **360**
Furrow Bee, Southern Bronze **324**

Furrow Bee, Squat **362**
Furrow Bee, Tufted **358**
Furrow Bee, Turquoise **338**
Furrow Bee, White-footed **348**
Furrow Bee, White-zoned **350**
Halictus confusus 7, **324**, 331
Halictus eurygnathus 4, 6, 11, 324, **326**, 328, 329, 415
Halictus maculatus 2, 14, 324, 395, **461**
Halictus quadricinctus 395
Halictus rubicundus 324, 327, **328**, 394, 395, 403
Halictus subauratus 331
Halictus tumulorum 137, 324, 325, **330**, 393, 403
Heriades truncorum **164**, 196, 197, 203
Honey Bee **40**
Hoplitis adunca 166
Hoplitis claviventris **166**, 195, 200, 201
Hoplitis leucomelana 166, 201
Hoplitis spinulosa 195
Hylaeus annularis **9**, **434**, 444, 445
Hylaeus brevicornis **436**, 445
Hylaeus communis **438**, 451
Hylaeus confusus **440**, 449, 455
Hylaeus cornutus **442**
Hylaeus dilatatus 435, **444**
Hylaeus hyalinatus **446**
Hylaeus incongruus 440, **448**
Hylaeus masoni 435
Hylaeus pectoralis 5, 7, 8, 11, 22, 438, **450**
Hylaeus pictipes **452**
Hylaeus punctulatissimus 455
Hylaeus signatus **454**
Hylaeus spilotus 435
Hylaeus, Spined **442**
Ivy Bee **426**
Lasioglossum aeratum 349
Lasioglossum albipes **332**, 336, 403
Lasioglossum angusticeps 369
Lasioglossum brevicorne **334**, 380, 381, 409
Lasioglossum calceatum 332, 333, **336**, 381, 389, 403
Lasioglossum cupromicans **338**, 349, 356, 357, 379
Lasioglossum fratellum **340**, 343, 381, 391, 396, 397
Lasioglossum fulvicorne 341, **342**, 391, 396, 397
Lasioglossum laevigatum **344**
Lasioglossum laticeps 389, 391, 403
Lasioglossum lativentre **346**, 372, 373, 389, 409
Lasioglossum leucopus 339, **348**, 356, 357, 379, 399
Lasioglossum leucozonium 345, **350**, 385, 389, 394, 403
Lasioglossum malachurum **352**, 364, 395, 403
Lasioglossum minutissimum 285, **354**, 399
Lasioglossum morio 339, 348, 349, **356**, 379, 393, 399, 401, 405, 462
Lasioglossum nitidiusculum 137, **358**, 361, 387, 393, 401
Lasioglossum parvulum 137, 358, 359, **360**, 386, 387, 393

Lasioglossum pauperatum 355, **362**
Lasioglossum pauxillum **364**, 387, 391, 401
Lasioglossum prasinum 7, **366**, 387, 411
Lasioglossum punctatissimum **368**, 387, 399
Lasioglossum puncticolle **370**
Lasioglossum quadrinotatum 347, **372**, 389, 409
Lasioglossum rufitarse 375, 391, 393
Lasioglossum semilucens 11, **374**, 399
Lasioglossum sexstrigatum 4, 137, 332, **376**, 399, 401
Lasioglossum smeathmanellum 137, 338, 339, 349, 356, 357, **378**
Lasioglossum villosulum 137, 335, **380**, 393, 409
Lasioglossum xanthopus **382**, 403, 416, 417
Lasioglossum zonulum 345, 351, **384**, 415
Leafcutter Bee, Black-headed **170**
Leafcutter Bee, Brown-footed **178**
Leafcutter Bee, Coast **176**
Leafcutter Bee, Patchwork **168**
Leafcutter Bee, Silvery **172**
Leafcutter Bee, Willughby's **180**
Leafcutter Bee, Wood-carving **174**
Long-horned Bee **86**
Long-horned Bee, Scarce **458**
Macropis europaea 7, 8, **22**, 450
Macropis fulvipes 23
Mason Bee, Blue **188**
Mason Bee, Cliff **460**
Mason Bee, Fringe-horned **192**
Mason Bee, Gold-fringed **182**
Mason Bee, Orange-vented **190**
Mason Bee, Red **186**
Mason Bee, Red-tailed **184**
Mason Bee, Spined **194**
Mason Bee, Welted **166**
Megachile centuncularis 156, 157, 159, 163, **168**, 179
Megachile circumcincta 11, 155, 159, 161, 163, **170**, 181
Megachile ericetorum 168
Megachile leachella 5, 9, 157, 159, 161, **172**
Megachile ligniseca 157, 163, **174**
Megachile maritima 7, 9, 17, 152, 153, 155, 159, 171, 175, **176**, 181
Megachile versicolor 157, 159, 169, **178**
Megachile willughbiella 155, 161, 163, 171, 175, 177, **180**
Melecta albifrons 35, 39, **88**, 459
Melecta luctuosa 2, 11, 39, 88, 89, **459**
Melitta dimidiata 24
Melitta haemorrhoidalis **24**, 107
Melitta leporina 25, **26**, 29, 106, 107, 463
Melitta tricincta 25, 26, **28**, 107
Melitta, Clover **26**
Melitta, Gold-tailed **24**
Mini-miner, Alfken's **208**
Mini-miner, Common **272**
Mini-miner, Impunctate **302**
Mini-miner, Long-fringed **284**
Mini-miner, Plain **274**
Mini-miner, Shiny-margined **298**
Mini-miner, Thick-margined **238**
Mining Bee, Ashy **226**

Mining Bee, Big-headed **222**
Mining Bee, Bilberry **268**
Mining Bee, Black **288**
Mining Bee, Black-headed **276**
Mining Bee, Blackthorn **316**
Mining Bee, Broad-faced **292**
Mining Bee, Broad-margined **304**
Mining Bee, Bryony **244**
Mining Bee, Buff-banded **300**
Mining Bee, Buffish **278**
Mining Bee, Buff-tailed **262**
Mining Bee, Carrot **282**
Mining Bee, Chocolate **296**
Mining Bee, Clarke's **228**
Mining Bee, Cliff **308**
Mining Bee, Coppice **260**
Mining Bee, Grey-backed **314**
Mining Bee, Grey-banded **234**
Mining Bee, Grey-gastered **310**
Mining Bee, Grey-patched **280**
Mining Bee, Groove-faced **210**
Mining Bee, Gwynne's **218**
Mining Bee, Hawk's-beard **250**
Mining Bee, Hawthorn **224**
Mining Bee, Heather **252**
Mining Bee, Large Gorse **220**
Mining Bee, Large Meadow **264**
Mining Bee, Large Sallow **212**
Mining Bee, Large Scabious **258**
Mining Bee, Long-fringed **232**
Mining Bee, Oak **240**
Mining Bee, Orange-tailed **256**
Mining Bee, Painted **246**
Mining Bee, Perkins' **460**
Mining Bee, Red-backed **294**
Mining Bee, Red-girdled **266**
Mining Bee, Sandpit **216**
Mining Bee, Short-fringed **236**
Mining Bee, Small Flecked **230**
Mining Bee, Small Gorse **286**
Mining Bee, Small Sallow **290**
Mining Bee, Small Sandpit **214**
Mining Bee, Small Scabious **270**
Mining Bee, Tawny **248**
Mining Bee, Tormentil **306**
Mining Bee, Trimmer's **312**
Mining Bee, White-bellied **254**
Mining Bee, Wilke's **318**
Mining Bee, Yellow-legged **242**
Mourning Bee, Common **88**
Mourning Bee, Square-spotted **459**
Nomad Bee, Bilberry **112**
Nomad Bee, Black-horned **134**
Nomad Bee, Blunthorn **106**
Nomad Bee, Blunt-jawed **140**
Nomad Bee, Broad-banded **138**
Nomad Bee, Cat's-ear **120**
Nomad Bee, Early **124**
Nomad Bee, Fabricius' **98**
Nomad Bee, Flat-ridged **128**
Nomad Bee, Flavous **102**
Nomad Bee, Fork-jawed **132**
Nomad Bee, Fringeless **96**
Nomad Bee, Gooden's **114**
Nomad Bee, Large Bear-clawed **90**
Nomad Bee, Lathbury's **122**
Nomad Bee, Little **104**

Nomad Bee, Long-horned **118**
Nomad Bee, Marsham's **126**
Nomad Bee, Orange-horned **110**
Nomad Bee, Painted **108**
Nomad Bee, Panzer's **130**
Nomad Bee, Sheppard's **136**
Nomad Bee, Short-spined **116**
Nomad Bee, Silver-sided **92**
Nomad Bee, Six-banded **459**
Nomad Bee, Small Bear-clawed **94**
Nomad Bee, Variable **142**
Nomad Bee, Yellow-shouldered **100**
Nomada alboguttata 4, 8, 9, **90**, 95, 217
Nomada argentata 6, 11, **92**, 271
Nomada armata 259
Nomada baccata 7, 91, **94**, 215
Nomada bifasciata 4, 108, 115, 135, 255
Nomada conjungens **96**, 293
Nomada errans 283
Nomada fabriciana **98**, 101, 133, 211, 219, 225, 279, 461
Nomada facilis 93, 120, 121, 251
Nomada ferruginata 4, 9, **100**, 291, 317
Nomada flava **102**, 115, 125, 127, 131, 132, 139, 241, 281, 297
Nomada flavoguttata 97, **104**, 137, 239, 273, 275, 299, 303
Nomada flavopicta 25, 27, 29, **106**, 110
Nomada fucata **108**, 115, 135, 243
Nomada fulvicornis **110**, 221, 281, 289, 309, 311
Nomada fusca 247
Nomada glabella 11, 102, **112**, 130, 131, 132, 139, 269
Nomada goodeniana 108, **114**, 127, 139, 227, 279, 281, 309, 311
Nomada guttulata 9, **116**, 141, 267
Nomada hirtipes **118**, 223
Nomada integra 9, 93, **120**, 263
Nomada lathburiana 4, **122**, 227, 315
Nomada leucophthalma **124**, 213, 229
Nomada marshamella 110, 115, **126**, 241, 281, 297, 313
Nomada obtusifrons 2, **128**, 231, 307
Nomada panzeri 102, 112, 125, **130**, 132, 139, 247, 249, 261, 269, 305, 317
Nomada roberjeotiana 231, 307
Nomada ruficornis 99, **132**, 257
Nomada rufipes 7, **134**, 235, 253, 277, 283, 301
Nomada sexfasciata 2, 11, 87, **459**
Nomada sheppardana **136**, 357, 359, 361, 377, 379
Nomada signata **138**, 249
Nomada striata 117, **140**, 241, 295, 319
Nomada subcornuta 110
Nomada succincta 108, 115, 265, 279, 281
Nomada zonata 4, **142**, 233, 237
Osmia aurulenta 9, **182**, 184
Osmia bicolor 9, 183, **184**, 187, 193
Osmia bicornis **186**, 203, 466
Osmia caerulescens 155, **188**, 191, 201, 203
Osmia cornuta 4, 182, 186, 187
Osmia leaiana 189, **190**, 195, 203
Osmia niveata 189, 191
Osmia pilicornis 2, 6, 9, 11, 162, 182, **192**

Osmia spinulosa 166, 167, **194**, 199, 203
Osmia xanthomelana 2, 162, 182, 192, **460**
Pantaloon Bee **20**
Panurgus banksianus **320**, 322, 323
Panurgus calcaratus 320, 321, **322**
Red Bartsia Bee **28**
Resin Bee, Small-headed **164**
Rophites quinquespinosus **461**
Rophites, Five-spined **461**
Scissor Bee, Large **150**
Scissor Bee, Small **148**
Sea Aster Bee **424**
Shaggy Bee, Large **320**
Shaggy Bee, Small **322**
Sharp-tail Bee, Dull-vented **154**
Sharp-tail Bee, Grooved **160**
Sharp-tail Bee, Large **152**
Sharp-tail Bee, Rufescent **162**
Sharp-tail Bee, Shiny-vented **156**
Sharp-tail Bee, Square-jawed **158**
Sphecodes albilabris 4, 419
Sphecodes crassus 359, 361, 365, 369, **386**, 390, 391
Sphecodes ephippius 215, 331, 337, 347, 351, 373, **388**, 392, 413
Sphecodes ferruginatus 343, 365, 386, **390**
Sphecodes geoffrellus 331, 339, 349, 357, 359, 361, 381, **392**, 401
Sphecodes gibbus 329, **394**, 403, 411
Sphecodes hyalinatus 341, 343, **396**
Sphecodes longulus 355, 375, **398**, 409
Sphecodes marginatus 335, 375, 381
Sphecodes miniatus 377, 392, 393, **400**
Sphecodes monilicornis 329, 333, 337, 353, 383, 385, 392, 395, **402**, 411
Sphecodes niger 357, 386, **404**
Sphecodes pellucidus 217, 351, 367, 394, **406**
Sphecodes puncticeps 335, 347, 373, 381, 399, **408**
Sphecodes reticulatus 215, 367, 395, 403, **410**
Sphecodes rubicundus 265, 388, 389, **412**, 417
Sphecodes scabricollis 8, 327, 385, **414**
Sphecodes spinulosus 383, 413, **416**
Stelis breviuscula 11, 165, **196**, 199, 203
Stelis odontopyga 4, 195, 197, **198**, 203
Stelis ornatula 167, 189, 196, **200**
Stelis phaeoptera 9, 11, 191, 197, 199, **202**, 466
Stelis punctulatissima 147, 191, **204**
Wool Carder Bee **146**
Xylocopa valga 145
Xylocopa violacea **144**
Xylocopa virginica 144
Yellow-face Bee, Chalk **444**
Yellow-face Bee, Common **438**
Yellow-face Bee, Hairy **446**
Yellow-face Bee, Large **454**
Yellow-face Bee, Little **452**
Yellow-face Bee, Reed **450**
Yellow-face Bee, Shingle **434**
Yellow-face Bee, Short-horned **436**
Yellow-face Bee, White-jawed **440**
Yellow-face Bee, White-lipped **448**
Yellow Loosestrife Bee **22**

About the author

James Power spent much of his childhood exploring the wide-open spaces of Salisbury Plain. The freedom to roam for hours on end and discover populations of Marsh Fritillary butterflies, Stone Curlews and Fairy Shrimps sparked an interest which ultimately led to a career in nature conservation. Regrettably, bees were not yet on his radar, but their time would come.

His career as a nature conservationist has lasted almost 40 years and has included spells with two different Wildlife Trusts, Defra, the Severn Gorge Countryside Trust, the National Trust, and even the National Parks department in Malawi. Within Britain, he's worked in Wiltshire, Yorkshire, Surrey, Shropshire, and now—the perfect place to land—Sussex.